Fourth Edition

Jones' Instrument Technology Volume 2

Measurement of Temperature and Chemical Composition

Edited by **B E Noltingk**

BUTTERWORTH HEINEMANN

Butterworth-Heinemann Ltd
Linacre House, Jordan Hill, Oxford OX2 8DP

PART OF REED INTERNATIONAL BOOKS

OXFORD LONDON BOSTON
MUNICH NEW DELHI SINGAPORE SYDNEY
TOKYO TORONTO WELLINGTON

First published 1985
Reprinted 1987, 1992

British Library Cataloguing in Publication Data
Jones, E. B. (Ernest Beachcroft)
Jones' instrument technology - 4th ed.
Vol 2: measurement of temperature and chemical composition
1. Measuring instruments
I. Title II. Noltingk, B. E. III. Jones, E. B. (Ernest Beachcroft). Instrument technology
620'.0044 QC100.5

ISBN 0 7506 0766 1

Library of Congress Cataloguing in Publication Data
Jones, E. B.
Jones' instrument technology - 4th ed.
Includes index.
ISBN 0 7506 0766 1
1. Mechanical measurements 2. Measurement of temperature and chemical composition
1. Engineering instruments. I. Noltingk, B. E.
II. Jones, E. B. (Ernest Beachcroft). Instrument technology
TA165.159674 1984 620'.0028 84-4273

Printed and bound in Great Britain by
Thomson Litho Ltd, East Kilbride, Scotland.

Contents

Contributors

Sir Claud Hagart-Alexander Bt., BA, MInstMC, DL, had a long experience in instrumentation with ICI Ltd. He is now a director of Instrumentation Systems Ltd.

The chapters on Chemical Analysis have been contributed by a team from Central Electricity Research Laboratories. **Mr W. G. Cummings** BSc, CChem, FRSC, MInstE, MInstMC, was until recently, head of the Analytical Chemistry Section, where **Mr A. C. Smith** BSc, CChem, FRSC, MInstP, is now in charge; **Dr C. K. Laird** BSc, PhD, CChem, MRSC and **Dr K. Torrance** BSc, PhD, also work in that Section. **Dr D. B. Meadowcroft** BSc, PhD, MInstP, MICorrST, is in the Materials Branch of C.E.R.L.

Preface

As in the first volume of this updated *Instrument Technology*, I must express my thanks to the contributors. They have, to a greater or lesser extent (but mainly greater), tolerated my cajoling them in two ways. First, to write at all. Secondly, to cover an agreed range of topics with an agreed thoroughness. I believe that a useful book has emerged. I might especially thank Mr Bill Cummings for applying his practical wisdom to helping define the areas of analytical chemistry that needed to be dealt with in a modern book on instrument technology.

I am picking up a frequent theme of E. B. Jones, the original author, in underlining both the value of technicians in today's technical industry and the need for more of them. We hope that our book will play its part in their training and indeed in the training and understanding of many folk who need to have some knowledge of instrumentation. We have tried to tread a balanced path between the expounding of fundamental science and the description of trivial practical details. But with emphasis on the practical. I was once told the story of a distant country where, yes, they were very keen on football; their ministry of education decreed that there should be football lessons. So every school put the subject on its curriculum – but only in the classrooms: there were no pitches to play on and no balls to play with. Neither instrument technology nor football can be well learned unless the experimental is blended with the theoretical.

Instrument technology has been divided into four volumes. It would have been impossibly bulky if it had not! In the first volume we grouped the measurements of mechanical quantities. Here, in the second, we have two broad subjects of very wide interest, temperature and chemical composition. They are both important for process control, but we have tried not to be tied too tightly to on-line instrumentation. Although the titles do not exactly correspond, it can be seen that between them Volumes 1 and 2 of the new edition cover most of the topics appearing in the old Volumes 1 and 2. The new Volume 3 has several new subjects while Volume 4 deals with matters of general interest that stretch across the measurement of many different quantities, justifying its title of Systems. More details can be seen in the list of contents printed on page v.

BEN
1985

1 Temperature measurement

C. HAGART-ALEXANDER

1.1 Temperature and heat

1.1.1 Application considerations

Temperature is one of the most frequently used process measurements. Almost all chemical processes and reactions are temperature-dependent. Not infrequently in chemical plant, temperature is the only indication of the progress of the process. Where the temperature is critical to the reaction, a considerable loss of product may result from incorrect temperatures. In some cases, loss of control of temperature can result in catastrophic plant failure with the attendant damage and possibly loss of life.

Another area where accurate temperature measurement is essential is in the metallurgical industries. In the case of many metal alloys, typically steel and aluminium alloys, the temperature of the heat treatment the metal receives during manufacture is a crucial factor in establishing the mechanical properties of the finished product.

There are many other areas of industry where temperature measurement is essential. Such applications include steam raising and electricity generation, plastics manufacture and moulding, milk and dairy products and many other areas of the food industries.

Then, of course, where most of us are most aware of temperature is in the heating and air-conditioning systems which make so much difference to people's personal comfort.

Instruments for the measurement of temperature, as with so many other instruments, are available in a wide range of configurations. Everyone must be familiar with the ubiquitous liquid-in-glass thermometer. There is then a range of dial thermometers with the dial attached directly to the temperature measuring element, i.e. local reading thermometers. Remote reading instruments are also available where the measuring system operates the dial directly through a length of metal capillary tubing. The distance between the sensing 'bulb' and the dial, or readout, of these instruments is limited to about thirty metres. Where the temperature readout is required at a long distance from the location of the sensing element there are two main options; either an electrical measuring technique such as a thermocouple or resistance thermometer can be used or where the distances between the plant measurement locations and the control room are very long it is usually better to use temperature transmitters. These instruments use the same types of probes as other temperature measuring instruments. The transmitting mechanism is normally attached directly to the probe. It may also have a local readout facility as well as its transmitting function which is to convert the measurement effect into a pneumatic or electrical signal suitable for transmission over long distances' (See Volume 4 on transmitters.)

Temperature measurement effects are also used directly for simple control functions such as switching on and off an electric heater or the direct operation of a valve, i.e. thermostats.

1.1.2 Definitions

For the understanding of temperature measurement it is essential to have an appreciation of the concepts of temperature and other heat-related phenomena.

1.1.2.1 Temperature

The first recorded temperature measurement was carried out by Galileo at the end of the sixteenth century. His thermometer depended on the expansion of air. Some form of scale was attached to his apparatus for he mentions 'degrees of heat' in his records.

As with any other measurement, it is necessary to have agreed and standardized units of measurement. In the case of temperature the internationally recognized units are the kelvin and the degree Celsius. The definitions of these units are set out in Section 1.2.

One must differentiate between heat and temperature. The effect of temperature is the state of agitation, both oscillation and rotation of molecules in a medium. The higher the temperature of a body the greater the vibrational energy of its molecules and the greater its potential to transfer this molecular kinetic energy to another body. Temperature is the potential to cause heat to move from a point of higher temperature to one of lower temperature. The rate of heat transfer is a function of that temperature difference.

1.1.2.2 Heat

Heat is thermal energy. The quantity of heat in a body is proportional to the temperature of that body, i.e. it is its heat capacity multiplied by its absolute temperature.

Heat is energy and as such is measured in units of energy. Heat is measured in joules. (Before the international agreements on the SI system of units heat was measured in calories. One calorie was approximately 4.2 joules.)

1.1.2.3 Specific heat capacity

Different materials absorb different amounts of heat to produce the same temperature rise. The specific heat capacity, or more usually the specific heat, of a substance is the amount of heat which, when absorbed by 1 kg of that substance, will raise its temperature by one kelvin

$$\text{Specific heat capacity} = \text{J kg}^{-1}\ \text{K}^{-1}$$

1.1.2.4 Thermal conductivity

The rate at which heat is conducted through a body depends upon the material of the body. Heat travels very quickly along a bar of copper, for instance, but more slowly through iron. In the case of non-metals, ceramics or organic substances, the thermal conduction occurs more slowly still. The heat conductivity is not only a function of the substance but also the form of the substance. Plastic foam is used for heat insulation because the gas bubbles in the foam impede the conduction of heat. Thermal conductivity is measured in terms of:

$$\frac{\text{energy} \times \text{length}}{\text{area} \times \text{time} \times \text{temperature difference}}$$

$$\text{thermal conductivity} = \frac{\text{J.m}}{\text{m}^2.\text{s.K}} = \text{J.m}^{-1}.\text{s}^{-1}.\text{K}^{-1}$$

1.1.2.5 Latent heat

When a substance changes state from solid to liquid or from liquid to vapour it absorbs heat without change of temperature. If a quantity of ice is heated at a constant rate its temperature will rise steadily until it reaches a temperature of 0 °C; at this stage the ice will continue to absorb heat with no change of temperature until it has all melted to water. Now as the heat continues to flow into the water the temperature will continue to rise but at a different rate from before due to the different specific heat of water compared to ice. When the water reaches 100 °C the temperature rise will again level off as the water boils, changing state from water to steam. Once all the water has boiled to steam the temperature will rise again but now at yet another rate dependent on the specific heat of steam. This is illustrated in Figure 1.1.

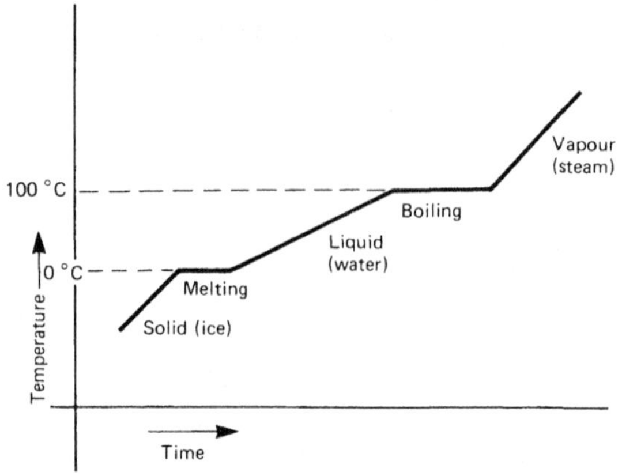

Figure 1.1 Increase of temperature during change of state of a mass of water under conditions of constant energy input.

The amount of heat required to convert a kilogram of a substance from solid state to liquid state is the 'latent heat of fusion'.

Likewise the 'latent heat of evaporation' is the amount of heat required to convert a kilogram of liquid to vapour.

This levelling of temperature rise during change of state accounts for the constant freezing temperatures and constant boiling temperatures of pure materials. The units of measurement of latent heat are joules per kilogram:

$$\text{latent heat} = \text{J.kg}^{-1}$$

1.1.2.6 Thermal expansion

Expansion of solids When a solid is heated, it increases in volume. It increases in length, breadth and thickness. The increase in length of any side of a solid will depend upon the original length l_0, the rise in temperature t, and the coefficient of linear expansion α.

The coefficient of linear expansion may be defined as the increase in length per unit length when the temperature is raised 1 K. Thus, if the temperature of a rod of length l_0, is raised from 0 °C to t °C, then the new length, l_t, will be given by:

$$l_t = l_0 + l_0 . \alpha t = l_0(1 + \alpha t) \tag{1.1}$$

The value of the coefficient of expansion varies from substance to substance and the coefficients of linear expansion of some common materials are given in Table 1.1.

Table 1.1 Coefficients of linear expansion of solids
Extracted from *Tables of Physical and Chemical Constants* by Kaye and Laby (Longmans). The values given are per kelvin and, except where some temperature is specified, for a range about 20 degrees

Substance	α (*ppm*)
Aluminium	25.5
Copper	16.7
Gold	13.9
Iron (cast)	10.2
Lead	29.1
Nickel	12.8
Platinum	8.9
Silver	18.8
Tin	21.4
Brass (typical)	18.9
Constantan (Eureka), 60 Cu, 40 Ni	17.0
Duralumin	22.6
Nickel steel,	
10% Ni	13.0
30% Ni	12.0
36% Ni (Invar)	−0.3 to +2.5
40% Ni	6.0
Steel	10.5 to 11.6
Phosphor bronze, 97.6 Cu, 2 Sn, 0.2 P	16.8
Solder, 2 Pb, 1 Sn	25
Cement and concrete	10
Glass (soda)	8.5
Glass (Pyrex)	3
Silica (fused) −80° to 0 °C	0.22
Silica (fused) 0° to 100 °C	0.50

The increase in area with temperature, i.e. the coefficient of superficial expansion is approximately twice the coefficient of linear expansion. The coefficient of cubic expansion is almost three times the coefficient of linear expansion.

In engineering practice it is necessary, especially in large structures, to make allowance for thermal expansion. For instance bridges are built with expansion joints. Many instruments are designed with temperature compensation to accommodate thermal expansion. Thermal expansion can be made use of for temperature measurement, as is dealt with in Section 1.3.

If great accuracy is required when measuring lengths with a scale made of metal, allowance should be made for the increase in length of the scale when its temperature is greater than that at which it was calibrated. Owing to the expansion of the scale, a length which was originally l_1 at the temperature t_1, at which the scale was calibrated, will have increased to l_2 where

$$l_2 = l_1[1 + \alpha(t_2 - t_1)] \tag{1.2}$$

Here t_2 is the temperature at which the measurement is made, and α the coefficient of expansion of the metal of the scale.

A 1 mm division on the scale will therefore now measure

$$1 + \alpha(t_2 - t_1)\,\text{mm} \tag{1.3}$$

An actual length l_2 mm will therefore measure

$$\frac{l_2}{1 + \alpha(t_2 - t_1)} \tag{1.4}$$

The length will therefore appear to be smaller than it actually is. To make this error negligibly small, secondary standards of length are made of Invar, a nickel steel alloy whose linear coefficient of expansion is nearly zero.

Expansion of liquids and gases In dealing with the expansion of liquids and gases it is necessary to consider the volume expansion, or cubical expansion. Both liquids and gases have to be held by a container, which will also expand, so that the apparent expansion of the liquid or gas will be less than the true or absolute expansion. The true coefficient of expansion of a liquid is equal to the coefficient of apparent expansion plus the coefficient of cubical expansion of the containing vessel. Usually the expansion of a gas is so much greater than that of the containing vessel that the expansion of the vessel may be neglected in comparison with that of the gas.

The coefficient of expansion of a liquid may be defined in two ways. First, there is the zero coefficient of expansion, which is the increase in volume per degree rise in temperature, divided by the volume at 0 °C, so that volume V_t at temperature t is given by:

$$V_t = V_0(1 + \beta t) \tag{1.5}$$

where V_0 is the volume at 0 °C and β is the coefficient of cubical expansion.

There is also the mean coefficient of expansion between two temperatures. This is the ratio of the increase in volume per degree rise of temperature, to the original volume. That is,

$$\beta = \frac{V_{t_2} - V_{t_1}}{V_{t_1}(t_2 - t_1)} \tag{1.6}$$

where V_{t_1} is the volume at temperature t_1, and V_{t_2} is the volume at temperature t_2.

This definition is useful in the case of liquids that do not expand uniformly, e.g. water.

1.1.2.7 Radiation

There are three ways in which heat may be transferred: conduction, convection, and radiation. Conduction is, as already covered, the direct transfer of heat through matter. Convection is the indirect transfer of heat by the thermally induced circulation of a liquid or gas; in 'forced convection', the circulation is increased by a fan or pump. Radiation is the direct transfer of heat (or other form of energy) across space. Thermal radiation is electromagnetic radiation and comes within the infrared, visible and ultraviolet regions of the electromagnetic spectrum. The demarcation between these three classes of radiation is rather indefinite but as a guide the wavelength bands are shown in Table 1.2.

Table 1.2 Wavelengths of thermal radiation

Radiation	*Wavelength* (μm)
Infrared	100–0.8
Visible light	0.8–0.4
Ultraviolet	0.4–0.01

So far as the effective transfer of heat is concerned the wavelength band is limited to about 10 μm in the infrared and to 0.1 μm in the ultraviolet. All the radiation in this band behaves in the same way as light. The radiation travels in straight lines, may be reflected or refracted and the amount of radiant energy falling on a unit area of a detector is inversely proportional to the square of the distance between the detector and the radiating source.

1.2 Temperature scales

To measure and compare temperatures it is necessary to have agreed scales of temperature. These temperature scales are defined in terms of physical phenomena which occur at constant temperatures. The temperatures of these phenomena are known as 'fixed points'.

1.2.1 Celsius temperature scale

The Celsius temperature scale is defined by international agreement in terms of two fixed points, the ice point and the steam point. The temperature of the ice point is defined as zero degrees Celsius and the steam point as one hundred degrees Celsius.

The ice point is the temperature at which ice and water exist together at a pressure of 1.0132×10^5 N.m^{-2} (originally one standard atmosphere = 760 mm of mercury). The ice should be prepared from distilled water in the form of fine shavings and mixed with ice-cold distilled water.

The steam point is the temperature of distilled water boiling at a pressure of 1.0132×10^5 N.m^{-2}. The temperature at which water boils is very dependent on pressure. At a pressure p, N.m^{-2} the boiling point of water t_p in degrees Celsius is given by

$$t_p = 100 + 2.795 \times 10^{-4}(p - 1.013 \times 10^{-5}) - 1.334 \times 10^{-9}(p - 1.013 \times 10^5)^2 \quad (1.7)$$

The temperature interval of 100 °C between the ice point and the steam point is called the fundamental interval.

1.2.2 Kelvin, absolute or thermodynamic temperature scale

The earlier scales of temperature depended upon the change with temperature of some property, such as size, of a substance. Such scales depended upon the nature of the substance selected. About the middle of the nineteenth century, Lord Kelvin defined a scale of temperature in terms of the mechanical work which may be obtained from a reversible heat engine working between two temperatures, and which, therefore, does not depend upon the properties of a particular substance. Kelvin divided the interval between the ice and steam points into 100 divisions so that one kelvin represents the same temperature interval as one Celsius degree. (The unit of the Kelvin or thermodynamic temperature scale is the 'kelvin'.) The definition of the kelvin is the fraction 1/273.16 of the thermodynamic temperature of the triple point of water. This definition was adopted by the thirteenth meeting of the General Conference for Weights and Measures in 1967 (13th CGPM, 1967).

It has also been established that an ideal gas obeys the gas law $PV = RT$, where T is the temperature on the absolute or kelvin scale and where P is the pressure of the gas, V is the volume occupied and R is the universal gas constant. Thus, the behaviour of an ideal gas forms a basis of temperature measurement on the absolute scale. Unfortunately the ideal gas does not exist, but the so-called permanent gases, such as hydrogen, nitrogen, oxygen and helium, obey the law very closely, provided the pressure is not too great. For other gases and for the permanent gases at greater pressures, a known correction may be applied to allow for the departure of the behaviour of the gas from that of an ideal gas. By observing the change of pressure of a given mass of gas at constant volume, or the change of volume of the gas at constant pressure, it is possible to measure temperatures on the absolute scale.

The constant-volume gas thermometer is simpler in form, and is easier to use, than the constant-pressure gas thermometer. It is, therefore, the form of gas thermometer which is most frequently used. Nitrogen has been found to be the most suitable gas to use for temperature measurement between 500 and 1500 °C, while at temperatures below 500 °C hydrogen is used. For very low temperatures, helium at low pressure is used.

The relationship between the kelvin and Celsius scales is such that zero degrees Celsius is equal to 273.15 K

$$t = T - 273.15 \tag{1.8}$$

where t represents the temperature in degrees Celsius and T is the temperature kelvin.

It should be noted that temperatures on the Celsius scale are referred to in terms of degrees Celsius, °C, temperatures on the absolute scale are in kelvins, K, no degree sign being used. For instance the steam point is written in Celsius, 100 °C, but on the Kelvin scale 373.15 K.

1.2.3 International Practical Temperature Scale of 1968 (IPTS-68)

The gas thermometer, which is the final standard of reference, is, unfortunately, rather complex and cumbersome, and entirely unsuitable for industrial use. Temperature measuring instruments capable of a very high degree of repeatability are available. Use of these instruments enables temperatures to be reproduced to a very high degree of accuracy, although the actual value of the temperature on the thermodynamic scale is not known with the same degree of accuracy. In order to take advantage of the fact that temperature scales may be reproduced to a much higher degree of accuracy than they can be defined, an International Practical Temperature Scale was adopted in 1929 and revised in 1948. The latest revision of the scale was in 1968 (IPTS-68) and this is the scale used in this book. The 1948 scale is still used in many places in industry. The differences between temperatures on the two scales are small, frequently within the accuracy of commercial instruments. Table 1.3 shows the deviation of the 1948 scale from the 1968 revision.

Table 1.3 Deviation of IPTS-68 from IPTS-48

t_{68} (°C)	$t_{68} - t_{48}$
−200	0.022
−150	−0.013
0	0.000
50	0.010
100	0.000
200	0.043
400	0.076
600	0.150
1000	1.24

The International Practical Temperature Scale is based on a number of defining fixed points each of which has been subject to reliable gas thermometer or radiation thermometer observations and these are linked by interpolation using instruments which have the highest degree of reproducibility. In this way the International Practical Temperature Scale is conveniently and accurately reproducible and provides means for identifying any temperature within much narrower limits than is possible on the thermodynamic scale.

The defining fixed points are established by realizing specified equilibrium states between phases of pure substances. These equilibrium states and the values assigned to them are given in Table 1.4.

Table 1.4 Defining fixed points of the IPTS-68[(1)]

Equilibrium state	*Assigned value of International Practical temperature* T_{68}	t_{68}
Triple point of equilibrium hydrogen	13.81 K	−259.34 °C
Boiling point of equilibrium hydrogen at pressure of 33 330.6 kN . m^{-2}	17.042 K	−256.108 °C
Boiling point of equilibrium hydrogen	20.28 K	−252.87 °C
Boiling point of neon	27.102 K	−246.048 °C
Triple point of oxygen	54.361 K	−218.789 °C
Boiling point of oxygen	90.188 K	−182.962 °C
Triple point of water[(3)]	273.16 K	0.01 °C
Boiling point of water[(2)(3)]	373.15 K	100 °C
Freezing point of zinc	692.73 K	419.58 °C
Freezing point of silver	1235.08 K	961.93 °C
Freezing point of gold	1337.58 K	1064.43 °C

(1) Except for the triple points and one equilibrium hydrogen point (17.042 K) the assigned values of temperature are for equilibrium states at a pressure $p_0 = 1$ standard atmosphere (101.325 kN . m^{-2}).

In the realization of the fixed points small departures from the assigned temperatures will occur as a result of the differing immersion depths of thermometers or the failure to realize the required pressure exactly. If due allowance is made for these small temperature differences, they will not affect the accuracy of realization of the Scale.

(2) The equilibrium state between the solid and liquid phases of tin (freezing point of tin has the assigned value of $t_{68} = 231.9681$ °C and may be used as an alternative to the boiling point of water.

(3) The water used should have the isotopic composition of ocean water.

The scale distinguishes between the International Practical Kelvin Temperature with the symbol T_{68} and the International Practical Celsius Temperature with the symbol t_{68} the relationship between T_{68} and t_{68} is

$$t_{68} = T_{68} - 273.15\ \text{K} \tag{1.9}$$

The size of the degree is the same on both scales, being 1/273.16 of the temperature interval between absolute zero and the triple point of water (0.01 °C). Thus, the interval between the ice point 0 °C and the boiling point of water 100 °C is still 100 Celsius degrees. Temperatures are expressed in kelvins below 273.15 K (0 °C) and degrees Celsius above 0 °C.

Temperatures between and above the fixed points given in Table 1.4 can be interpolated as follows.

From 13.81 K to 630.74 °C the standard instrument is the platinum resistance thermometer. The thermometer resistor must be strain-free, annealed pure platinum having a resistance ratio $W(T_{68})$ defined by

$$W(T_{68})=\frac{R(T_{68})}{R(273.15\ \mathrm{K})} \qquad (1.10)$$

where R is the resistance, which must not be less than 1.392 50 ohms at $T_{68}=373.15$ K, i.e. the resistance ratio

$$\frac{R(100\ ^\circ\mathrm{C})}{R(0\ ^\circ\mathrm{C})}$$

is greater than the ratio 1.3920 of the 1948 scale, i.e. the platinum must be purer.

Below 0 °C the resistance–temperature relationship of the thermometer is found from a reference function and specified deviation equations. From 0 °C to 630.74 °C two polynomial equations provide the resistance temperature relationship. This will be discussed further in the section on resistance thermometers.

From 630.74 °C to 1064.43 °C the standard instrument is the platinum 10 per cent rhodium/platinum thermocouple, the electromotive force–temperature relationship of which is represented by a quadratic equation and is discussed in the appropriate section.

Above 1337.58 K (1064.43 °C) the scale is defined by Planck's law of radiation with 1337.58 K as the reference temperature and the constant c_2 has a value 0.014 388 metre kelvin. This will be discussed in the section on radiation thermometers.

In addition to the defining fixed points the temperatures corresponding to secondary points are given. These points, particularly the melting or freezing points of metals, form convenient workshop calibration points for temperature measuring devices (see Table 1.5).

Table 1.5 Secondary reference points (IPTS-68)

Substance	*Equilibrium state*	*Temperature (K)*
Normal hydrogen	TP	13.956
Normal hydrogen	BP	20.397
Neon	TP	24.555
Nitrogen	TP	63.148
Nitrogen	BP	77.342
Carbon dioxide	Sublimation point	194.674
Mercury	FP	234.288
Water	Ice point	273.15
Phenoxy benzine	TP	300.02
Benzoic acid	TP	395.52
Indium	FP	429.784
Bismuth	FP	544.592
Cadmium	FP	594.258
Lead	FP	600.652
Mercury	BP	629.81
Sulphur	BP	717.824
Copper/aluminium eutectic	FP	821.38
Antimony	FP	903.89
Aluminium	FP	933.52
Copper	FP	1357.6
Nickel	FP	1728
Cobalt	FP	1767
Palladium	FP	1827
Platinum	FP	2045
Rhodium	FP	2236
Iridium	FP	2720
Tungsten	FP	3660

TP: triple point; FP: freezing point; BP: boiling point.

Table 1.6 Comparison of temperature scales

	K	°C	°F	°R
Absolute zero	0	−273.15	−523.67	0
Boiling point O_2	90.19	−182.96	−361.33	162.34
Zero Fahrenheit	255.37	−17.78	0	459.67
Ice point	273.15	0	32	491.67
Steam point	373.15	100	212	671.67
Freezing point of silver	1235.08	961.93	1763.47	2223.14

1.2.4 Fahrenheit and Rankine scales

These two temperature scales are now obsolescent in Britain and the United States, but as a great deal of engineering data, steam tables etc. have been published using the Fahrenheit and Rankine temperature a short note for reference purposes is relevant.

Fahrenheit This scale was proposed in 1714. Its original fixed points were the lowest temperature obtainable with ice and water which was taken as zero. Human blood heat was made 96 degrees (98.4 on the modern scale). On this scale the ice point is at 32 °F and the steam point at 212 °F. There does not appear to be any formal definition of the scale.

To convert from the Fahrenheit to Celsius scale, if t is the temperature in Celsius and f the temperature in Fahrenheit

$$t=\tfrac{5}{9}(f-32) \qquad (1.11)$$

Rankine The Rankine scale is the thermodynamic temperature corresponding to Fahrenheit. Zero in Rankine is of course the same as zero kelvin. On the

Rankine scale the ice point is at 491.67 °R. Zero Fahrenheit is 459.67 °R. To convert temperature from Fahrenheit to Rankine, where R is the Rankine temperature

$$R = f + 459.67 \tag{1.12}$$

Table 1.6 illustrates the relationship between the four temperature scales. It has to be emphasized again that the Fahrenheit and Rankine scales are essentially obsolete.

1.2.5 Realization of temperature measurement

Techniques for temperature measurement are very varied. Almost any temperature-dependent effect may be used for temperature measurement. Sections 1.3–1.6 describe the main techniques for temperature measurement used in industry. However, in laboratories or under special industrial conditions, a wider range of instruments is available. In Table 1.7 is a summary of the more usually used measuring instruments in the range quoted. All measuring instruments require to be calibrated against standards. In the case of temperature the standards are the defining fixed points on the IPTS-68. These fixed points are not particularly easy to achieve in workshop conditions. Although the secondary points are intended as workshop standards it is more usual, in most instrument workshops, to calibrate against high grade instruments whose calibration is traceable to the IPTS-68 fixed points.

1.3 Measurement techniques – direct effects

Instruments for measuring temperature described in this section are classified according to the nature of the change in the measurement probe produced by the change of temperature. They have been divided into four classes: liquid expansion, gas expansion, change of state, and solid expansion.

1.3.1 Liquid-in-glass thermometers

The glass thermometer must be the most familiar of all thermometers. Apart from its industrial and laboratory use it finds application in both domestic and medical fields.

1.3.1.1 Mercury-filled glass thermometer

The coefficient of cubical expansion of mercury is about eight times greater than that of glass. If, therefore, a glass container holding mercury is heated, the mercury will expand more than the container. At a high temperature, the mercury will occupy a greater fraction of the volume of the container than at a low temperature. If, then, the container is made in the form of a bulb with a capillary tube attached, it can be so arranged that the surface of the mercury is in the capillary tube, its position along the tube will change with temperature and the assembly used to indicate temperature. This is the principle of the mercury-in-glass thermometer.

The thermometer, therefore, consists simply of a

Table 1.7 Temperature measurement techniques

Range (K)	*Technique*	*Application*	*Resolution* (K)
0.01–1.5	Magnetic susceptance of paramagnetic salt	Laboratory	0.001
0.1–50	Velocity of sound in acoustic cavity	Laboratory standard	0.0001
0.2–2	Vapour pressure	Laboratory standard	0.001
1.5–100	Germanium resistance thermometer	Laboratory standard	0.0001
1.5–100	Carbon resistance thermometer	Laboratory	0.001
1.5–1400	Gas thermometer	Laboratory	0.002
		Industrial	1.0
210–430	Silicon P–N junction	Laboratory	0.1
		Industrial	—
4–500	Thermistor	Laboratory	0.001
		Industrial	0.1
11–550	Quartz crystal oscillator	Laboratory	0.001
		Industrial	—
15–1000	Platinum resistance thermometer	Standard	0.000 01
		Industrial	0.1
20–2700	Thermocouple	General-purpose	1.0
30–3000	Sound velocity in metal rod	Laboratory	1%
130–950	Liquid-in-glass	General-purpose	0.1
130–700	Bimetal	Industrial	1–2
270–5000	Total radiation thermometer	Industrial	10
270–5000	Spectrally selective radiation thermometer	Industrial	2

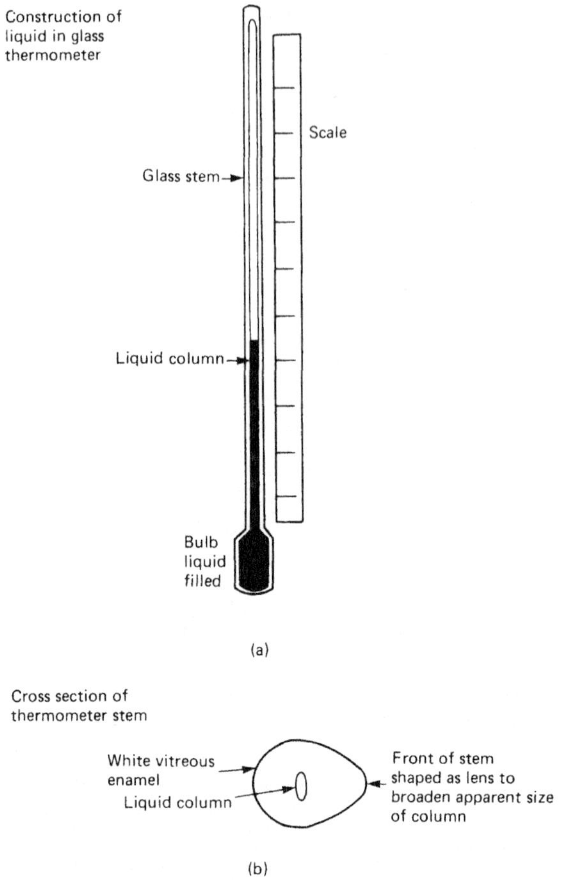

Figure 1.2 Mercury-in-glass thermometer: (a) thermometer and scale, (b) cross-section of thermometer stem.

stem of suitable glass tubing having a very small, but uniform, bore. At the bottom of this stem there is a thin-walled glass bulb. The bulb may be cylindrical or spherical in shape, and has a capacity very many times larger than that of the bore of the stem. The bulb and bore are completely filled with mercury, and the open end of the bore sealed off either at a high temperature, or under vacuum, so that no air is included in the system. The thermometer is then calibrated by comparing it with a standard thermometer in a bath of liquid whose temperature is carefully controlled.

When the standard thermometer and the thermometer to be calibrated have reached equilibrium with the bath at a definite temperature, the point on the glass of the thermometer opposite the top of the mercury meniscus is marked. The process is repeated for several temperatures. The intervals between these marks are then divided off by a dividing machine. In the case of industrial thermometers, the points obtained by calibration are transferred to a brass plate, which is then fixed with the tube into a suitable protecting case to complete the instrument.

The stem of the thermometer is usually shaped in such a way that it acts as a lens, magnifying the width of the mercury column. The mercury is usually viewed against a background of glass which has been enamelled white. Figure 1.2 shows the typical arrangement for a liquid-in-glass thermometer.

Mercury-in-glass thermometers are available in three grades. The limits of error of grades A and B are specified by the National Physical Laboratory, and are given in the tables in the British Standards Code No. 1041 on 'Temperature measurement'. Grade C is a commercial grade of thermometer, and no limits of accuracy are specified. Thermometers of this grade are, of course, cheaper than those of the other grades and their price varies, to a certain degree, according to their accuracy. They should be compared from time to time during use with thermometers of known accuracy. See section on errors due to ageing.

Whenever possible, thermometers should be calibrated, standardized, and used, immersed up to the reading (i.e. totally immersed) as this avoids errors that are due to the fact that the emergent column of mercury and the glass stem are at a lower temperature than that of the bulb, and are therefore not expanded by the same amount. Errors introduced in this way should be allowed for if accurate readings are required, particularly at high temperatures. Some thermometers are, however, calibrated for 'partial immersion', and should be used immersed to the specified depth.

When reading a thermometer an observer should keep his eye on the same level as the top of the mercury column. In this way errors due to parallax will be avoided.

Figure 1.3 shows the effect of observing the thermometer reading from the wrong position. When viewed from (a) the reading is too high. Taken from (b) the reading is correct, but from (c) it is too low.

A mercury-in-glass thermometer has a fairly large thermal capacity (i.e. it requires quite an appreciable amount of heat to change its temperature by one degree), and glass is not a very good conductor of heat. This type of thermometer will, therefore, have a definite thermal lag. In other words, it will require a definite time to reach the temperature of its surroundings. This time should be allowed for before any reading is taken. If there is any doubt as to whether the thermometer has reached equilibrium with a bath of liquid having a constant temperature, then readings should be taken at short intervals of time. When the reading remains constant the thermometer must be in equilibrium with the bath. If the temperature is varying rapidly the thermometer may never indicate the temperature accurately, particularly if the tested medium is a gas.

Glass thermometers used in industry are usually

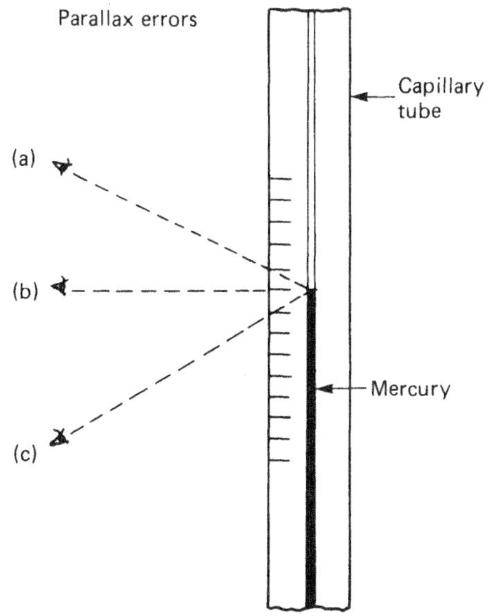

Figure 1.3 Parallax errors when reading glass thermometer.

protected by metal sheaths. These sheaths may conduct heat to or from the neighbourhood of the thermometer bulb and cause the thermometer to read either high or low according to the actual conditions prevailing. A thermometer should, therefore, be calibrated, whenever possible, under the conditions in which it will be used, if accurate temperature readings are required. If, however, the main requirement is that the temperature indication be consistent for the same plant temperature, then an error introduced is not so important, so long as the conditions remain the same, and the error is constant.

Errors due to ageing It is often assumed that provided a mercury-in-glass thermometer is in good condition it will always give an accurate reading. This is not always so, particularly with cheap thermometers. A large error may be introduced by changes in the size of the bulb due to ageing. When glass is heated to a high temperature, as it is when a thermometer is made, it does not, on cooling, contract to its original volume immediately. Thus, for a long time after it has been made the bulb continues to contract very slowly so that the original zero mark is too low on the stem, and the thermometer reads high. This error continues to increase over a long period, and depends upon the type of glass used in the manufacture of the thermometer. In order to reduce to a minimum the error due to this cause, during manufacture thermometers are annealed by baking for several days at a temperature above that which they will be required to measure, and then cooled slowly over a period of several days.

Another error due to the same cause is the depression of the zero when a thermometer is cooled rapidly from a high temperature. When cooled, the glass of the thermometer bulb does not contract immediately to its original size so that the reading on the thermometer at low temperature is too low, but returns to normal after a period of time. This period depends upon the nature of the glass from which the bulb is made.

High temperature thermometers Mercury normally boils at 357 °C at atmospheric pressure. In order to extend the range of a mercury-in-glass thermometer beyond this temperature, the top end of the thermometer bore is enlarged into a bulb having a capacity of about 20 times that of the bore of the stem. This bulb, together with the bore above the mercury, is then filled with nitrogen or carbon dioxide at a sufficiently high pressure to prevent the mercury boiling at the highest temperature at which the thermometer will be used. In order to extend the range to 550 °C, a pressure of about 20 bar is required. In spite of the existence of this gas at high pressure, there is a tendency for mercury to vaporize from the top of the column at high temperatures and to condense on the cooler portions of the stem in the form of minute globules which will not join up again with the main bulk of the mercury. It is, therefore, inadvisable to expose a thermometer to high temperatures for prolonged periods.

At high temperatures the correction for the temperature of the emergent stem becomes particularly important, and the thermometer should, if possible, be immersed to the top of the mercury column. Where this is not possible, the thermometer should be immersed as far as conditions permit, and a correction made to the observed reading, for the emergent column. To do this, the average temperature of the emergent column should be found by means of a short thermometer placed in several positions near to the stem. The emergent column correction may then be found from the formula:

$$\text{correction} = 0.0016(t_1 - t_2)n \text{ on Celsius scale}$$

where t_1 is the temperature of the thermometer bulb, t_2 is the average temperature of the emergent column, and n is the number of degrees exposed. The numerical constant is the coefficient of apparent expansion of mercury in glass.

1.3.1.2 Use of liquids other than mercury

In certain industrial uses, particularly in industries where the escape of mercury from a broken bulb might cause considerable damage to the products, other liquids are used to fill the thermometer. These liquids are also used where the temperature range of the mercury-in-glass thermometer is not suitable. Table 1.8 lists them together with their range of usefulness.

Table 1.8 Liquids used in glass thermometers (expansion type)

Liquid	*Temperature range* (°C)
Mercury	−35 to +510
Alcohol	−80 to +70
Toluene	−80 to +100
Pentane	−200 to +30
Creosote	−5 to +200

1.3.1.3 Mercury-in-glass electric contact thermometer

A mercury-in-glass thermometer can form the basis of a simple on/off temperature controller which will control the temperature of an enclosure at any value between 40 °C and 350 °C.

Mercury is a good electrical conductor. By introducing into the bore of a thermometer two platinum contact wires, one fixed at the lower end of the scale and the other either fixed or adjustable from the top of the stem, it is possible to arrange for an electrical circuit to be completed when a predetermined temperature is reached. The current through the circuit is limited to about 25 mA. This current is used to operate an electronic control circuit. Contact thermometers find applications in laboratories for the temperature control of water baths, fluidized beds and incubators. With careful design, temperature control to 0.1 K can be attained.

Formerly, fixed temperature contact thermometers were used for the temperature control of quartz crystal oscillator ovens, but now this duty is more usually performed by thermistors or semiconductor sensors which can achieve better temperature control by an order of magnitude.

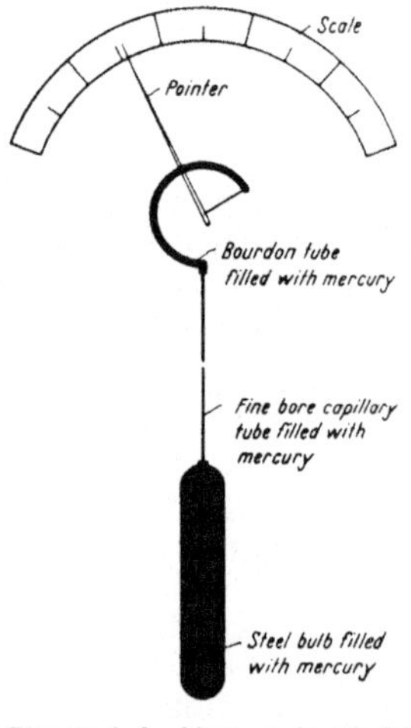

Figure 1.4 Mercury-in-steel thermometer.

1.3.2 Liquid-filled dial thermometers

1.3.2.1 Mercury-in-steel thermometer

Two distinct disadvantages restrict the usefulness of liquid-in-glass thermometers in industry – glass is very fragile, and the position of the thermometer for accurate temperature measurement is not always the best position for reading the scale of the thermometer.

These difficulties are overcome in the mercury-in-steel thermometer shown in Figure 1.4. This type of thermometer works on exactly the same principle as the liquid-in-glass thermometer. The glass bulb is, however, replaced by a steel bulb and the glass capillary tube by one of stainless steel. As the liquid in the system is now no longer visible, a Bourdon tube is used to measure the change in its volume. The Bourdon tube, the bulb and the capillary tube are completely filled with mercury, usually at a high pressure. When suitably designed, the capillary tube may be of considerable length so that the indicator operated by the Bourdon tube may be some distance away from the bulb. In this case the instrument is described as being a 'distant reading' or 'transmitting' type.

When the temperature rises, the mercury in the bulb expands more than the bulb so that some mercury is driven through the capillary tube into the Bourdon tube. As the temperature continues to rise, increasing amounts of mercury will be driven into the Bourdon tube, causing it to uncurl. One end of the Bourdon tube is fixed, while the motion of the other end is communicated to the pointer or pen arm. As there is a large force available the Bourdon tube may be made robust and will give good pointer control and reliable readings.

The Bourdon tube may have a variety of forms, and the method of transmitting the motion to the pointer also varies. Figure 1.5 shows one form of Bourdon tube in which the motion of the free end is transmitted to the pointer by means of a segment and pinion. The free end

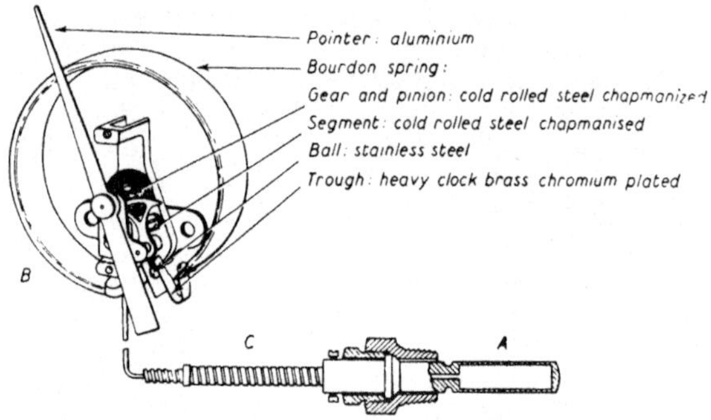

Figure 1.5 Construction of mercury-in-steel thermometer (courtesy The Foxboro Company).

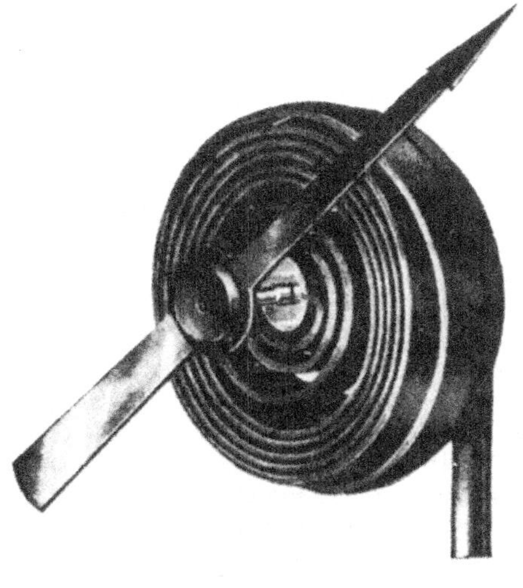

Figure 1.6 Multi-turn Bourdon tube.

of the tube forms a trough in which a stainless steel ball at the end of the segment is free to move. The ball is held against the side of the trough by the tension in the hair-spring. By using this form of construction lost motion and angularity error are avoided, and friction reduced to a minimum. Ambient-temperature compensation may be obtained by using a bimetallic strip, or by using twin Bourdon tubes in the manner described under the heading of capillary compensation.

Figure 1.6 shows a Bourdon tube having a different form, and a different method of transmitting the motion to the pen arm. This Bourdon tube is made of steel tube having an almost flat section. A continuous strip of the tubing is wound into two coils of several turns. The coils are arranged one behind the other so that the free end of each is at the centre while the outer turn of the coils is common to both, as can be seen in the illustration. One end of the continuous tube – the inner end of the back coil – is fixed and leads to the capillary tube; while the other end – the inner end of the front coil – is closed, and is attached to the pointer through a small bimetallic coil which forms a continuation of the Bourdon tube. This bimetallic coil compensates for changes brought about in the elastic properties of the Bourdon tube and in the volume of the mercury within the Bourdon tube due to ambient temperature changes.

This particular formation of the tube causes the pointer to rotate truly about its axis without the help of bearings, but bearings are provided to keep the pointer steady in the presence of vibration. The friction at the bearings will, therefore, be very small as there is little load on them. As the end of the Bourdon tube rotates the pointer directly, there will be no backlash.

Thermometer bulbs The thermometer bulb may have a large variety of forms depending upon the use to which it is put. If the average temperature of a large enclosure is required, the bulb may take the form of a considerable length of tube of small diameter either arranged as a U or wound into a helix. This form of bulb is very useful when the temperature of a gas is being measured, for it presents a large surface area to the gas and is therefore more responsive than the forms having a smaller surface area for the same cubic capacity.

In the more usual form, the bulb is cylindrical in shape and has a robust wall: the size of the cylinder depends upon many factors, such as the filling medium and the temperature range of the instrument, but in all cases, the ratio of surface area to volume is kept at a maximum to reduce the time lag in the response of the thermometer.

The flange for attaching the bulb to the vessel in which it is placed also has a variety of forms depending upon whether the junction has to be gas-tight or not, and upon many other factors. Figure 1.7 shows some forms of bulbs.

The capillary tube and its compensation for ambient temperature The capillary tube used in the mercury-in-steel thermometer is usually made from stainless steel, as mercury will amalgamage with other metals. Changes of temperature affect the capillary and the mercury it contains, and hence the thermometer reading; but if the capillary has a very small capacity, the error owing to changes in the ambient temperature will be negligible.

Where a capillary tube of an appreciable length is used, it is necessary to compensate for the effects brought about by changes in the temperature in the

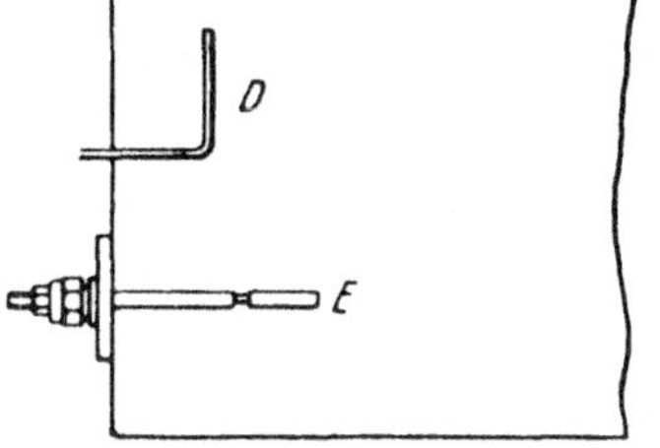

Figure 1.7 Forms for bulbs for mercury-in-steel thermometers.

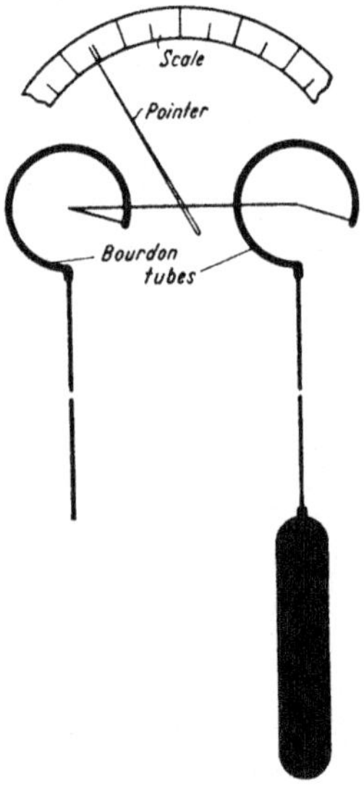

Figure 1.8 Ambient temperature compensation of mercury-in-steel thermometer.

neighbourhood of the tube. This may be done in a number of ways. Figure 1.8 illustrates a method which compensates not only for the changes of temperature of the capillary tube, but also for the changes of temperature within the instrument case. In order to achieve complete temperature compensation two thermal systems are used, which are identical in every respect except that one has a bulb and the other has not. The capillary tubes run alongside each other, and the Bourdon tubes are in close proximity within the same case. If the pointer is arranged to indicate the difference in movement between the free ends of the two Bourdon tubes, then it will be indicating an effect which is due to the temperature change in the bulb only. If compensation for case temperature only is required, then the capillary tube is omitted in the compensating system, but in this case the length of capillary tube used in the uncompensated system should not exceed about 8 metres.

Another method of compensating for temperature changes in the capillary tube is to use a tube of comparatively large bore and to insert into the bore a wire made of Invar, or other alloy with a very low coefficient of expansion. Mercury has a coefficient of cubical expansion about six times greater than that of stainless steel. If the expansion of the Invar wire may be regarded as being negligibly small, and the wire is arranged to fill five-sixths of the volume of the capillary bore, then the increase in the volume of the mercury which fills the remaining one-sixth of the bore will exactly compensate for the increase in volume of the containing capillary tube. This method requires the dimensions both of the bore of the capillary tube and of the diameter of the wire insert to be accurate to within very narrow limits for accurate compensation. The insert may not necessarily be continuous, but may take the form of short rods, in which case it is, however, difficult to eliminate all trapped gases.

Compensation for changes in the temperature of the capillary tube may also be achieved by introducing compensating chambers, of the form shown in Figure 1.9, at intervals along the length of the capillary tube. These chambers operate on exactly the same principle as the Invar-wire-insert type of capillary tube, but the proportion of the chamber occupied by the Invar is now arranged to compensate for the relative increase in volume of the mercury within the chamber and in the intervening length of capillary tube.

1.3.2.2 Other filling liquids

Admirable though mercury may be for thermometers, in certain circumstances it has its limitations, particularly at the lower end of the temperature scale. It is also very expensive to weld mercury systems in stainless steel. For these and other reasons, other liquids are used in place of mercury. Details of the liquids used in liquid-in-metal thermometers, with their usual temperature ranges, are given in Table 1.9. Comparison with Table 1.8 shows that liquids are used for different temperature ranges in glass and metal

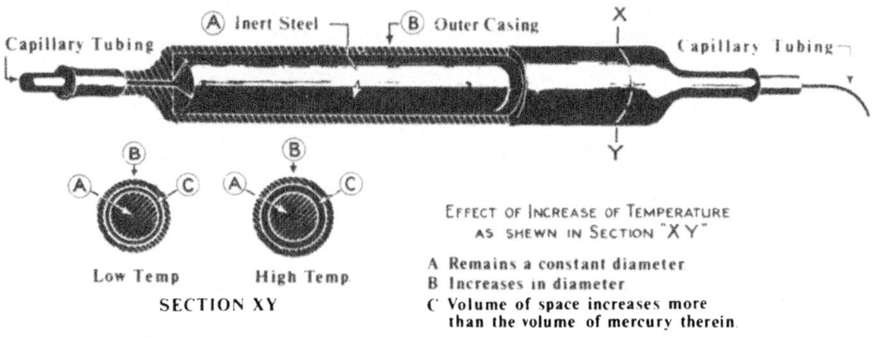

Figure 1.9 Ambient temperature compensation chamber.

Table 1.9 Liquids used in metal thermometers (expansion type)

Liquid	*Temperature range* (°C)
Mercury	−39 to +650
Xylene	−40 to +400
Alcohol	−46 to +150
Ether	+20 to +90
Other organic liquids	−87 to +260

thermometers. In general, in metal thermometers, liquids can be used up to higher temperatures than in glass thermometers as they can be filled to higher pressures.

When liquids other than mercury are used, the bulb and capillary tube need no longer be made of steel. The material of the bulb may, therefore, be chosen from a wide range of metals and alloys, and is selected to give the maximum resistance to any corrosive action which may be present where the bulb is to be used.

The capillary tube, too, may be made from a variety of materials, although copper and bronze are the most common. When capillary tubes are made from materials other than stainless steel, it may be necessary to protect them from corrosion or mechanical damage. This may be done by covering the tube with thermal insulation material – formerly asbestos was used – and winding the whole in a heavy spiral of bronze. In cases where a bronze outer casing is likely to be damaged either by acid fumes or mechanically, it may be replaced by a stainless steel spiral which results in a much stronger but slightly less flexible construction. For use in damp places, or where the tube is liable to be attacked by acid fumes, the capillary and bronze spiral may be protected by a covering of moulded rubber, polyvinyl chloride, or rubber-covered woven-fabric hose. For use on chemical plants such as sulphuric acid plants both the capillary tube and the bulb are protected by a covering of lead.

The construction of the liquid-in-metal thermometer is the same as that of the mercury-in-steel thermometer, and compensation for changes in ambient temperature may be achieved in the same ways.

Further facts about liquid-in-metal thermometers will be found in Table 1.11 on p. 17 showing the comparison of the various forms of non-electrical dial thermometers.

In installations where liquid-filled instruments with very long capillaries are used, care must be taken to see that there is not a significant height difference between the bulb location and that of the instrument. If there is a large height difference, the pressure due to the column of liquid in the capillary will be added to (or subtracted from) the pressure due to the expansion of the liquid in the bulb resulting in a standing error in the temperature reading. This problem is at its worst with mercury-filled instruments. Instruments with double capillary ambient temperature compensation, Figure 1.7, are, of course, also compensated for static head errors.

1.3.3 Gas-filled instruments

The volume occupied by a given mass of gas at a fixed pressure is a function of both the molecular weight of the gas and its temperature. In the case of the 'permanent gases' provided the temperature is significantly above zero kelvin the behaviour of a gas is represented by the equation

$$pv = RT \tag{1.13}$$

where p is pressure in $N.m^{-2}$, v is volume in m^3, T is the temperature in K and R is the gas constant with a value of $8.314\ J.mol^{-1}.K^{-1}$.

If, therefore, a certain volume of inert gas is enclosed in a bulb, capillary and Bourdon tube, and most of the gas is in the bulb, then the pressure as indicated by the Bourdon tube, may be calibrated in terms of the temperature of the bulb. This is the principle of the gas-filled thermometer.

Since the pressure of a gas maintained at constant volume increases by 1/273 of its pressure at 0 °C for every degree rise in temperature, the scale will be linear provided the increase in volume of the Bourdon tube, as it uncurls, can be neglected in comparison with the total volume of gas.

An advantage of the gas-filled thermometer is that the gas in the bulb has a lower thermal capacity than a similar quantity of liquid, so that the response of the thermometer to temperature changes will be more rapid than that for a liquid-filled system with a bulb of the same size and shape.

The coefficient of cubical expansion of a gas is many times larger than that of a liquid or solid (air, 0.0037; mercury, 0.00018; stainless steel, 0.00003). It would therefore appear at first sight, that the bulb for a gas-filled system would be smaller than that for a liquid-filled system. The bulb must, however, have a cubical capacity many times larger than that of the capillary tube and Bourdon tube, if the effects of ambient temperature changes upon the system are to be negligible.

It is extremely difficult to get accurate ambient temperature compensation in any other way. The change in dimensions of the capillary tube due to a temperature change is negligible in comparison with the expansion of the gas. Introducing an Invar wire into the capillary bore would not be a solution to the problem, because the wire would occupy such a large proportion of the bore that extremely small variations in the dimensions of the bore or wire would be serious.

Placing an exactly similar capillary and Bourdon

tube alongside that of the measuring system, and measuring the difference in the change of the two Bourdons does not give accurate compensation. This can be seen from the following example. Suppose the capillary and Bourdon tube have a capacity equal to 1/100 part of the capacity of the whole system. Let the ambient temperature rise by 10 °C, while the bulb remains at the same temperature. In the compensating system the pressure will increase by 10/273 of the pressure at 0 °C. In the measuring system this temporary increase in pressure in the capillary tube will soon be reduced by gas flowing into the bulb from the capillary tube until the pressures in the bulb and capillary are the same. Thus, the increase in pressure in the measuring system will only be about one-hundredth of the pressure increase in the compensating system.

More accurate compensation would be obtained by having a compensating system which also included a bulb maintained at ambient temperature, but this again would not give the completely accurate compensation.

Further facts about gas expansion thermometers will be found in Table 1.11, in which certain forms of dial thermometers are compared.

1.3.4 Vapour pressure thermometers

Suppose a container is partially filled with liquid and the space above the liquid is completely evacuated. The molecules of liquid will be in motion, and will be moving in an entirely random manner. From time to time, molecules having a vertical component of velocity will reach the surface of the liquid. If this vertical component of velocity is great enough, a molecule will be able to leave the liquid in spite of the fact that other molecules are attracting it back into the liquid. Thus, after a time, a number of liquid molecules will occupy the space above the liquid. These molecules, too, will be in a state of random motion, and from time to time molecules will leave the vapour and pass back into the liquid.

At first, the rate at which molecules are returning to the liquid will be less than the rate at which they leave, and the vapour above the liquid is said to be 'unsaturated'. Eventually, however, as the number of molecules in the vapour state increases, the rate at which the molecules leave the liquid will be exactly equal to the rate at which they return, and the quantity – and, therefore, the pressure – of the vapour in the space will remain constant. When this is so, the vapour is said to be 'saturated'.

If the temperature of the container is raised, the velocity of the molecules of liquid will be increased. The number of molecules now having sufficient energy to leave the liquid will be increased, and a greater number will leave the liquid. The rate at which molecules leave the liquid will now be greater than the rate at which they return so that the vapour pressure in the space will increase. It will continue to increase until the saturated vapour pressure for the new temperature is reached, when the molecules of vapour will again be returning to the liquid at the same rate as that at which liquid molecules leave.

If, instead of having a fixed top, the container has a movable piston, then, if the volume of the space is increased without the temperature changing, the vapour pressure will temporarily fall as the same number of vapour molecules are occupying a larger space. Now, however, the rate at which the molecules return to the liquid will be reduced, and will not be as great as the rate at which they leave the liquid. The pressure of the vapour will therefore increase, until the rate at which molecules return to the liquid again balances the rate at which they leave. The vapour pressure returns, therefore, to the 'saturated' vapour pressure for the particular temperature.

Table 1.10 Liquids used in vapour pressure thermometers

Liquid	*Critical temperature* (°C)	*Boiling point* (°C)	*Typical ranges available* (°C)
Argon	−122	−185.7	Used for measuring very low temperatures down to −253 °C in connection with the liquefaction of gases
Methyl chloride	143	−23.7	0 to 50
Sulphur dioxide	157	−10	30 to 120
Butane (n)	154	−0.6	20 to 80
Methyl bromide		4.6	30 to 85
Ethyl chloride	187	12.2	30 to 100
Diethyl ether	194	34.5	60 to 160
Ethyl alcohol	243	78.5	30 to 180
Water	375	100	120 to 220
Toluene	321	110.5	150 to 250

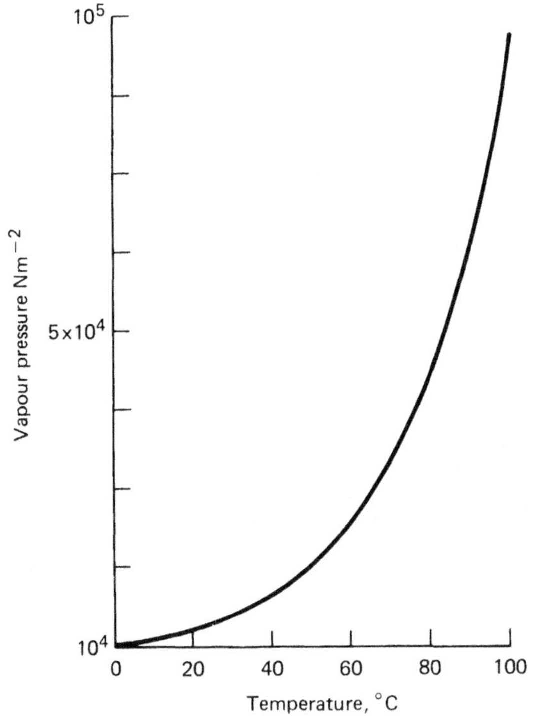

Figure 1.10 Saturated vapour pressure of water.

In the same way, if the volume of the space is reduced, molecules will leave the vapour at a greater rate than that at which they leave the liquid, so that the vapour pressure, which was temporarily increased, will fall until it is again the saturated vapour pressure of the liquid at the particular temperature.

Thus, provided there is always liquid and vapour present, the saturated vapour pressure of the liquid depends only upon its temperature, and is independent of the size of the container.

If a thermometer system similar to that described for gas expansion thermometers is arranged so that the system contains both liquid and vapour and the interface between liquid and vapour is in the bulb, that is, at the temperature whose value is required, then the vapour pressure as measured by the Bourdon tube will give an indication of the temperature. This indication will be completely independent of the volume of the bulb, the capillary and the Bourdon tube and therefore independent of expansion due to ambient temperature changes.

The saturated vapour pressure of a liquid is not linear with temperature. Figure 1.10 shows the temperature—vapour pressure relationship for a typical liquid. The form of the vapour pressure graphs for other volatile liquids is of a similar form. It will be seen that pressure versus temperature is non-linear. A thermometer based on vapour pressure will have a scale on which the size of the divisions increases with increasing temperature.

The realization of a vapour instrument is essentially the same as a gas-filled instrument except that in the latter the whole instrument is filled with a permanent gas while in the former the bulb is filled partly with liquid and partly with gas. This arrangement is shown diagrammatically in Figure 1.11(a).

Many liquids are used for vapour-pressure-actuated thermometers. The liquid is chosen so as to give the required temperature range, and so that the usual operating temperature comes within the widely spaced graduations of the instrument. In some forms of the instrument, a system of levers is arranged to give a linear portion to the scale over a limited portion of its range. By suitable choice of filling liquid, a wide variety of ranges is available, but the range for any particular filling liquid is limited. The choice of material for bulb construction is also very wide. Metals – such as copper, steel, Monel metal, tantalum – may be used. Table 1.10 shows a number of liquids commonly used for vapour-pressure thermometers together with their useful operating ranges.

In the instrument, shown diagrammatically in Figure 1.11(a) a quantity of liquid partially fills the bulb. The surface of the liquid in the bulb should be at the temperature which is being measured. The method by which the vapour pressure developed in the bulb is transmitted to the Bourdon tube will depend upon whether the temperature of the capillary tube and Bourdon tube is above or below that of the bulb.

If the ambient temperature of the capillary and Bourdon tube is above that of the bulb, then they will be full of vapour, which will transmit the vapour pressure, as shown in Figure 1.11(a). When the ambient temperature increases, it will cause the vapour in the capillary and Bourdon tube to increase in pressure temporarily, but this will cause vapour in the bulb to condense until the pressure is restored to the saturated vapour pressure of liquid at the temperature of the bulb.

Vapour pressure instruments are not usually satisfactory when the temperature being measured at the bulb is near the ambient temperature of the capillary and the Bourdon tube. In particular, significant measurement delays occur as the measured temperature crosses the ambient temperature. These delays are caused by the liquid distilling into or out of the gauge and capillary, Figure 1.11(b).

If there is a significant level difference between the bulb and the gauge, an error will be produced when liquid distils into the capillary due to the pressure head from the column of liquid.

When rapid temperature changes of the bulb occur passing through ambient temperature the movement of the instrument pointer may be quite erratic due to

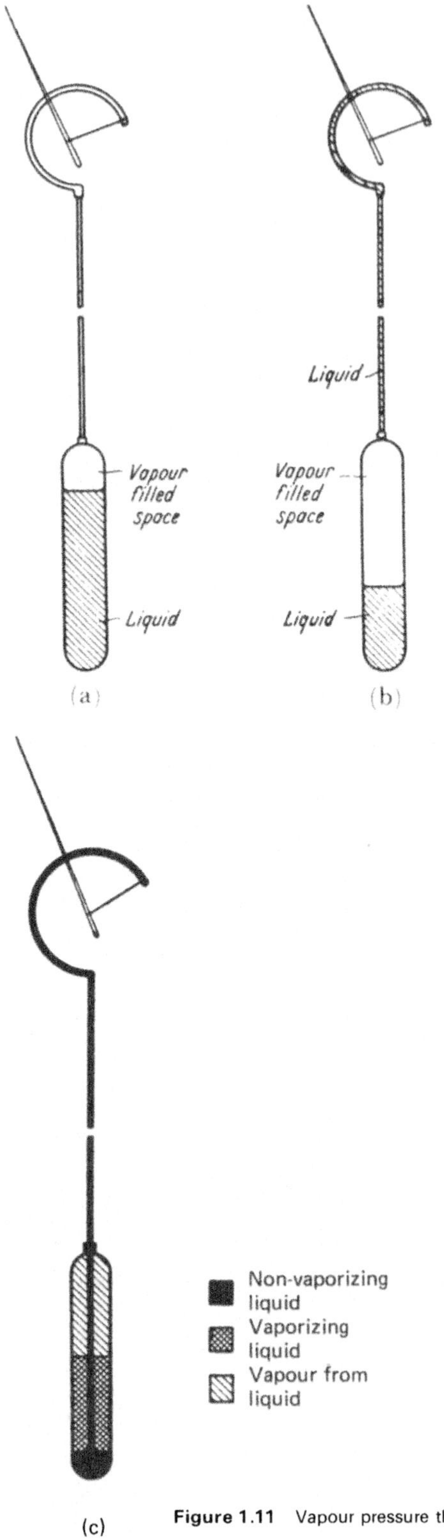

Figure 1.11 Vapour pressure thermometer.

the formation of bubbles in the capillary.

In order to overcome the defects brought about by distillation of the liquid into, and out of, the capillary and Bourdon tubes, these tubes may be completely filled with a non-vapourizing liquid which serves to transmit the pressure of the saturated vapour from the bulb to the measuring system. To prevent the non-vaporizing liquid from draining out of the capillary tube, it is extended well down into the bulb, as shown in Figure 1.11(c), and the bulb contains a small quantity of the non-vaporizing fluid. The non-vaporizing fluid will still tend to leave the capillary tube unless the bulb is kept upright.

Vapour pressure thermometers are very widely used because they are less expensive than liquid- and gas-filled instruments. They also have an advantage in that the bulb can be smaller than for the other types.

The range of an instrument using a particular liquid is limited by the fact that the maximum temperature for which it can be used must be well below the critical temperature for that liquid. The range is further limited by the non-linear nature of the scale.

In Table 1.11 the three types of fluid-filled thermometers are compared.

1.3.5 Solid expansion

Thermal expansion of solids, usually metals, forms the basis of a wide range of inexpensive indicating and control devices. These devices are not particularly accurate, typically errors of as much as ± 5 K or more may be expected, but due to their low cost they find wide application especially in consumer equipment. As indicated earlier in this section this technique is also used to provide temperature compensation in many instruments.

The temperature sensitive elements using solid expansion fall into two groups: rod sensing probes and bimetal strips.

There are so many applications that only one or two examples will be given to illustrate the techniques.

1.3.5.1 Rod sensing probes

The widest application of this technique is for immersion thermostats for use in hot water temperature control. Figure 1.12 shows diagrammatically the operation of an immersion thermostat. The micro-switch is operated by the thermal expansion of the brass tube. The reference length is provided by a rod of low thermal expansion such as Invar. These thermostats, though not particularly accurate and having a switching differential of several kelvin, provide a very rugged and reliable control system for a non-critical application such as domestic hot water

Table 1.11 Comparison of three types of dial thermometers

	Liquid-in-metal	*Gas expansion (constant volume)*	*Vapour pressure*
Scale	Evenly divided.	Evenly divided.	Not evenly divided. Divisions increase in size as the temperature increases. Filling liquid chosen to give reasonably uniform scale in the neighbourhood of the operating temperatures.
Range	Wide range is possible with a single filling liquid, particularly with mercury. By choice of suitable filling liquid, temperatures may be measured between −200 °C and 570 °C, but not with a single instrument.	Usually has a range of at least 50 °C between −130 °C and 540 °C. Can be used for a lower temperature than mercury in steel.	Limited for a particular filling liquid, but with the choice of a suitable liquid almost any temperature between −50 °C and 320 °C may be measured. Instrument is not usually suitable for measuring temperatures near ambient temperatures owing to the lag introduced when bulb temperature crosses ambient temperature.
Power available to operate the indicator.	Ample power is available so that the Bourdon tube may be made robust and arranged to give good pointer control.	Power available is very much less than that from liquid expansion.	Power available is very much less than that from liquid expansion.
Effect of difference in level of bulb and Bourdon tube.	When the system is filled with a liquid at high pressure, errors due to difference of level between bulb and indicator will be small. If the difference in level is very large, a correction may be made.	No head error, as the pressure due to difference in level is negligible in comparison with the total pressure in the system.	Head error is not negligible, as the pressure in the system is not large. Error may be corrected over a limited range of temperature if the ratio pressure to deflection of the pointer can be considered constant over that range. In this case the error is corrected by resetting the pointer.
Effect of changes in barometric pressure.	Negligible.	May produce a large error. Error due to using the instrument at a different altitude from that at which it was calibrated may be corrected by adjusting the zero. Day to day variations in barometric pressure may be corrected for in the same way.	Error may be large, but may be corrected by resetting the pointer as for head error. Day to day errors due to variation in barometric pressure may be corrected by zero adjustment.
Capillary error.	Compensation for change in ambient temperature obtained as described in text (page 12).	Difficult to eliminate (see page 13).	No capillary error.
Changes in temperature at the indicator.	Compensation obtained by means of a bimetallic strip.	Compensation obtained by means of bimetallic strip.	Errors due to changes in the elasticity of the Bourdon tube are compensated for by means of a bimetallic strip.
Accuracy.	$\pm\frac{1}{2}\%$ of range to 320 °C $\pm 1\%$ of range above 320 °C.	$\pm 1\%$ of differential range of the instrument if the temperature of the capillary and Bourdon tube does not vary too much.	$\pm 1\%$ of differential range even with wide temperature variation of the capillary and Bourdon tube.

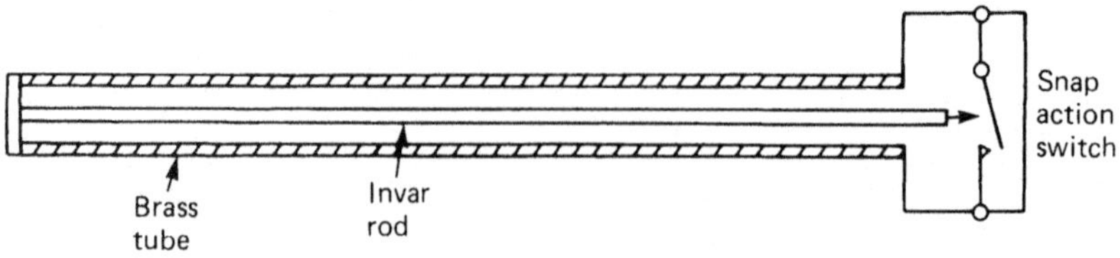

Figure 1.12 Rod thermostat.

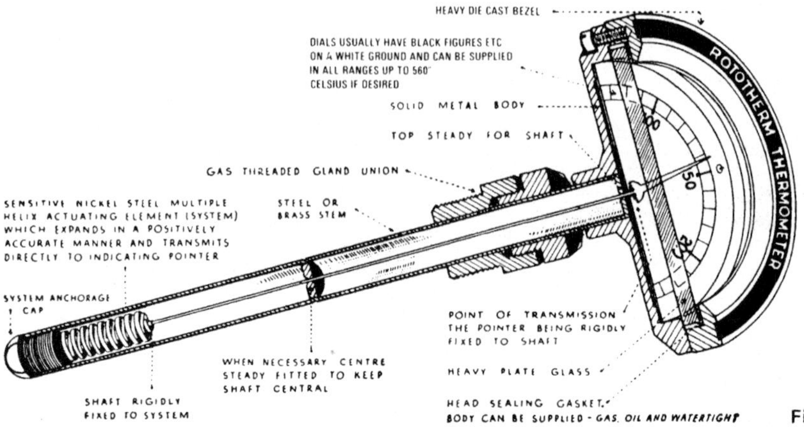

Figure 1.13 Dial thermometer.

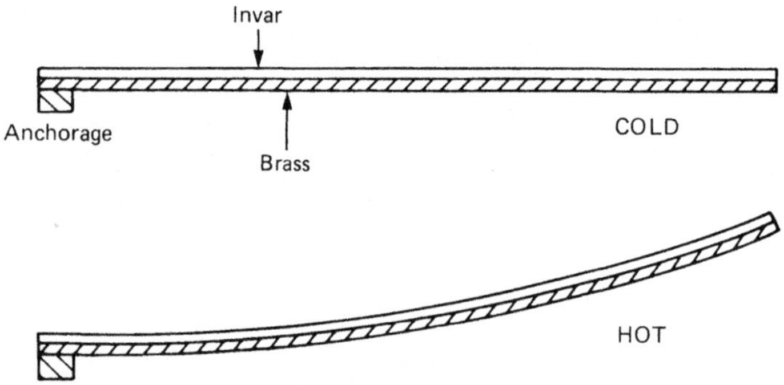

Figure 1.14 Action of bimetal strip.

control. Life spans of fifty years continuous operation are fairly typical.

Figure 1.13 shows another rod application. In this case to achieve greater sensitivity the expanding component is coiled.

1.3.5.2 Bimetal strip thermometer

Bimetal strips are fabricated from two strips of different metals with different coefficients of thermal expansion bonded together to form, in the simplest case, a cantilever. Typical metals are brass and Invar. Figure 1.14 illustrates this principle. As the temperature rises the brass side of the strip expands more than the Invar side, resulting in the strip curling, in this case upwards.

In this 'straight' form a bimetal strip can form part of a micro-switch mechanism thus forming a temperature-sensitive switch or thermostat.

To construct a thermometer the bimetal element is coiled into a spiral or helix. Figure 1.15 shows a typical coiled thermometer element.

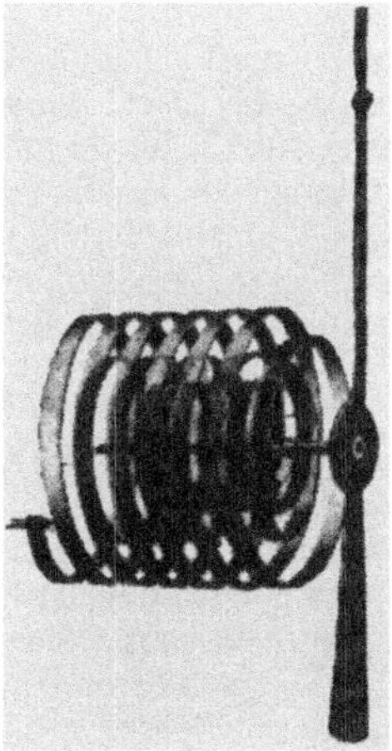

Figure 1.15 Helical bimetal strip.

A long bimetal strip, consisting of an Invar strip welded to a higher expansion nickel–molybdenum alloy wound around without a break into several compensated helices, arranged coaxially one within the other forms the temperature-sensitive element of an instrument which may be designed to measure temperature. This method of winding the strip enables a length, sufficient to produce an appreciable movement of the free end, to be concentrated within a small space. It also makes it possible to keep the thermal capacity of the element and its stem at a low value, so the instrument will respond rapidly to small temperature changes.

The helices in the winding are so compensated that any tendency towards lateral displacement of the spindle in one helix is counteracted by an opposite tendency on the part of one or more of the other helices. Thus, the spindle of the instrument is fully floating, retaining its position at the centre of the scale without the help of bearings. The instrument is, therefore, not injured by mechanical shocks which would damage jewelled bearings.

This particular design also results in the angular rotation of the spindle being proportional to the change in temperature for a considerable temperature range. The instrument has a linear temperature scale, and can be made to register temperatures up to 300 °C to within ±1 per cent of the scale range.

Due to its robust construction, this instrument is used on many industrial plants, and a slightly modified form is used in many homes and offices to indicate room temperature. It can be made for a large variety of temperature ranges and is used in many places where the more fragile mercury-in-glass thermometer was formerly used.

1.4 Measurement techniques – electrical

1.4.1 Resistance thermometers

All metals are electrical conductors which at all but very low temperatures offer resistance to the passage of electric current. The electrical resistance exhibited by a conductor is measured in ohms. The proportional relationship of electrical current and potential difference is given by Ohm's law:

$$R = E/I \qquad (1.14)$$

where R is resistance in ohms, E is potential difference in volts, and I is current in amperes. Different metals show widely different resistivities. The resistance of a conductor is proportional to its length and inversely proportional to its cross-sectional area, i.e.

$$R = \rho \frac{L}{A} \qquad (1.15)$$

or

$$\rho = R \frac{A}{L} \qquad (1.16)$$

where R is resistance of the conductor, ρ is resistivity of the material, L is length of the conductor, and A is cross-sectional area of the conductor. The units of resistivity are ohms . metre.

The resistivity of a conductor is temperature-dependent. The temperature coefficient of resistivity is positive for metals, that is, the resistance increases with temperature, and for semiconductors the temperature coefficient is negative. As a general guide at normal ambient temperatures the coefficient of resistivity of most elemental metals lies in the region of 0.35 per cent to 0.7 per cent per kelvin.

Table 1.12 Resistivities of different metals

Metal	*Resistivity at 293 K microhms . metre*	*Temperature coefficient of resistivity* (K^{-1})
Aluminium	282.4	0.0039
Brass (yellow)	700	0.002
Constantan	4900	10^{-5}
Copper (annealed)	172.4	0.003 93
Gold	244	0.0034
Iron (99.98%)	1000	0.005
Mercury	9578	0.000 87
Nichrome	10 000	0.0004
Nickel	780	0.0066
Platinum (99.85%)	11 060	0.003 927
Silver	159	0.0038
Tungsten	560	0.0045

1 Resistivities of metals are dependent on the purity or exact composition of alloys. Some of the above figures represent average values.
2 Temperature coefficients of resistivity vary slightly with temperature. The above values are for 20 °C.

Table 1.12 shows the resistivity and temperature coefficients for a number of common metals: both elements and alloys.

The metals most used for resistance measurement are platinum, nickel and copper. These metals have the advantage that they can be manufactured to a high degree of purity and consequently they can be made with very high reproducibility of resistance characteristics. Copper has the disadvantage of a low resistivity resulting in inconveniently large sensing elements and has the further disadvantage of poor resistance to corrosion resulting in instability of electrical characteristics. The main area of application of copper for resistance thermometers is in electronic instrumentation where it is in a controlled environment and where an essentially linear temperature characteristic is required.

1.4.1.1 Platinum resistance thermometers

Platinum is the standard material used in the resistance thermometer which defines the International Practical Temperature Scale, not because it has a particularly high coefficient of resistivity, but because of its stability in use. In fact, a high coefficient is not, in general, necessary for a resistance thermometer material as resistance values can be determined with a high degree of accuracy using suitable equipment and taking adequate precautions.

Platinum, having the highest possible coefficient of resistivity, is considered the best material for the construction of thermometers. A high value of this coefficient is an indication that the platinum is of high purity. The presence of impurities in resistance thermometer material is undesirable, as diffusion, segregation and evaporation may occur in service, resulting in a lack of stability of the thermometer. The temperature coefficient of resistivity is also sensitive to internal strains so that it is essential that the platinum should be annealed at a temperature higher than the maximum temperature of service. The combination of purity and adequate annealing is shown by a high value of the ratio of the resistances at the steam and ice points. To comply with the requirements of the International Practical Temperature Scale of 1968 this ratio must exceed 1.392 50.

It is essential that the platinum element is mounted in such a way that it is not subject to stress in service.

Platinum is used for resistance thermometry in industry for temperatures up to 800 °C. It does not oxidize, but must be protected from contamination. The commonest cause of contamination of platinum resistance thermometers is contact with silica, or silica-bearing refractories, in a reducing atmosphere. In the presence of a reducing atmosphere, silica is reduced to silicon which alloys with platinum making it brittle. Platinum resistance thermometers may be used for temperatures down to about 20 K.

For measuring temperatures between 1 K and 40 K doped germanium sensors are usually used, while carbon resistors are used between 0.1 K and 20 K. Above 20 K platinum has a greater temperature coefficient of resistivity and has a greater stability. Between 0.35 K and 40 K a new resistance thermometer material (0.5 atomic % iron–rhodium) is also used.

Calibration of resistance thermometers To conform with IPTS-68 the resistance of the thermometer at temperatures below 0 °C is measured at a number of defining points and the calibration is obtained by difference from a reference function W which is defined and tabulated in the scale. The differences from the function ΔW are expressed by polynomials, the coefficients of which are obtained from calibration at fixed points for each of the ranges 13.81–20.28 K, 20.28–54.361 K, 54.361–90.188 K and 90.188–273.15 K. The last mentioned range was formerly defined by the Callendar–Van Dusen equation but the difference from the reference function given by the equation

$$\Delta W = A + B\left[\frac{t_{68}}{100\,^\circ\mathrm{C}} - 1\right]\frac{t_{68}}{100} \quad (1.17)$$

is now used, where t_{68} is the temperature in °C and the constants A and B are determined by measurements of W at 100 °C and −182.962 °C (90.188 K).

For the range 0 °C to 630.74 °C the Callendar equation is still used but a correction term is added so that the calibration procedure is to measure the resistance of the thermometer at 0 °C (obtained by way of the triple point of water), the boiling point of water (100 °C) and the freezing point of zinc 419.58 °C on the 1968 scale (formerly 419.505 °C on the 1948 scale). The Callendar equation is then used to determine the intermediate value of t':

$$t' = \frac{1}{\alpha}(W(t') - 1) + \delta\left(\frac{t'}{100\,^\circ\mathrm{C}}\right)\left(\frac{t'}{100\,^\circ\mathrm{C}} - 1\right) \quad (1.18)$$

The procedure is then to correct t' by an amount which varies with temperature but is the same for all thermometers which meet the specification of the scale:

$$t_{68} = t' + 0.45\left(\frac{t'}{100\,^\circ\mathrm{C}}\right)\left(\frac{t'}{100\,^\circ\mathrm{C}} - 1\right) \times \left(\frac{t'}{419.58\,^\circ\mathrm{C}} - 1\right)\left(\frac{t'}{630.74\,^\circ\mathrm{C}} - 1\right)\,^\circ\mathrm{C} \quad (1.19)$$

The value of α for a given specimen of platinum is the same on the 1948 and 1968 scales but the value of δ changes because of the change in the assigned zinc point; for example, a δ coefficient of 1.492 on the old scale becomes 1.497 on the new.

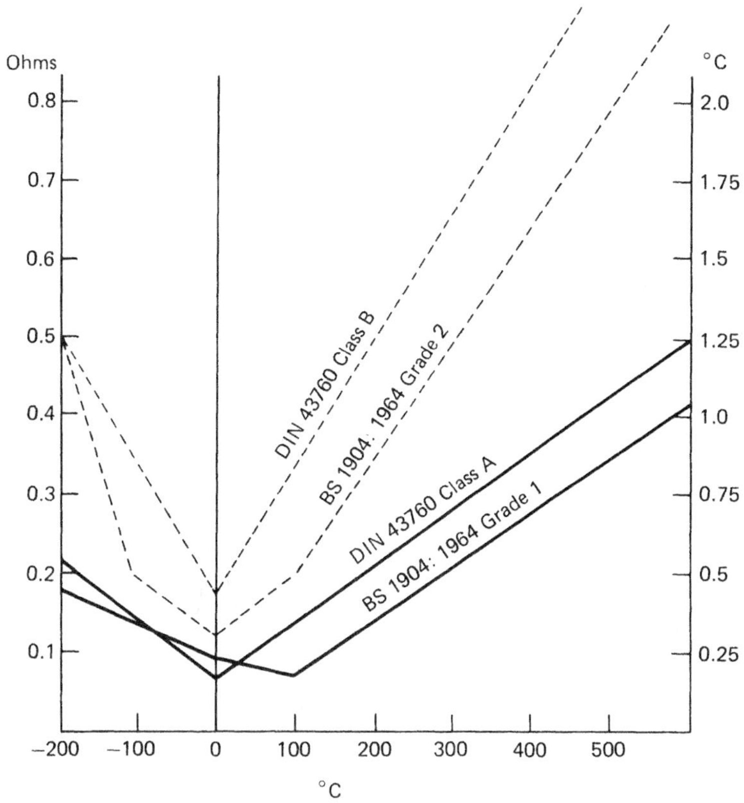

Figure 1.16 BS and DIN specifications for 100 ohm platinum resistance thermometers (courtesy Kent Industrial Instruments Ltd.)

Industrial resistance thermometers Industrial platinum resistance thermometers manufactured in Britain and Europe in general conform to BS 1904, 1964 or DIN 43 760 (1980). BS 1904 is currently being revised to conform with DIN 43 760 (1980). Figure 1.16 shows the tolerances in ohms and kelvin permitted by these two standards. Thermometers are usually produced to Grade 2 or Class B for general industrial use but they are available to Grade 1 or Class A. The normal resistance value is 100 ohms and the fundamental interval (the increase of resistance between 0 °C and 100 °C) is 38.5 ohms.

The calibration of a resistance thermometer is based on IPTS-68 and is usually carried out by comparison with a standard resistance thermometer. Platinum resistance sensors may be designed for any range within the limits of 15 K and 800 °C and may be capable of withstanding pressures up to 600 bar and vibration up to 60 g's, or more, at frequencies up to 2000 Hz. The size of the sensitive element may be as small as 2 mm diameter by 8 mm long, in the case of the miniature fast response elements, to 6 mm diameter by 50 mm long in the more rugged types. A wide range of sensor designs is available, the form used depending upon the duty and the speed of response required. Some typical forms of construction are illustrated in Figure 1.17. Figure 1.17(a) shows a high temperature form in which the spiral platinum coil is bonded at one edge of each turn with high temperature glass inside cylindrical holes in a ceramic rod. In the high accuracy type, used mainly for laboratory work, the coil is not secured at each turn but is left free to ensure a completely strain-free mounting, Figure 1.17(b). Where a robust form, suitable for use in aircraft and missiles or any severe vibration condition is required, the ceramic is in solid rod form and the bifilar wound platinum coil is sealed to the rod by a glass coating as shown in Figure 1.17(c). Where the sensor is intended for use for measuring surface temperatures, the form shown in Figure 1.17(d) is used. In all forms, the ceramic formers are virtually silica-free and the resistance element is sealed in with high temperature glass to form an impervious sheath which is unaffected by most gases and hydrocarbons. The external leads, which are silver or platinum of a diameter much larger than the wire of the resistance element, are welded to the fine

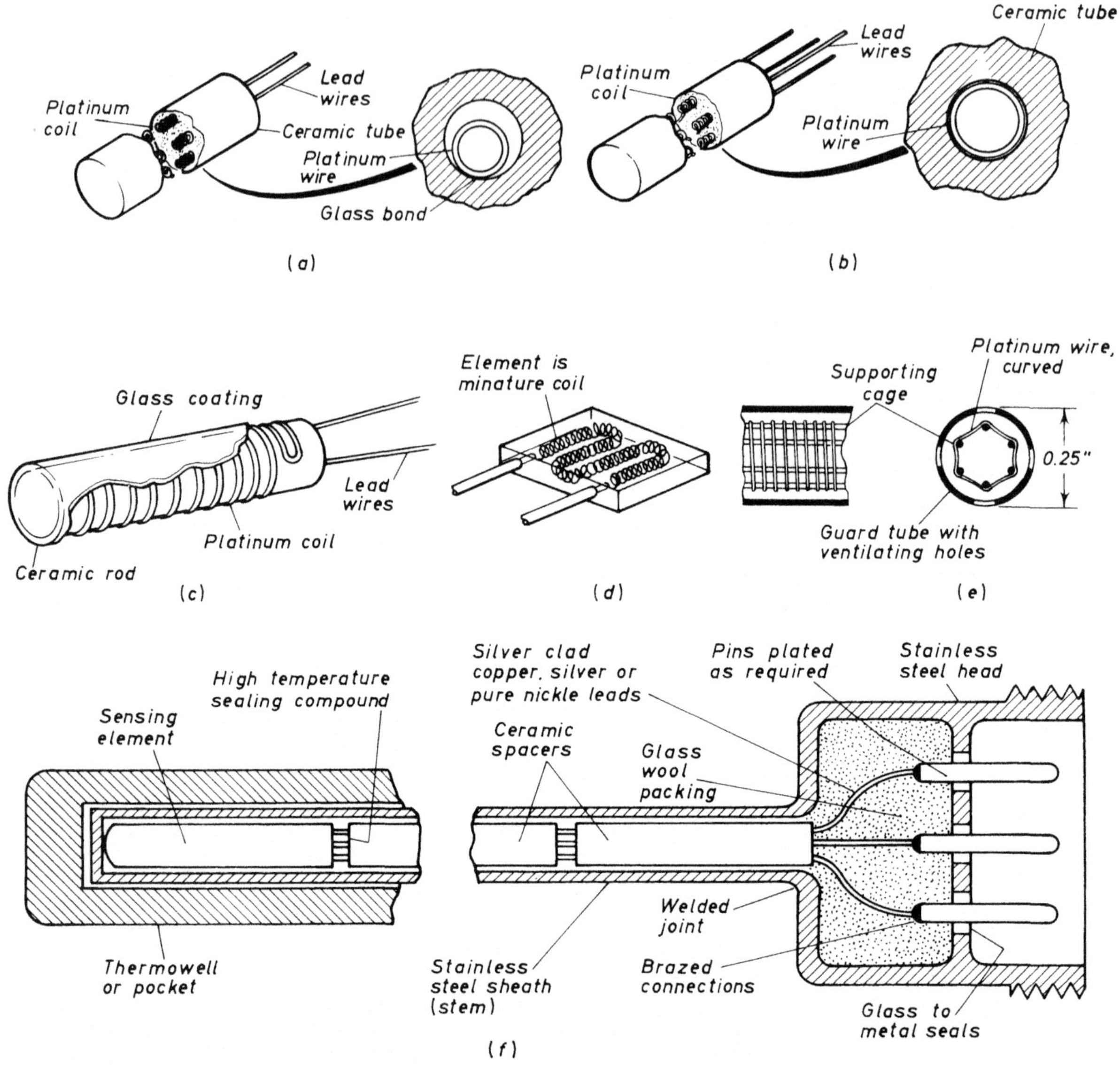

Figure 1.17 Construction of resistance thermometers (courtesy Rosemount Engineering).

platinum wire wholly inside the glass seal.

The inductance and capacitance of elements are made as low as possible in order to allow their use with a.c. measuring instruments. Typically the elements shown will have self-inductance of 2 μH per 100 Ω and the element self-capacitance will not exceed 5 pF. The current passed through a resistance thermometer to measure the resistance must be limited to minimize errors by self-heating of the resistance element. Typical maximum acceptable current is 10 mA for a 100 ohm thermometer.

The rate of response of a resistance thermometer is a function of its construction and encapsulation. A heavy industrial type may have a response of one or two minutes when plunged into water while a naked type, like that shown in Figure 1.17(e), will be only a few milliseconds under the same conditions. Figure 1.17(f) shows the cross-section of a resistance thermometer encapsulated in a metal tube. Figure 1.18 shows a range of typical industrial resistance thermometers.

A more recent development of resistance ther-

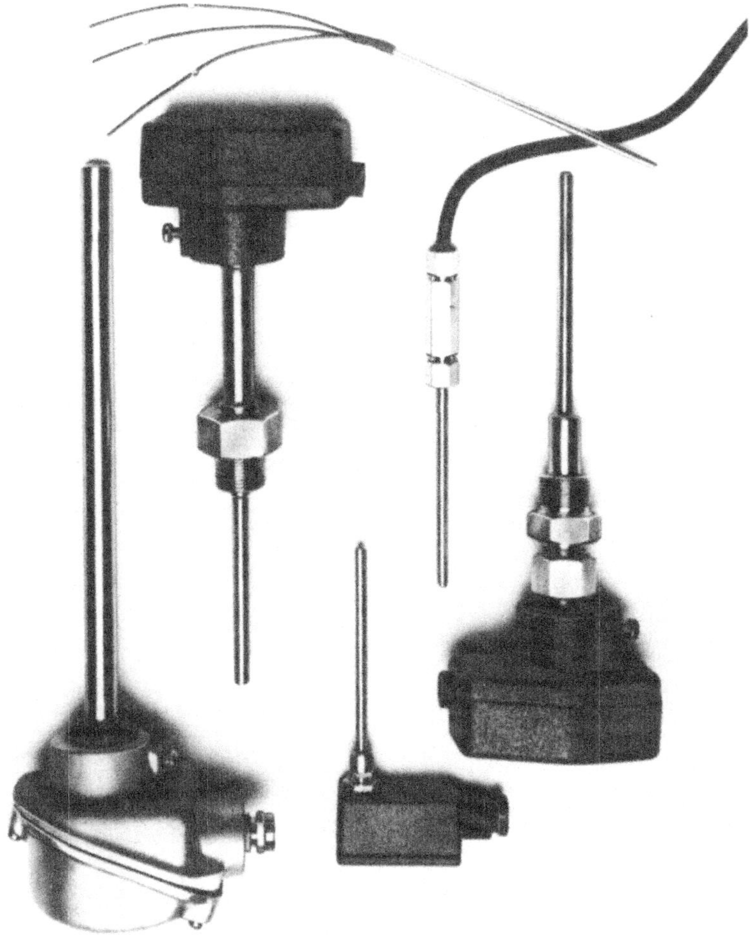

Figure 1.18 Typical industrial resistance thermometers (courtesy Kent Industrial Instruments Ltd.).

mometers has been the replacement of the wire-wound element of the conventional resistance thermometer by a metallized film track laid down on a glass or ceramic substrate. These thermometer elements are made by similar techniques to those used for making hybrid integrated electronic circuits. After the laying down of the metallized film the film is trimmed by a laser to achieve the required parameters.

1.4.1.2 Nickel resistance thermometers

Nickel forms an inexpensive alternative to platinum for resistance thermometers. The usable range is restricted to $-200\,^{\circ}$C to $+350\,^{\circ}$C. But the temperature coefficient of resistivity of nickel is 50 per cent higher than that of platinum which is an advantage in some instruments. Nickel resistance thermometers find wide use in water-heating and air-conditioning systems.

As mentioned above, the current through a resistance thermometer sensor must be kept low enough to limit self-heating. However in some applications such as flow meters, anemometers and psychrometers, the self-heating effect is used, the final temperature of the sensor being a function of the flow rate of the process fluid or air. See also Chapter 1 of Volume 1.

1.4.1.3 Resistance thermometer connections

When resistance thermometers are located at some distance from the measuring instrument the electrical resistance of the connecting cables will introduce errors of reading. This reading error will, of course, vary as the temperature of the cables changes. However, this error can be compensated by the use of extra conductors. Since, normally, the change of resistance of a resistance thermometer is measured in a Wheatstone bridge circuit or a modified Wheatstone

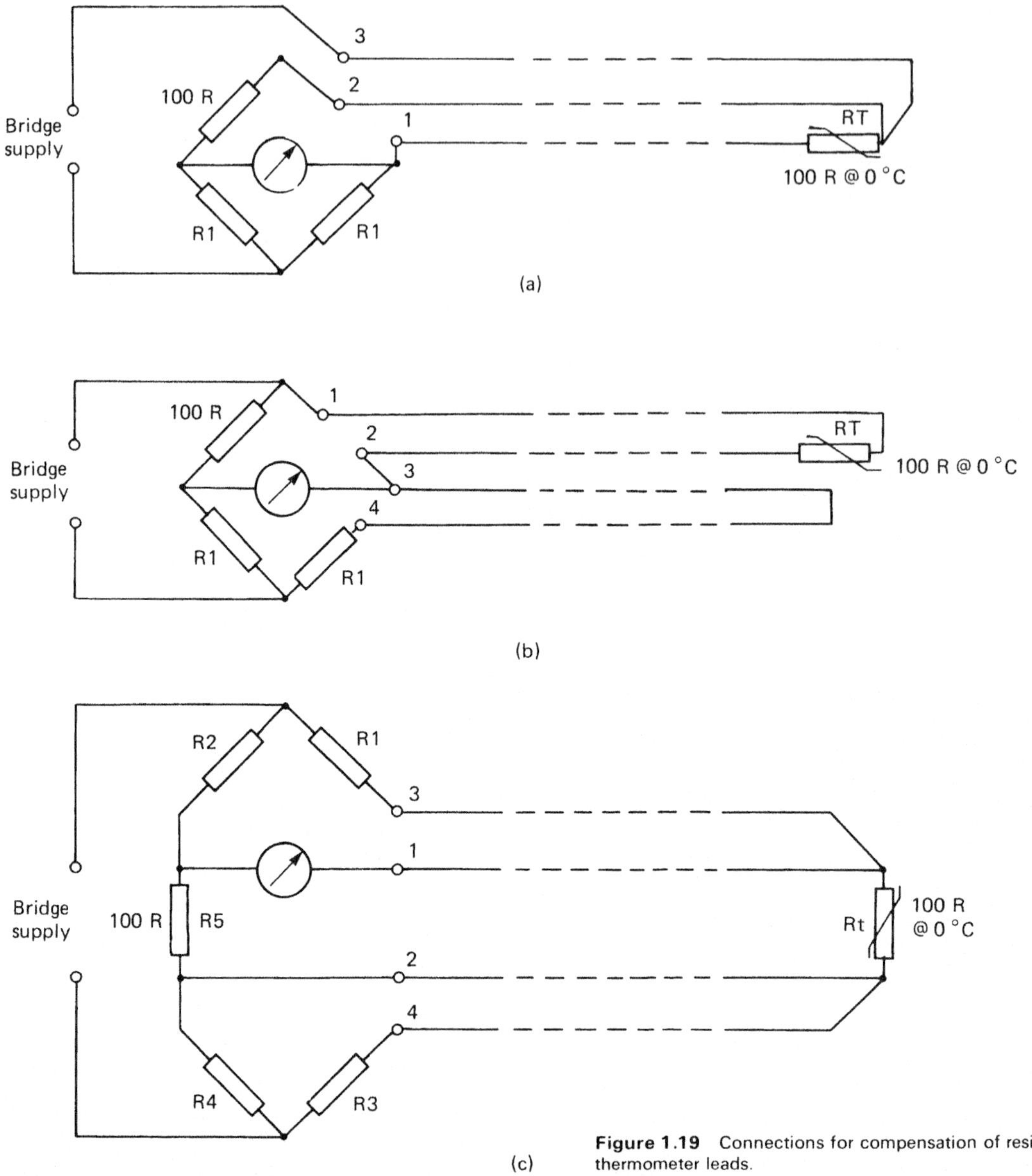

Figure 1.19 Connections for compensation of resistance thermometer leads.

bridge, the compensating conductors can be connected in the opposite side of the bridge. In this way bridge unbalance is only a function of the change of resistance of the thermometer element. Figure 1.19(a) shows three-wire compensation. The resistance of wire 1 is added to that of the resistance thermometer but is balanced by wire 2 in the reference side of the bridge. Wire 3 supplies the power to the bridge. In Figure 1.19(b) four-wire compensation is shown. The resistance of wires 1 and 2 which connect to the resistance thermometer are compensated by the resistance of wires 3 and 4 which are connected together at the resistance thermometer and are again in the opposite arm of the bridge. A Kelvin double bridge is illustrated in Figure 1.19(c). Resistors R1 and R3 set up a constant current through the resistance thermometer. Resistors R2 and R4 set up a constant current in the reference resistor R5 such that the voltage V_R is equal to the voltage V_t across the resistance thermometer when it is at 0 °C. At any other temperature $V_t = I_t R_t$ and the meter will indicate the difference between V_t and V_R which will be proportional to the temperature. The indicator must have a very high resistance so that the current in conductors 1 and 2 is essentially zero. Refer to Volume 3.

1.4.2 Thermistors

1.4.2.1 Negative temperature coefficient thermistors

An alternative to platinum or nickel for resistance thermometer sensing elements is a semiconductor composed of mixed metal oxides. The composition of these materials depends on the particular properties required. Combinations of two or more of the following oxides are used: cobalt, copper, iron, magnesium, manganese, nickel, tin, titanium, vanadium and zinc. Devices made of these materials are called thermistors. They consist of a piece of the semiconductor to which two connecting wires are attached at opposite sides or ends. Thermistors have a negative temperature coefficient; that is, as the temperature rises the electrical resistance of the device falls. This variation of resistance with temperature is much higher than in the case of metals. Typical resistance values are 10 kilohms at 0 °C and 200 ohms at 100 °C. This very high sensitivity allows measurement or control to a very high resolution of temperature differences. The accuracy is not as good as for a metallic resistance thermometer owing to the difficulty in controlling the composition of the thermistor material during manufacture. The resolution differs across the usable span of the devices due to their non-linearity. With the right choice of device characteristics it is nevertheless possible to control a temperature to within very close limits: 0.001 degree Celsius temperature change is detectable.

The total range that can be measured with thermistors is from −100 °C to +300 °C. However, the span cannot be covered by one thermistor type – four or five types are needed.

The physical construction of thermistors covers a wide range. The smallest are encapsulated in glass or epoxy beads of 1–2.5 mm diameter, bigger ones come as discs 5–25 mm diameter or rods 1–6 mm diameter and up to 50 mm length. The bigger devices are able to pass quite high currents and so operate control equipment directly without need of amplifiers. Thermistors are also available in metal encapsulations like those used for platinum resistance thermometers.

The big disadvantage of thermistors is that their characteristics are non-linear. The temperature coefficient of resistivity α at any temperature within the range of a sensor is given by:

$$\alpha = -B/T^2 \tag{1.20}$$

where B is the characteristic temperature constant for that thermistor and T is temperature in kelvin. The units of α are ohms . K^{-1}.

Most thermistors have a specified resistance at 20 °C or 25 °C. To determine the resistance at any other temperature equation (1.21) is used:

$$R_2 = R_1 \exp\left(\frac{B}{t_2} - \frac{B}{t_1}\right) \tag{1.21}$$

where R_1 is resistance of thermistor at temperature t_1(°C) and R_2 is resistance of thermistor at temperature t_2(°C).

Thermistors are available described as curve-matched. These devices are manufactured to fine tolerances and are interchangeable with an error of less than ±0.2 per cent. However, they are expensive and are only available in a limited range of formats.

In general most thermistors are manufactured with tolerances of 10 to 20 per cent. Instrumentation for use with these devices must have provision for trimming out the error. Thermistors do not have the stability of platinum resistance thermometers. Their characteristics tend to drift with time. Drifts of up to 0.1 °C or more can be expected from some types over a period of some months.

1.4.2.2 Positive temperature coefficient thermistors

Positive temperature coefficient (PTC) thermistors are manufactured from compounds of barium, lead and strontium titanates. PTC thermistors are primarily designed for the protection of wound equipment such as transformers and motors. The characteristics of these devices have the general shape shown in Figure 1.20. The resistance of PTC thermistors is low and

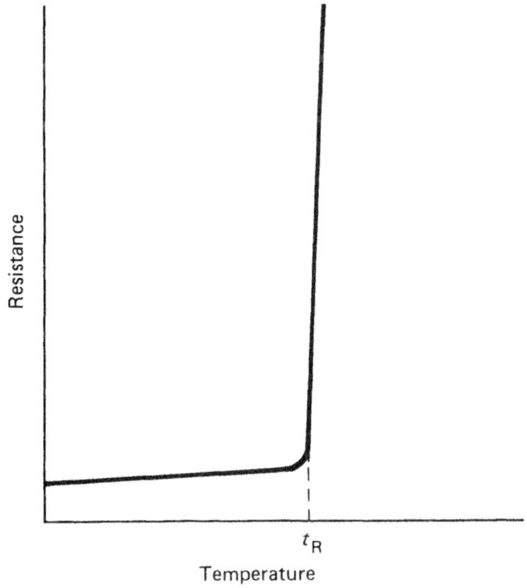

Figure 1.20 Resistance temperature characteristic for PTC thermistor.

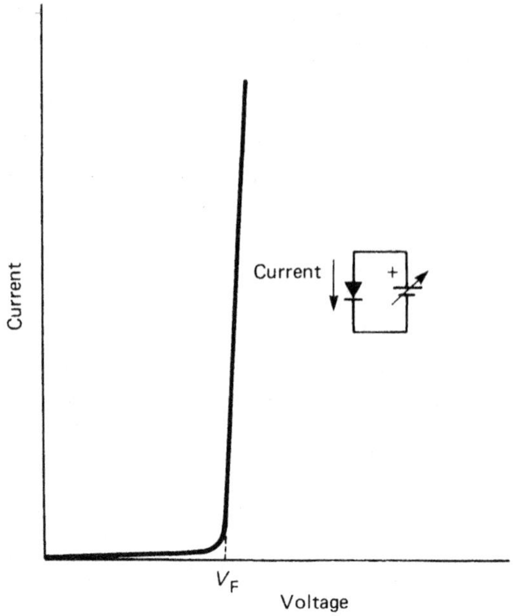

Figure 1.21 Forward bias characteristic of silicon diode.

relatively constant with temperature at low temperature. At temperature T_R the increase of resistance with temperature becomes very rapid. We refer to T_R as the reference or switching temperature.

In use PTC thermistors are embedded in the windings of the equipment to be protected. They are connected in series with the coil of the equipment contactor or protection relay. If the temperature of the windings exceeds temperature T_R the current becomes so small that power is effectively disconnected from the equipment.

1.4.3 Semiconductor temperature measurement

1.4.3.1 Silicon junction diode

Recently silicon semiconductors have been entering the field of temperature measurement. Figure 1.21 shows the forward bias characteristic of a silicon diode. At voltages below V_f, the forward conduction voltage, virtually no current flows. Above V_f the diode passes current. The voltage V_f represents the energy required by current carriers to cross the junction space charge. The value of V_f varies between diode types. It is typically 700 millivolts at 20 °C. The voltage V_f has a temperature coefficient which is essentially the same for all silicon devices of −2 mV per degree Celsius. The forward voltage against temperature characteristic is very nearly linear over the temperature range of −50 °C to +150 °C. This voltage change with temperature is substantial and as the characteristic is linear it makes a very useful measurement or control signal. There are two principal disadvantages to silicon diodes as control elements. The negative coefficient Figure 1.22 is not fail-safe. If the control loop is controlling a heater, breakage of the diode wires would be read by the controller as low temperature and full power would be applied to the heaters. The second disadvantage is the rather limited temperature range. Also if a silicon diode is heated above about 200 °C it is completely destroyed, effectively becoming a short circuit.

1.4.3.2 Temperature-sensing integrated circuits

The temperature characteristic of a silicon junction can be improved if the measuring diode is incorporated in an integrated circuit containing an amplifier. Devices are available either to provide an output current proportional to temperature or an output voltage proportional to temperature. Figure 1.23(a) shows the basis of such a device. Figure 1.23(b) shows the circuit of the Analog Devices temperature sensor type AD 590. The operating range of this device is −55 °C to +150 °C. The temperature is sensed by the emitter–base junctions of two transistors. If two identical transistors are operated at a constant ratio r of collector current densities then the difference in V_t in their base emitter voltages is given by equation (1.22):

$$V_t = \frac{KT}{q} . \ln r \qquad (1.22)$$

where K is Boltzmann's constant ($1.380\,66 \times 10^{-23}$ J.K^{-1}), q is the electron charge ($1.602\,19 \times 10^{-19}$ coulomb) and T is temperature in kelvins. It can be seen that V_t is directly proportional to temperature in kelvins. The voltage is converted to a temperature-

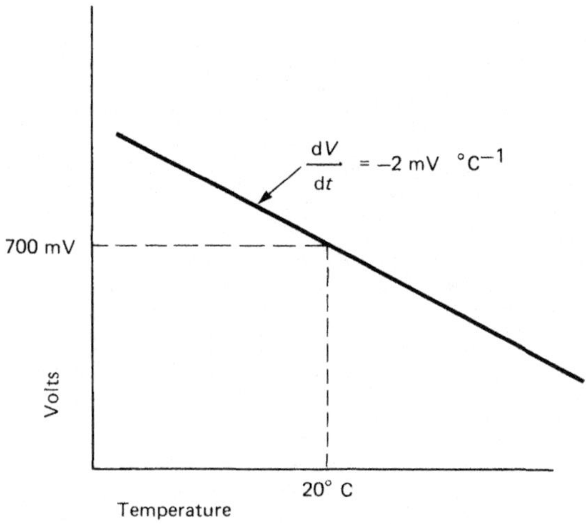

Figure 1.22 Temperature characteristic of silicon diode.

Figure 1.23 Semiconductor temperature sensors: (a) diode and amplifier, (b) Analog Devices I.C. temperature sensor circuit Type AD 590.

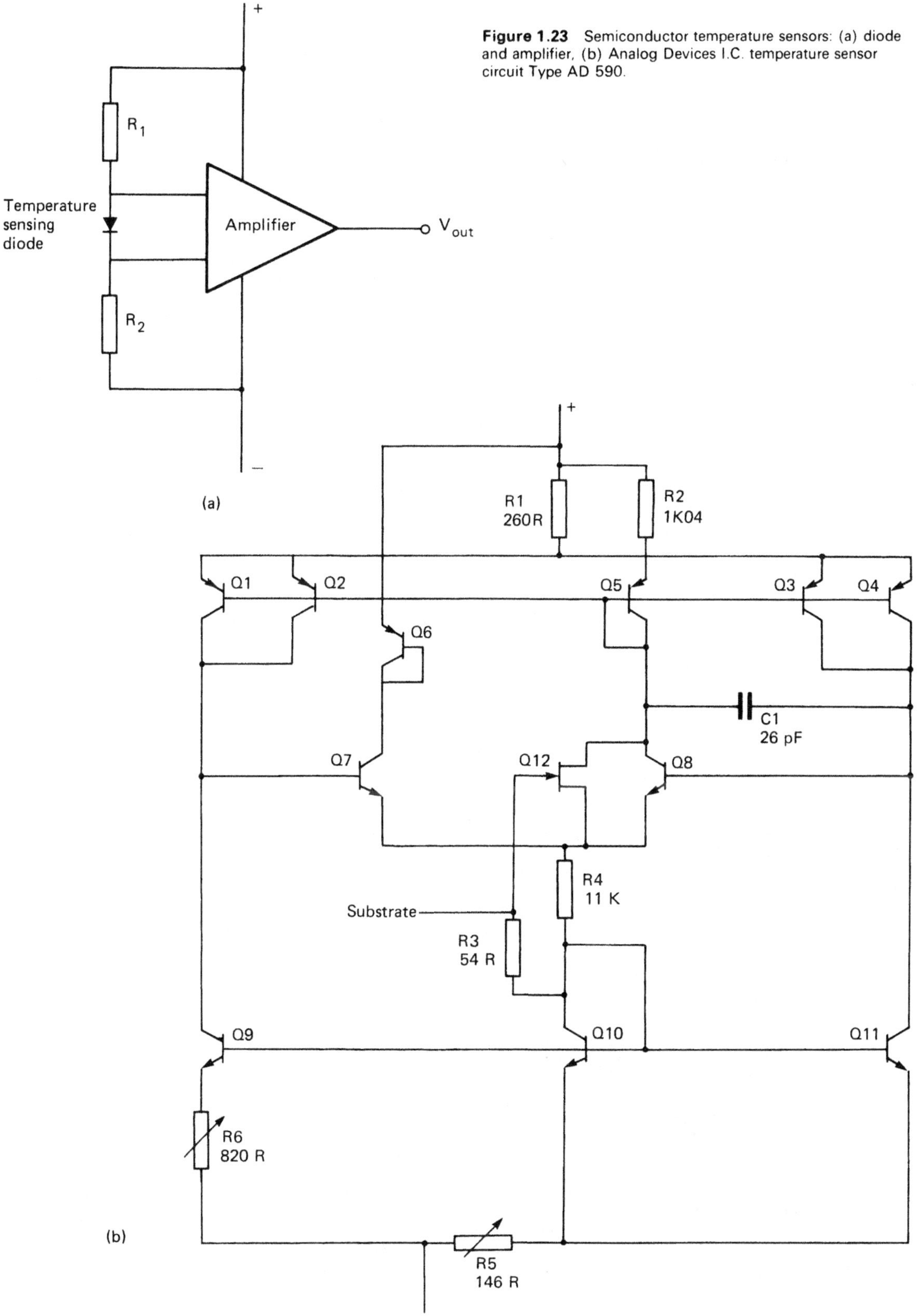

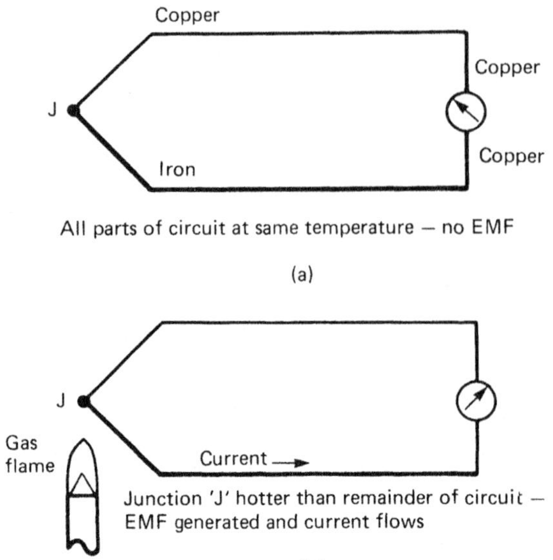

Figure 1.24 Basic thermocouple circuit.

dependent current I_t by low temperature coefficient thin film resistors R5 and R6. These resistors are laser-trimmed to give the required tolerance at 25 °C. Transistors Q_8 and Q_{11} provide the temperature-dependent voltage V_t. The remaining transistors provide the amplification to give the output current of one microampere per kelvin. The transistor Q_{10} supplies the bias and substrate leakage currents for the circuit. The device is packaged in a transistor can or ceramic capsule or it can be supplied as the naked chip for encapsulation into other equipment.

1.5 Measurement techniques – thermocouples

1.5.1 Thermoelectric effects

If an electrical circuit consists of entirely metallic conductors and all parts of the circuit are at the same temperature there will be no electromotive force in the circuit and therefore no current flows. However, if the circuit consists of more than one metal and if junctions between two metals are at different temperatures then there will be an e.m.f. in the circuit and a current will flow. Figure 1.24 illustrates this effect. The e.m.f. generated is called a thermoelectric e.m.f. and the heated junction is a thermocouple.

1.5.1.1 Seebeck effect

In 1821 Seebeck discovered that if a closed circuit is formed of two metals, and the two junctions of the metals are at different temperatures, an electric current will flow round the circuit. Suppose a circuit is formed by twisting or soldering together at their ends, as shown in Figure 1.25, wires of two different metals such as iron and copper. If one junction remains at room temperature, while the other is heated to a higher temperature, a current is produced, which flows from copper to iron at the hot junction, and from iron to copper at the cold one.

Seebeck arranged a series of 35 metals in order of their thermoelectric properties. In a circuit made up of any two of the metals, the current flows across the hot junction from the earlier to the later metal of the series. A portion of his list is as follows: Bi—Ni—Co—Pd—Pt—U—Cu—Mn—Ti—Hg—Pb—Sn—Cr—Mo—Rh—Ir—Au—Zn—W—Cd—Fe—As—Sb—Te.

1.5.1.2 Peltier effect

In 1834 Peltier discovered that when a current flows across the junction of two metals heat is absorbed at the junction when the current flows in one direction and liberated if the current is reversed. Heat is absorbed when a current flows across an iron–copper junction from copper to iron, and liberated when the current flows from iron to copper. This heating effect should not be confused with the Joule heating effect, which being proportional to I^2R, depends only upon the size of the current and the resistance of the conductor and does not change to a cooling effect when the current is reversed. The amount of heat liberated, or absorbed, is proportional to the quantity of electricity which crosses the junction, and the amount liberated, or absorbed, when unit current passes for a unit time is called the Peltier coefficient.

As heat is liberated when a current does work in overcoming the e.m.f. at a junction, and is absorbed when the e.m.f. itself does work, the existence of the Peltier effect would lead one to believe that the junction of the metals is the seat of the e.m.f. produced in the Seebeck effect. It would appear that an e.m.f. exists across the junction of dissimilar metals, its direction being from copper to iron in the couple considered. The e.m.f. is a function of the conduction

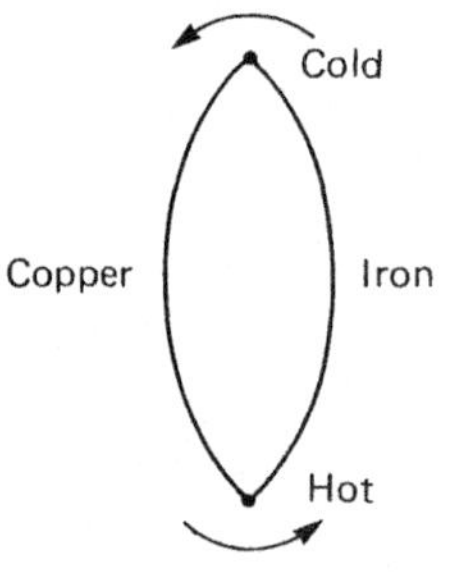

Figure 1.25 Simple thermocouple.

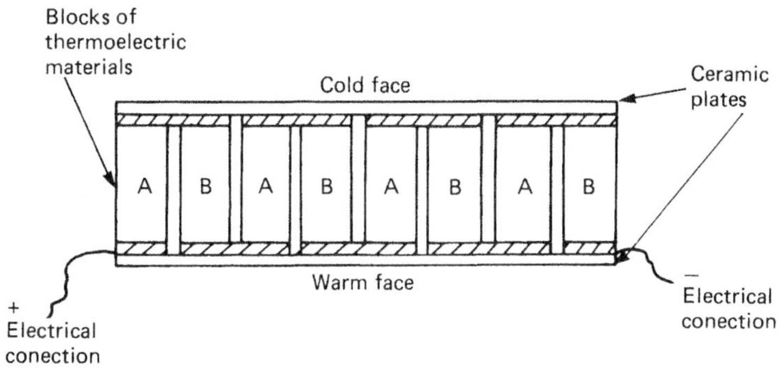

Figure 1.26 Peltier cooler.

electron energies of the materials making up the junction. In the case of metals the energy difference is small and therefore the e.m.f. is small. In the case of semiconductors the electron energy difference may be much greater, resulting in a higher e.m.f. at the junction. The size of this e.m.f. depends not only on the materials making up the junction but also upon the temperature of the junction. When both junctions are at the same temperature, the e.m.f. at one junction is equal and opposite to that at the second junction, so that the resultant e.m.f. in the circuit is zero. If, however, one junction is heated, the e.m.f. across the hot junction is greater than that across the cold junction, and there will be a resultant e.m.f. in the circuit which is responsible for the current:

$$\text{e.m.f. in the circuit} = P_2 - P_1$$

where P_1 is the Peltier e.m.f. at temperature T_1, and P_2 is the Peltier e.m.f. at temperature T_2 where $T_2 > T_1$. Peltier cooling is used in instrumentation where a small component is required to be cooled under precise control. Figure 1.26 shows diagrammatically the construction of such a cooler. The conductors and junctions have a big cross-section to minimize IR heating. The warmer face is clamped to a suitable heat sink while the cold face has the component to be cooled mounted in contact with it. Typical size for such a unit is of the order of 5–25 mm. The conductors in Peltier coolers may be either metals or semiconductors; in the latter case they are called Frigistors.

1.5.1.3 Thomson effect

Reasoning on the basis of the reversible heat engine, Professor William Thomson (later Lord Kelvin) pointed out that if the reversible Peltier effect was the only source of e.m.f., it would follow that if one junction was maintained at a temperature T_1, and the temperature of the other raised to T_2, the available e.m.f. should be proportional to $(T_2 - T_1)$. It may be easily shown that this is not true. If the copper–iron thermocouple, already described, is used, it will be found that on heating one junction while the other is maintained at room temperature, the e.m.f. in the circuit increases at first, then diminishes, and passing through zero, actually becomes reversed. Thomson, therefore, concluded that in addition to the Peltier effects at the junctions there were reversible thermal effects produced when a current flows along an unequally heated conductor. In 1856, by a laborious series of experiments, he found that when a current of electricity flows along a copper wire whose temperature varies from point to point, heat is liberated at any point P when the current at P flows in the direction of the flow of heat at P, that is when the current is flowing from a hot place to a cold place, while heat is absorbed at P when the current flows in the opposite direction. In iron, on the other hand, the heat is absorbed at P when the current flows ih the direction of the flow of heat at P, while heat is liberated when the current flows in the opposite direction from the flow of heat.

1.5.1.4 Thermoelectric diagram

It will be seen that the Seebeck effect is a combination of the Peltier and Thomson effects and will vary according to the difference of temperature between the two junctions, and with the metals chosen for the couple. The e.m.f. produced by any couple with the junctions at any two temperatures may be obtained from a thermoelectric diagram suggested by Professor Tait in 1871. On this diagram the thermoelectric line for any metal is a line such that the ordinate represents the thermoelectric power (defined as the rate of change of e.m.f. acting round a couple with the change of temperature of one junction) of that metal with a standard metal at a temperature represented by the abscissa. Lead is chosen as the standard metal as it does not show any measurable Thomson effect. The ordinate is taken as positive when, for a small difference of temperature, the current flows from lead to the metal at the hot junction. If lines a and b (Figure 1.27) represent the thermoelectric lines for two metals A and B then the e.m.f. round the circuit formed by the two metals, when the temperature of the cold junction

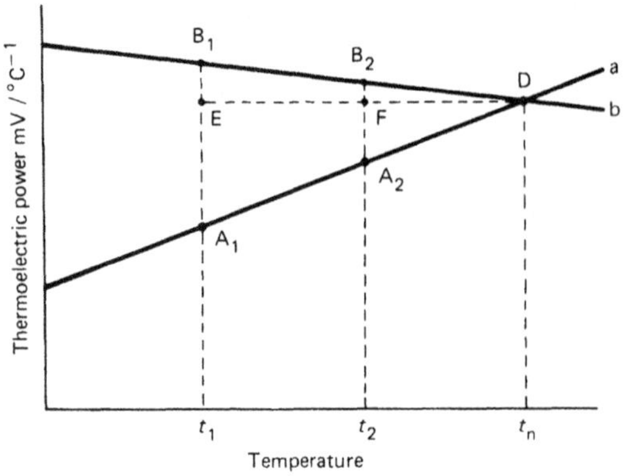

Figure 1.27 Thermoelectric diagram of two metals.

is t_1 and that of the hot junction is t_2, will be the difference in the areas of triangles A_1B_1D and A_2B_2D. Now the area of the triangle is

$$A_1B_1D = \tfrac{1}{2}(A_1B_1 \times ED) \quad (1.23)$$

and area

$$A_2B_2D = \tfrac{1}{2}(A_2B_2 \times FD) \quad (1.24)$$

$$\text{The e.m.f.} = \tfrac{1}{2}(A_1B_1 \times ED) - \tfrac{1}{2}(A_2B_2 \times FD) \quad (1.25)$$

Since triangles A_1B_1D and A_2B_2D are similar triangles the sides A_1B_1 and A_2B_2 are proportional to ED and FD respectively.

Therefore: e.m.f. $\propto ED^2 - FD^2$

But: $ED = t_n - t_1$ and $FD = t_n - t_2$

So: e.m.f. $\propto (t_n - t_1)^2 - (t_n - t_2)^2$

$$\propto (t_1 - t_2)\left(\frac{t_1 + t_2}{2} - t_n\right)$$

$$\text{Or e.m.f.} = K(t_1 - t_2)\left(\frac{t_1 + t_2}{2} - t_n\right) \quad (1.26)$$

where K is a constant which together with t_n must be obtained experimentally for any pair of metals. The temperarure t_n is called the neutral temperature. Equation (1.26) shows that the e.m.f. in any couple is proportional to the difference of temperature of the junctions and also to the difference between the neutral temperature and the average temperature of the junctions. The e.m.f. is zero either if the two junctions are at the same temperature or if the average of the temperature of the two junctions is equal to the neutral temperature. Figure 1.28 shows the graph of the e.m.f. of a zinc–iron thermocouple with temperature.

1.5.1.5 Thermoelectric inversion

This reversal of the thermoelectric e.m.f. is 'thermoelectric inversion'.

Figure 1.29 shows the thermoelectric lines for several common materials. It will be seen that the lines for iron and copper cross at a temperature of 275 °C. If the temperature of the cold junction of iron and copper is below 270 °C and the temperature of the other junction is raised, the thermoelectric e.m.f. of the circuit (represented by a trapezium) will increase until the temperature of the hot junction reaches 275 °C (when the e.m.f. is represented by a triangle). Further increase in the temperature of the hot junction will result in a decrease in the thermoelectric e.m.f. (the e.m.f. represented by the second triangle will be in the opposite sense). When the average temperature of the two junctions is 275 °C, or what comes to the same thing, the sum of the two temperatures is 550 °C, the areas of the two triangles will be equal and there will be no thermoelectric e.m.f.: 275 °C is the 'neutral temperature' for the copper–iron couple. With circuits of other materials, the neutral point will occur at different temperatures. Further increase in the temperature of the hot junction will produce a thermoelectric e.m.f. in the opposite direction: from iron to copper at the hot junction, which will again increase with increasing temperature of the hot junction as was seen with zinc and iron in Figure 1.28.

In choosing two materials to form a thermocouple to measure a certain range of temperature, it is very important to choose two which have thermoelectric lines which do not cross within the temperature range, that is, the neutral temperature must not fall within the range of temperature to be measured. If the neutral temperature is within the temperature range, there is some ambiguity about the temperature indicated by a certain value of the thermoelectric e.m.f., for there will be two values of the temperature of the hot junction for

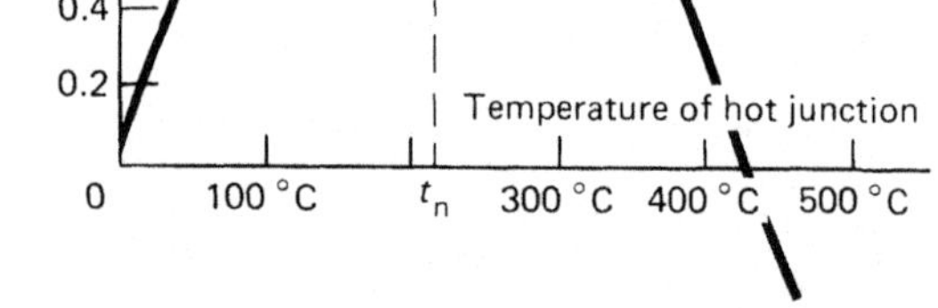

Figure 1.28 Temperature/emf curve for zinc/iron couple.

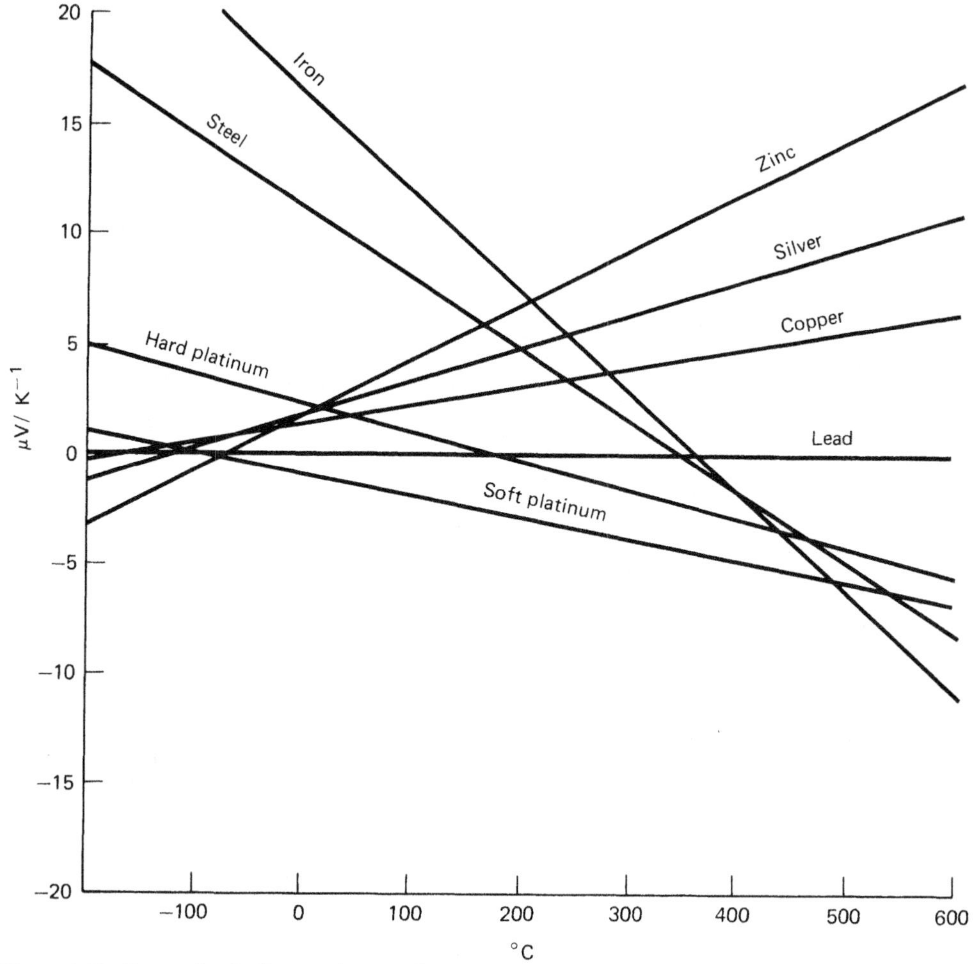

Figure 1.29 Thermoelectric diagrams for several metals.

which the thermoelectric e.m.f. will be the same. For this reason tungsten–molybdenum thermocouples must not be used at temperatures below 1250 °C.

1.5.1.6 Addition of thermoelectric e.m.f.s

In measuring the e.m.f. in any circuit due to thermoelectric effects, it is usually necessary to insert some piece of apparatus, such as a millivoltmeter, somewhere in the circuit, and since this generally involves the presence of junctions other than the two original junctions, it is important to formulate the laws according to which the e.m.f.s produced by additional junctions may be dealt with. These laws, discovered originally by experiment, have now been established theoretically.

Law of intermediate metals In a thermoelectric circuit composed of two metals A and B with junctions at temperatures t_1 and t_2 the e.m.f. is not altered if one or both the junctions are opened and one or more other metals are interposed between metals A and B, provided that all the junctions by which the single junction at temperature t_1 may be replaced are kept at t_1, and all those by which the junction at temperature t_2 may be replaced are kept at t_2.

This law has a very important bearing on the application of thermocouples to temperature measurement, for it means that, provided all the apparatus for measuring the thermoelectric e.m.f., connected in the circuit at the cold junction, is kept at the same temperature, the presence of any number of junctions of different metals will not affect the total e.m.f. in the circuit. It also means that if another metal is introduced into the hot junction for calibration purposes it does not affect the thermoelectric e.m.f., provided it is all at the temperature of the hot junction.

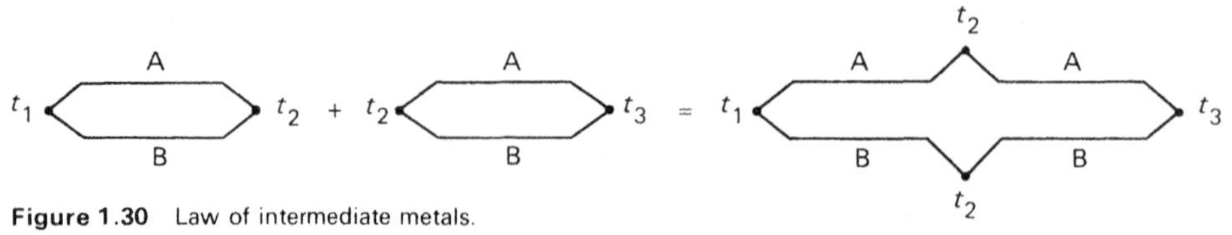

Figure 1.30 Law of intermediate metals.

Law of intermediate temperatures The e.m.f. E_{1-3} of a thermocouple with junctions at temperatures t_1 and t_3 is the sum of the e.m.f.s of two couples of the same metals, one with junctions at temperatures t_1 and t_2 (e.m.f. $= E_{1-2}$), and the other with junctions at t_2 and t_3 (e.m.f. $= E_{2-3}$), see Figure 1.30:

$$E_{1-2} + E_{2-3} = E_{1-3} \qquad (1.27)$$

This law is the basis upon which thermocouple measuring instruments can be manufactured.

1.5.1.7 Cold junction compensation

It is not normally practical in industrial applications to have thermocouple cold junctions maintained at 0 °C, but with the cold junctions at ambient temperature cold junction compensation is required. To achieve cold junction compensation consider a thermocouple with its hot junction at t °C and its cold junction at ambient, its e.m.f. being E_{a-t}. The instrument must indicate an e.m.f. equivalent to having the cold junction at 0 °C, i.e. an e.m.f. of E_{0-t}. This requires that an e.m.f. must be added at E_{a-t} to provide the required signal:

$$E_{0-t} = E_{a-t} + E_{0-a} \qquad (1.28)$$

The voltage E_{0-a} is called the cold junction compensation voltage.

This cold junction compensation e.m.f. can be provided automatically by the use of a temperature-sensitive element such as a resistance thermometer, thermistor or semiconductor sensor in the thermocouple circuit. Figure 1.31 shows such a circuit. In this circuit R_1, R_2 and R_3 are temperature-stable resistors and R_t is a resistance thermometer. The bridge is balanced when all components are at 0 °C and the e.m.f. appearing between points A and B is zero. As the temperature changes from 0 °C an e.m.f., which is the unbalance voltage of the bridge, exists across AB. This voltage is scaled by setting R_4 such that the e.m.f. AB is equal to E_{0-a} in equation (1.28).

Mechanical cold junction compensation An alternative cold junction compensation technique is used when a simple non-electronic thermometer is required. In this technique the thermocouple is connected directly to the terminals of a moving-coil galvanometer. A bimetal strip is connected mechanically to the mechanical zero adjustment of the instrument in such a way that the instrument zero is offset to indicate the ambient temperature. The e.m.f. E_{a-t} is then sufficient to move the pointer upscale to indicate the true temperature of the thermocouple.

1.5.1.8 Thermocouple circuit considerations

Galvanometer instruments A thermocouple circuit is like any other electrical circuit. There are one or more sources of e.m.f., which can be batteries, a generator or in this case the hot and cold junctions. There is a load, the indicator and there are electrical conductors, which have resistance, to connect the circuit together. The current in this circuit is, as always, governed by Ohm's law:

$$I = E/R \qquad (1.29)$$

where I is the current, E is the e.m.f. and R is the total circuit resistance.

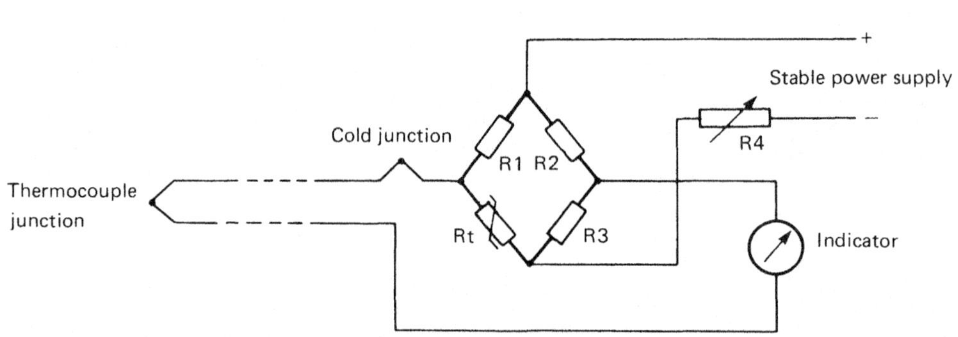

Figure 1.31 Bridge circuit to provide cold junction compensation.

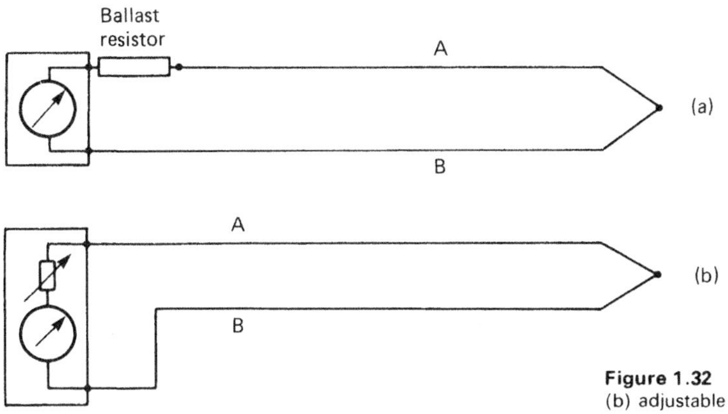

Figure 1.32 Use of ballast resistor: (a) external to instrument, (b) adjustable ballast mounted inside instrument.

In a practical thermocouple thermometer the resistance consists of the sum of the resistances of the thermocouple, the compensating cable (see Section 1.5.3.9) and the indicating instrument. Galvanometer-type thermocouple indicators with mechanical cold junction compensation as described in the previous section, are designed either to be used with an external circuit of stated resistance (this resistance value is usually marked on the dial) or they have an internal adjustable resistor. In the latter case the resistance of the external circuit must not exceed a stated maximum value and the adjustable resistor is adjusted to give the specified total circuit value. Where no internal resistance adjustment is provided the instrument must be used together with an external ballast resistor, see Figure 1.32(a). This resistor must be mounted as near as possible to the indicating instrument to ensure its being at the same temperature as the cold junction compensating mechanism. The usual practice when installing one of these instruments is to wind the ballast resistor with constantan wire on a small bobbin. The length of constantan wire is chosen to make up the required total resistance. On some instruments the bobbin is made integral with one of the indicator terminals. Figure 1.32(b) shows the arrangement with the ballast resistor integral with the indicating instrument.

Potentiometric instruments One way in which to circumvent the critical external resistor is to use a potentiometric indicating device. In a potentiometric device the thermocouple e.m.f. is opposed by an equal and opposite potential from the potentiometer; there is then no current in the circuit and therefore the circuit resistance value is irrelevant.

Potentiometric thermocouple indicators used to be quite common but are now not met so often. However, if the thermocouple indicator is, as it frequently is, a strip chart recorder, it is almost certain to be a potentiometric instrument. Figure 1.33(a) shows the potentiometric arrangement diagrammatically.

Electronic instruments In modern electronic instruments for thermocouple indication, whether they be analogue or digital devices, the input circuit 'seen' by the thermocouple is a high impedance amplifier. Again there is negligible current in the thermocouple circuit and as the resistance of the thermocouple circuit is of the order of 100 ohms while the amplifier input is likely to be a megohm or more the effect of the external circuit resistance is negligible. Electronic instruments allow their designer much more versatility for cold junction compensation. Instead of the bridge circuit of Figure 1.31 it is possible to arrange the cold junction correction after the input amplifier. This has the advantage that the voltage levels being worked with may be of the order of several volts amplitude instead of a few millivolts, making it easier to get a higher degree of accuracy for compensation. Figure 1.33(b) shows a block diagram of such an arrangement. Thermocouple input circuits are available as encapsulated electronic modules. These modules contain input amplifier and cold junction compensation. Since the cold junction consists of the input connections of the module, the connections and the cold junction sensor can be accurately maintained at the same temperature by encapsulation, giving very accurate compensation. These modules can be very versatile. Many are available for use with any of the normal thermocouples. The cold junction compensation is set to the thermocouple in use by connecting a specified value resistor across two terminals of the module. These modules are very convenient to use: they mount like other components directly on to an instrument circuit board. Typical module size is a 25 mm cube. Where the thermocouple instrument is based on a microcomputer the cold junction compensation can be done by software, the microcomputer being pro-

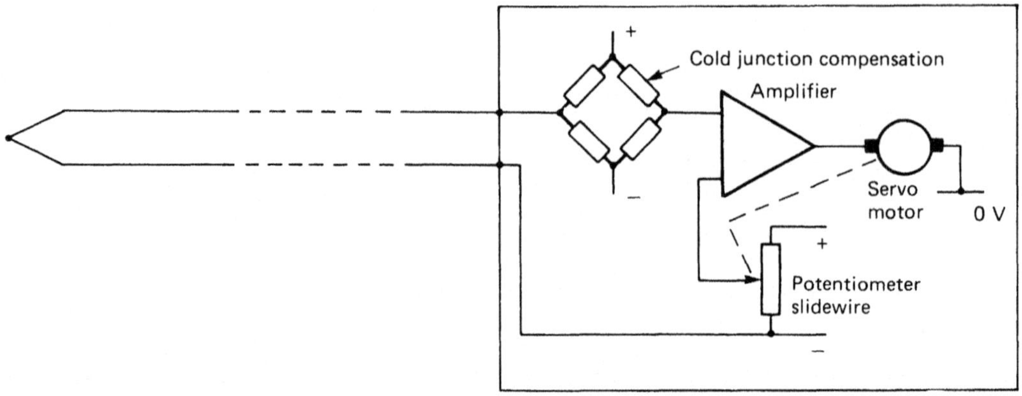

(a)

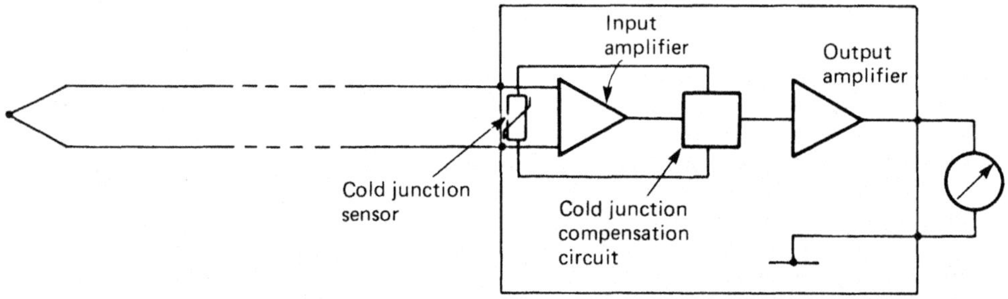

(b)

Figure 1.33 Cold junction compensation: (a) in conjunction with potentiometric indicating instrument, (b) alternative arrangement for cold junction compensation.

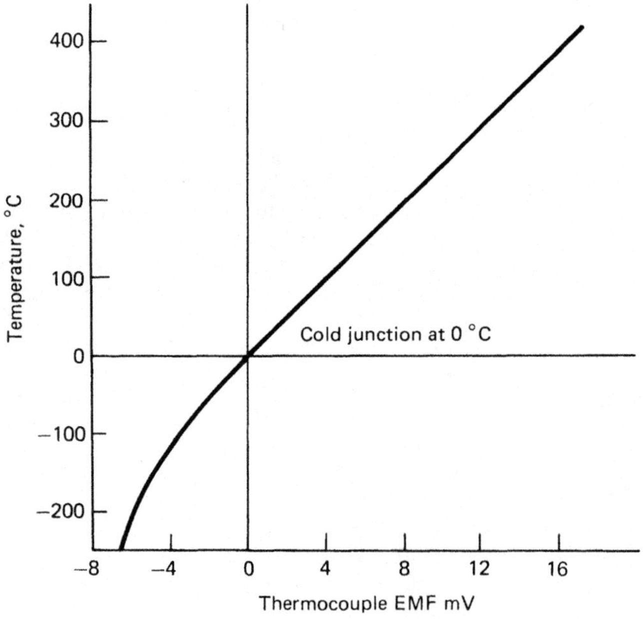

Figure 1.34 Type K thermocouple characteristic.

grammed to add the compensation value to the thermocouple output. In all electronic equipment for thermocouple signal processing the location of the sensor for cold junction temperature sensing is critical. It must be very close to the cold junction terminals and preferably in physical contact with them.

1.5.2 Thermocouple materials

Broadly, thermocouple materials divide into two arbitrary groups based upon cost of the materials, namely, base metal thermocouples and precious metal thermocouples.

1.5.2.1 Base metal thermocouples

The most commonly used industrial thermocouples are identified for convenience by type letters. The main types, together with the relevant British Standard specification and permitted tolerance on accuracy, are shown in Table 1.13. Also shown are their output e.m.f.s with the cold junction at 0 °C. These figures are given to indicate the relative sensitivities of the various couples. Full tables of voltages against hot junction temperatures are published in BS 4937. The standard also supplies the equations governing the thermocouple e.m.f.s for convenience for computer programming purposes. These equations are essentially square law; however, provided a thermocouple is used at temperatures remote from the neutral temperature its characteristic is very nearly linear. Figure 1.34 shows a plot of the characteristic for type K thermocouple. It can be seen that for temperatures in the range −50 °C to 400 °C the characteristic is approximately linear. The commonly used base metal thermocouples are types E, J, K, and T. Of these J and K are probably the most usual ones. They have a high e.m.f. output and type K is reasonably resistant to corrosion. Type T has a slight advantage, where the temperature measurement points are very remote from the instrumentation, that as one conductor is copper the overall resistance of the circuit can be lower than for other types. Table 1.14 shows the accuracy limits and recommended temperature measurement ranges as specified by the United States Standard ANSI MC 96 1 (1975).

1.5.2.2 Precious metal thermocouples

Thermocouples types B, R, and S clearly carry a considerable cost penalty and normally are only used when essential for their temperature range or their relatively high resistance to chemical attack. Their temperature top limit is 1500 °C for continuous use or 1650 °C for intermittent, spot reading, applications. This compares with 1100 °C continuous and 1300 °C intermittent for type K.

Errors in type R and S thermocouple readouts result from strain, contamination and rhodium drift.

The effect of strain is to reduce the e.m.f. resulting in low readings. The effect of strain may be removed by annealing the thermocouple. Installations should be designed to minimize strain on the thermocouple wires.

Contamination is by far the most common cause of thermocouple error and often results in ultimate mechanical failure of the wires. Elements such as Si, P, Pb, Zn, and Sn combine with platinum to form low melting point eutectics and cause rapid embrittlement and mechanical failure of the thermocouple wires. Elements such as Ni, Fe, Co, Cr, and Mn affect the e.m.f. output of the thermocouple to a greater or lesser degree, but contamination by these elements does not result in wire breakage and can only be detected by regular checking of the accuracy of the thermocouple. Contamination can be avoided by careful handling of the thermocouple materials before use and by the use of efficient refractory sheathing. Care should be taken to prevent dirt, grease, oil or soft solder coming into contact with the thermocouple wires before use. If the atmosphere surrounding the thermocouple sheath contains any metal vapour, the sheath must be impervious to such vapours.

Rhodium drift occurs if a rhodium–platinum limb is maintained in air for long periods close to its upper temperature limit. Rhodium oxide will form and volatilize, and some of this oxide can settle on, and react with, the platinum limb causing a fall in e.m.f. output. This is a comparatively slow process and is therefore only of significance in installations where the maximum stability and repeatability are required. Type B thermocouples are less susceptible to rhodium drift than types R or S, but type B has a lower e.m.f. than R and S and is subject to higher errors.

To overcome these disadvantages Pallador I which has a thermal e.m.f. comparable with iron/constantan, and Pallador II which provides a noble metal alternative to nickel–chromium/nickel–aluminium have been introduced. The positive limb of Pallador I is 10 per cent iridium–platinum and the negative limb is 40 per cent palladium–gold, while the positive limb of Pallador II is $12\frac{1}{2}$ per cent platinum–palladium and the negative 46 per cent palladium–gold. The maximum operating range for Pallador I thermocouples is 1000 °C and the maximum operating temperature for Pallador II is 1200 °C when protected by a 10 per cent rhodium–platinum sheath. (Pallador is the registered trade mark of Johnson Matthey Metals Ltd.)

The corresponding base metal thermocouple wires may be used as the compensating lead and an accuracy

Table 1.13 Thermocouples to British Standards

Type	*Conductors (positive conductor first)*	*BS 4937 Part No.*	*BS 1041, Part 4: 1966 Tolerance on temperature*	*Output for indicated temperature Cold junction at 0 °C*	*Service temperature. Max intermittant service in brackets*
B	Platinum: 30% Rhodium Platinum: 6% Rhodium	Part 7: 1974	0 to 1100 °C ± 3 °C 1100 to 1550 °C ± 4 °C	1.241 mV at 500 °C	*0 to 1 1500 °C (1700 °C)* Better life expectancy at high temperature than types R & S
K	Nickel: Chromium/Constantan (Chromel/Constantan) (Chromel/Advance)	Part 6: 1974	0 to 400 °C ± 3 °C	6.317 mV at 100 °C	*−200 to 850 °C (1100 °C)* Resistant to oxidizing atmospheres
J	Iron/Constantan	Part 3: 1973	0 to 300 °C ± 3 °C 300 to 850 °C ± 1%	5.268 mV at 100 °C	*−200 to 850 °C (1100 °C)* Low cost, suitable for general use
K	Nickel:Chromium/Nickel:Aluminium (Cromel/Alumel), (C/A), (T1/T2)	Part 4: 1973	0 to 400 °C ± 3 °C 400 to 1100 °C ± 0.75%	4.095 mV at 100 °C	*−200 to 1100 °C (1300 °C)* Good general purpose, best in oxidizing atmosphere
R	Platinum:13% Rhodium/Platinum	Part 2: 1973 }	0 to 1100 ± 1 °C 1100 to 1400 °C ± 2 °C 1400 °C ± 3 °C	{ 4.471 mV at 500 °C }	*0 to 1500 °C (1700 °C)* High temperature corrosion resistant
S	Platinum:10% Rhodium/Platinum	Part 1: 1973 }		{ 4.234 mV at 500 °C }	
T	Copper/Constantan; (Copper/Advance) (Cu/Con)	Part 5: 1974	0 to 100 °C ± 1 °C 100 to 400 °C ± 1%	4.277 mV at 100 °C	*−250 °C to 400 °C (500 °C)* High resistance to corrosion by water
	Rhodium:Iridium/Rhodium	Composition and accuracy to be agreed with manufacturer		Typically 6.4 mV at 1200 °C	0 to 2000 °C (2100 °C)
	Tungsten:Rhenium 5%/Tungsten: Rhenium 26%	Accuracy to be agreed with manufacturer		8.890 mV at 500 °C	0 to 2300 °C (2600 °C)
	Tungsten/Molybdenum	Composition and accuracy to be agreed with manufacturer		—	1250 to 2600 °C

Table 1.14 Thermocouples to American Standard ANSI MC 96 (1975)

Type	*Temperature range* (°C)	*Standard quality (whichever is greater)*	*Special quality (whichever is greater)*
B	800 to 1700	±0.5%	–
E	0 to 900	±1.7 °C or ±0.5%	±1.0 °C or ±0.4%
	−200 to 0	±1.7 °C or ±1%	–
J	0 to 750	±2.2 °C or ±0.75%	±1.1 °C or ±0.4%
K	0 to 1.200	±2.2 °C or ±0.75%	±1.1 °C or ±0.4%
	−200 to 0	±2.2 °C or ±2%	–
R, S	0 to 1450	±1.5 °C or ±2.5%	±0.6 °C or ±0.1%
T	0 to 350	±1.0 °C or 0.75%	±0.5 °C or ±0.4%
	−22 to 0	±1.0 °C or 1.5%	–

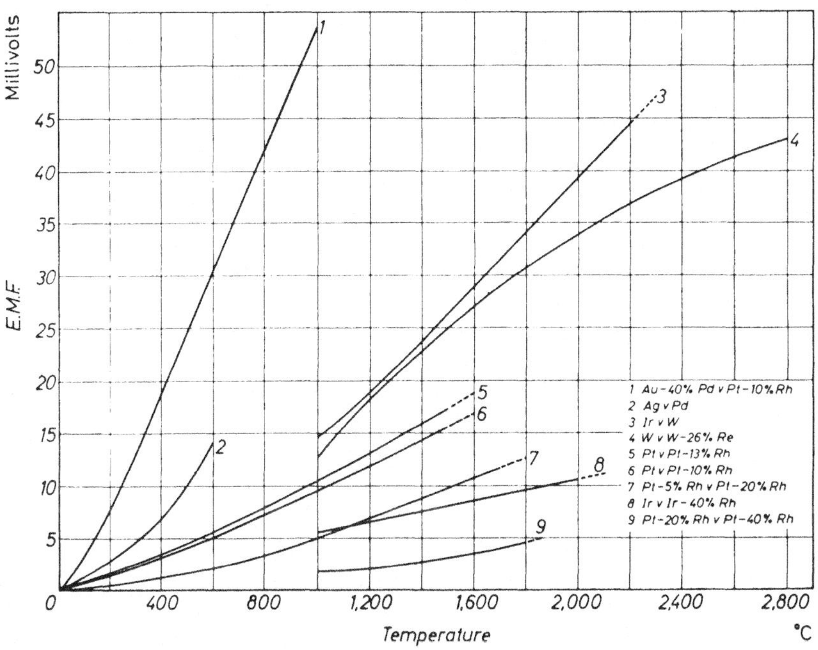

Figure 1.35 Summary of thermoelectric properties of precious metal thermocouples. Broken lines indicate areas for intermittent service.

of ±1 per cent will be attained on an instrument calibrated on the base metal characteristics. When the instrument is calibrated on the Pallador temperature–e.m.f. relationship an accuracy of ±2 K over the whole operating range is attainable.

Noble metal thermocouples may also be used for measuring cryogenic temperatures. Iron–gold/nickel–chromium or iron–gold/silver (normal silver with 0.37 atomic per cent gold) may be used for temperatures from 1 K to above 300 K.

Noble metal thermocouples are often used in the 'metal-clad' form with magnesia or alumina powder as the insulant. This form of construction is described in Section 1.5.3.2.

The following sheath materials are used: nickel, stainless steel, inconel in 1.6 and 3.2 mm sizes, and 5 per cent rhodium-plated and 10 per cent rhodium–platinum both in 1.0 mm sizes. For high temperature work other special thermocouples have been developed, tungsten 5 per cent rhenium/tungsten 20

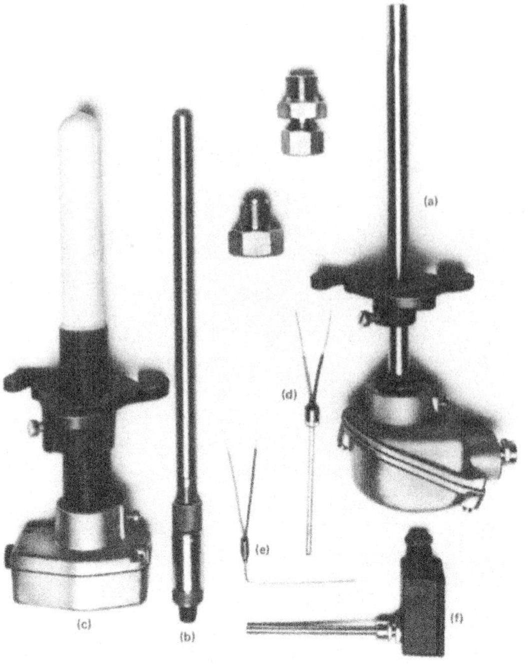

Figure 1.36 Examples of industrial thermocouple probes (courtesy Kent Industrial Measurements Ltd.).

per cent rhenium for use in hydrogen, vacuum and inert gas atmospheres up to 2320 °C and tungsten/molybdenum and tungsten/iridium for temperatures up to 2100 °C.

There is quite a wide range of precious metal thermocouples available. Types B, R and S are specified in BS 4937. These three are based only on platinum and rhodium. Gold, iridium, other 'platinum metals' and silver are also not uncommonly used. Figure 1.35 shows the characteristics of some of the options available.

1.5.3 Thermocouple construction

Thermocouples, like resistance thermometers and other temperature sensors, are available in a wide range of mechanical constructions.

1.5.3.1 Plain wire thermocouples

For use in protected environments, such as for laboratory use or inside otherwise enclosed equipment, plain wire thermocouples can be used. They are also used in plant where the fastest possible response is required. However, they suffer from the obvious disadvantage that they are both fragile and liable to chemical attack. The wires are available insulated with PVC or glass fibre sleeving, or for use with higher temperatures the wires can be insulated with refractory ceramic beads or sleeves.

1.5.3.2 Sheathed thermocouples

Thermocouples for use in plant situation, where robust construction is required or where they need to be interchangeable with other types of temperature measurement equipment, are available sheathed in steel or stainless steel designed for direct insertion into process vessels or for use in a thermometer pocket. Figure 1.36(a) and (b) show typical insertion probes. Where thermocouples are to be immersed in very corrosive process fluids or into very high temperature locations they are available constructed in ceramic sheaths as in Figure 1.36(c). Sheathed thermocouples, especially the ceramic ones, suffer from a slow response time, typically a minute or more. However, the locations where they are essential for their mechanical properties are usually in heavy plant where temperatures do not normally move fast in any case.

1.5.3.3 Mineral-insulated thermocouples

Probably the most versatile format for thermocouples is the mineral-insulated (MI) construction. In this form the thermocouples are made from mineral-insulated cable similar in concept to the MI cable used for electrical wiring applications. It differs however in that the conductors are of thermocouple wire and the sheath is usually stainless steel. The insulation, however, is similar, being in the form of finely powdered and densely compacted ceramic, usually aluminium oxide or magnesium oxide. Figure 1.36 shows MI thermocouples at (d), (e) and (f).

They are available in diameters from 1 millimetre up to 6 millimetres and can be supplied in any length required. The junction can be either insulated (a) or welded (b) to the tip of the sheath as shown in Figure 1.37. The latter arrangement has the advantage of very quick response. For some applications the junction being connected to the plant earth via the sheath tip can be unacceptable so in such cases insulated thermocouples must be used. The principal advantages are their quick response and mechanical flexibility, being able to be bent into almost any shape. Care must be taken if re-using MI thermocouples, for though they can be straightened or re-bent to a new shape this cannot be done too often. Either the wires break or the insulation gets displaced and the thermocouple becomes short-circuited. As shown in Figures 1.36 and

1.38, MI thermocouples can be supplied fitted with a variety of terminations. A further useful advantage of MI thermocouples is that the cable can be bought in rolls together with suitable terminations and the thermocouples can be made up to the required specifications on site. Also in situations where robust cabling is required, MI thermocouple cable can be used in lieu of compensating cable (see Section 1.5.3.9).

1.5.3.4 Surface contact thermocouples

Thermocouples for the measurement of the surface temperature of objects such as pipes or other components or plant items are available. On pipes a surface measurement makes a simple but not very accurate non-invasive temperature measurement. For higher temperatures or more rugged applications thermocouples are available embedded in a metal plate designed to be clamped or welded to the component to be measured. For lower temperature applications, below about 200 °C, or for use in protected environments, self-adhesive contact surface thermocouples are supplied. In these probes the thermocouple is embedded in a small plastic pad coated on one face with a suitable contact adhesive.

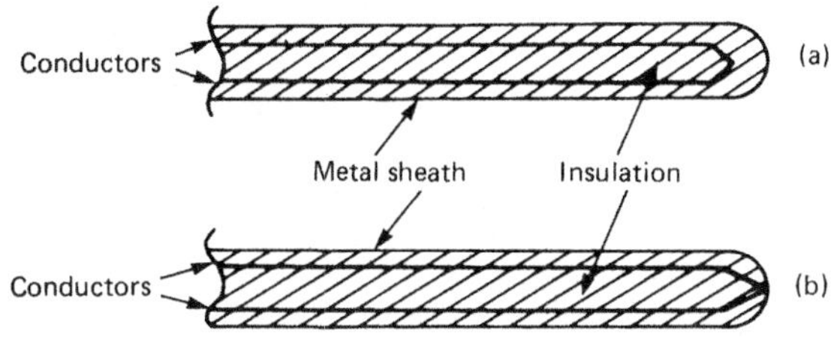

Figure 1.37 Mineral insulated thermocouples: (a) insulated junction, (b) junction welded to sheath.

1.5.3.5 Hot metal thermocouples

Where it is necessary to make spot measurements of the temperature of hot metal billets, very simple test prods are available which consist of a two-pronged 'fork'. The two prongs are made of the two thermocouple metals with sharpened points. When both prongs are in contact with the hot metal two junctions are formed, metal A to the billet and the billet to metal B. If the billet is large and enough time is allowed for the tips of the prongs to reach the temperature of the

Crimp on threaded seal type
Stainless steel 8 mm x 1.0 – 6g ISO metric externally threaded.
Sealing: epoxy resin
Max. operating temperature: 105° C

Pot seal type
Stainless steel screw-on pot
Sealing: plastic sealing compound
Max. operating temperature: 135°C

Connector head type
Die cast aluminium alloy connector head
Cable entry gland: 6 mm nylon, 10 mm nylon or 6/10 mm brass
Max. operating temperature: 105 ° C
See photograph on page 12.

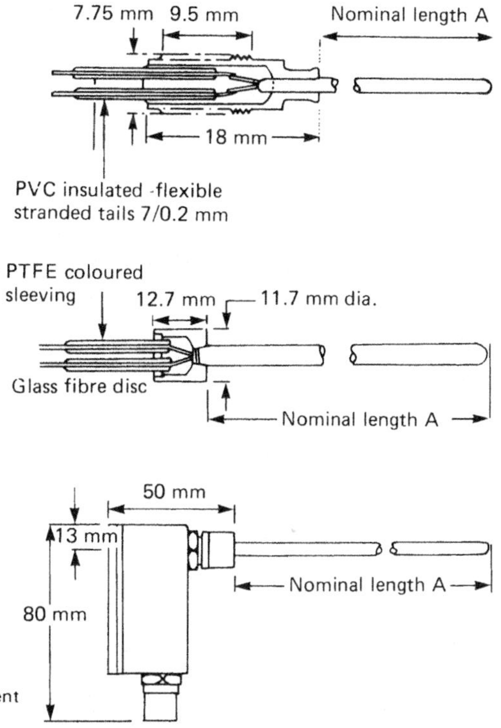

Figure 1.38 MI thermocouple terminations (courtesy Kent Industrial Measurements Ltd.).

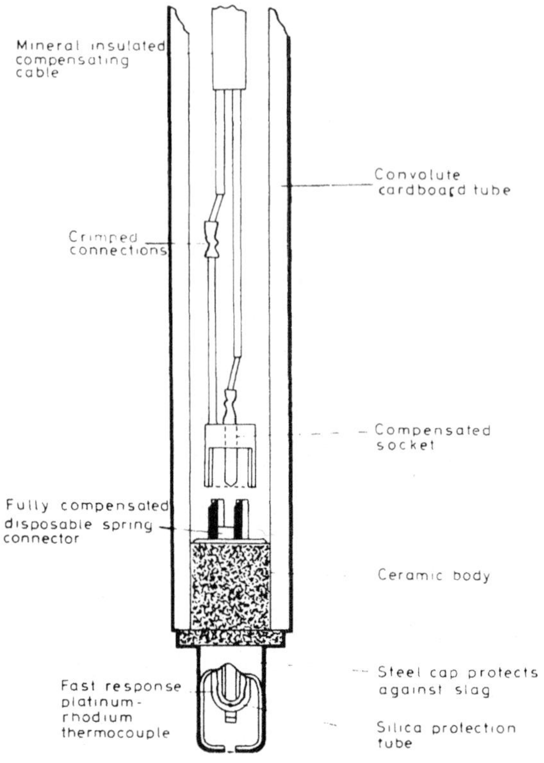

Figure 1.39 Liquid metal thermocouple.

billet then both junctions will be at the same temperature and the error thermal e.m.f.s cancel. This makes a simple, quick and very inexpensive way of measuring hot metal temperatures. The points of the prongs are screwed to the main assembly and are expendable. As soon as they lose their sharpness or begin to get corroded they can be changed.

1.5.3.6 Liquid metal thermocouples

When measuring the temperature of liquid metals such as steel it is desirable to use an expendable probe. The cost of a fully protected probe would be very high and the response time slow. For checking the temperature of liquid steel a dipstick probe can be used. The probe itself is robust and constructed with a socket of thermocouple material in the end. A disposable platinum–rhodium/platinum thermocouple itself lasts in the molten metal for a few seconds, long enough to take a temperature measurement. Figure 1.39 shows this arrangement.

1.5.3.7 Thermopiles

Where a very small temperature rise is to be measured many thermocouples may be connected in series. All the hot junctions are on the object whose temperature is to be measured and all the cold junctions are kept at a constant and known temperature. Where a quick temperature response is required these thermocouples can be of very thin wire of about 25 μm diameter. A speed of response of the order of 10 milliseconds can be achieved. Typical applications of thermopiles are to be found in infrared radiation measurement. This subject is dealt with in Section 1.6.

1.5.3.8 Portable thermocouple instruments

With the development, over the last decade of micro-electronic equipment, portable electrical thermometers have become very popular. They are available with either analogue or digital readouts. The analogue instruments are about the size of an analogue multi-meter, the digital instruments are about the size of a pocket calculator. While most of these instruments use type K thermocouples they are available for use with other thermocouple materials. There are also portable thermometers available using resistance thermometer or thermistor sensors. However, the thermocouple instruments are on the whole the most popular. The more sophisticated instruments have the option to use more than one type of thermocouple: a switch on the instrument sets it for the type in use. They are also available with a switched option to read out in Celsius or Fahrenheit. A range of hand-held probes are supplied for use with these instruments. Figure 1.40 shows some of the options available. The spring-loaded thermocouples are for surface contact measurements, hypodermic probes are supplied for such applications as temperature measurements in food such as meat where it may be an advantage to know the internal temperature of the material. Figure 1.41 shows typical analogue and digital readout instruments.

1.5.3.9 Thermocouple compensating cable

Ideally a thermocouple connects back to the reading instrument with cables made of the same metals as the thermocouple. This does however have two disadvantages in industrial conditions. Firstly many thermocouple metals have high electrical resistance. This means that on long runs, which on a big plant may be up to 1000 metres or more, heavy gauge conductors must be used. This is not only expensive but also makes the cables difficult to handle. Secondly in the case of precious metal thermocouples, types B, R and S for instance, the cost would be very high indeed. To overcome these problems compensating cables are used, see Figure 1.42. These cables are made of base metal and are of lower resistivity than the thermocouple material. The alloys used have thermoelectric properties that essentially match the thermocouples

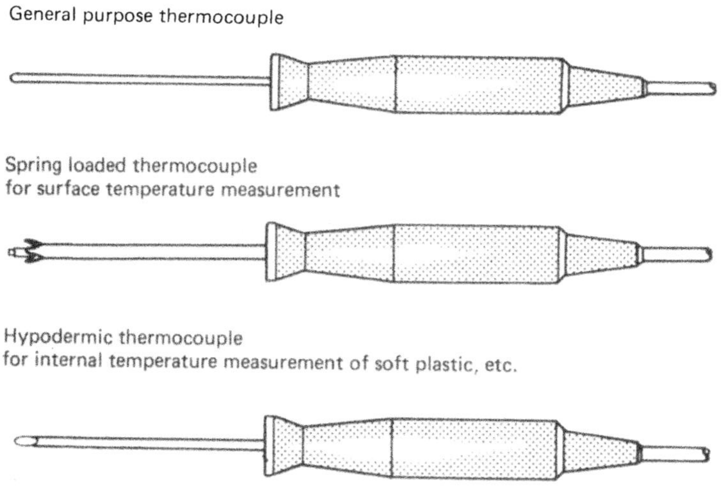

Figure 1.40 Hand-held thermocouple probes.

Figure 1.41 Portable analogue and digital thermocouple instruments (courtesy Comark Electronics Ltd.).

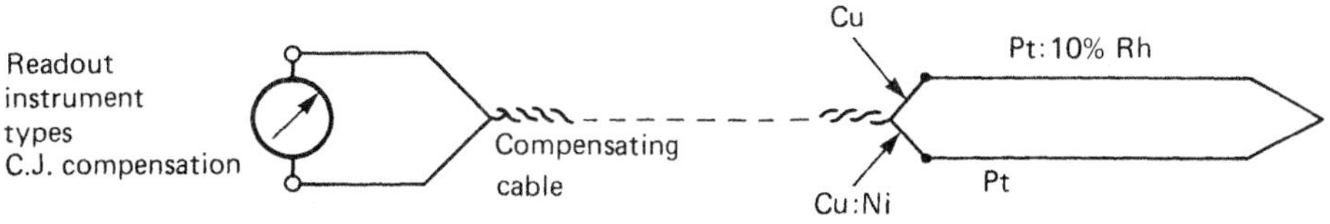

Figure 1.42 Thermocouple compensating cable.

themselves over a limited ambient temperature range.

Examples of compensating cables are:

Type	*Composition*	*Thermo-couples compensated*	*Temperature limitations*
U	Copper/copper–nickel	R and S	0–50 °C
Vx	Copper/Constantan	K	0–80 °C

Other base metal thermocouples such as types J and T comprise relatively inexpensive and low resistance metals. They are therefore normally installed using cables consisting of the same metals as the thermocouples themselves.

1.5.3.10 Accuracy consideration

The very extensive use of thermocouples stems from their great versatility combined with their low cost. However, as seen in Tables 1.13 and 1.14, thermocouples have a fairly wide permitted tolerance. This is due to the fact that most metals used for thermocouples are alloys and it is not possible to manufacture alloys to the same reproducibility as pure metals. It must be said that, in general, manufacturers do manufacture their thermocouples to better tolerance than BS 4937 demands. But, where the highest accuracy is required, it is essential to calibrate thermocouples on installation and to recalibrate them at regular intervals to monitor any deterioration due to corrosion or diffusion of foreign elements into the hot junction.

Where high accuracy is required it is necessary to calibrate first the thermocouple readout instrument and then the thermocouple itself in conjunction with the instrument.

The calibration of instruments can be done with a precision millivolt source which injects a signal equivalent to the temperature difference between the ambient or cold junction temperature and a temperature in the region in which the thermocouple is to be used.

To calibrate or check thermocouples the hot junction must be kept at an accurately known temperature. This can be done by inserting it into a heated isothermal block. An isothermal block is a block of metal, large compared with the thermocouple being measured and made of copper or aluminium. The block has provision for heating it and in some cases cooling. It is well insulated from the environment and is provided with suitable holes for inserting various sizes of thermocouple. Where not so high precision is required the thermocouple can be immersed in a heated fluidized sand bath. This consists of an open vessel fitted with a porous bottom (usually made of sintered metal). Heated air is forced up through the bottom. The vessel is filled with carefully graded sand. With the air coming up through it the sand behaves like a liquid. It takes up the temperature of the air. The sand is a good heat transfer medium. The apparatus makes a most convenient way of calibrating temperature probes. Where maximum accuracy is essential the thermocouple should be calibrated against one of the IPTS-68 secondary reference points. Table 1.5 shows some of the points.

In carrying out these calibrations the whole installation needs to be calibrated: thermocouple readout instrument together with compensating cable. In cases where very high accuracy is required, compensating cable should not be used, the conductors should be thermocouple metal for the full length of the installation.

There is now on the market some very versatile equipment for thermocouple calibration. Figure 1.43 shows a microprocessor-controlled calibrator. Its facilities include thermocouple simulation for types E, J, K, R, S and T, thermocouple output measurement with cold junction compensation and resistance thermometer simulation. Tests can be static or dynamic using ramp functions. Resolution is to 0.1 K and accuracy is ± 0.01 per cent of reading.

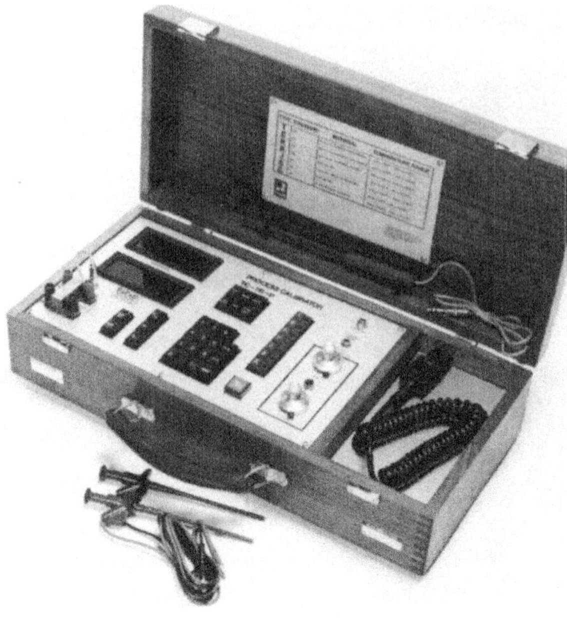

Figure 1.43 Thermocouple calibration equipment (courtesy Haven Automation Ltd.).

As with any other type of temperature measurement the location of the thermocouple junctions is critical. This is just as important for the cold junction as for the hot junction. It must be remembered that there may well be a temperature gradient over quite short distances in an instrument and unless the cold junction temperature sensor is in close thermal contact with the cold junction itself a reading error of several degrees Celsius may result. This problem is at its worst with mains electricity powered measuring instruments where there is a certain amount of heat liberated by the power unit.

The point to remember is that it is not usually adequate to measure the air temperature in the vicinity of the cold junctions. The sensor should be in good thermal contact with them.

An obvious point, but one which surprisingly often causes trouble, is the mismatch between the thermocouple and the measuring instrument. The obvious mismatch is using the wrong type of thermocouple or compensating cable.

In the case of galvanometric instruments inaccuracies occur if sufficient care has not been taken in the winding of the make-up resistor or if the thermocouple has been changed and the new external circuit resistance not checked. Careless location or make-up of the ballast resistor so that one of the cold junction terminals is too remote from the cold junction compensating element causes variable errors of several degrees as the ambient temperature changes. Where the ballast resistor required is of a low value, 10 ohms or so, the best arrangement may well be to use a coil of compensating cable of the right resistance.

1.6 Measurement techniques – radiation thermometers

1.6.1 Introduction

As was mentioned in Section 1.1, thermal energy may be transferred from one body to another by radiation as well as by conduction. The amount of thermal energy or heat leaving a body by radiation and the wavelength of that radiation are functions of the temperature of the body.

This dependence on temperature of the characteristics of radiation is used as the basis of temperature measurement by radiation thermometers.

1.6.1.1 Black body radiation

An ideal black body is one that at all temperatures will absorb all radiation falling on it without reflecting any whatever in the direction of incidence. The absorptive power of the surface, being the proportion of incident radiation absorbed, will be unity. Most surfaces do not absorb all incident radiation but reflect a portion of it. That is, they have an absorptive power of less than unity.

A black body is also a perfect radiator. It will radiate more radiation than a body with an absorptive power of less than unity. The emissive power is called the 'emissivity' of a surface. The emissivity is the ratio of the radiation emitted at a given temperature compared to the radiation from a perfectly black body at the same temperature.

The total emissivity of a body is the emissive power over the whole band of thermal radiation wavelengths and is represented by ε_t. When only a small band of wavelengths is considered the term 'spectral emissivity' is used and a subscript is added defining the wavelength band, e.g., $\varepsilon_{1.5}$ indicates the emissivity at 1.5 μm wavelength.

The emissivity of surfaces is not usually the same over all wavelengths of the spectrum. In general the emissivity of metals is greater at shorter wavelengths and the emissivity of oxides and refractory materials is greater at longer wavelengths. Some materials may have a very low emissivity at a particular wavelength band and higher emissivities at shorter and longer wavelength. For instance, glass has an emissivity of almost zero at 0.65 μm.

Realization of a black body radiator A black body radiator is achieved in practice by an enclosure, A in Figure 1.44, having a relatively small orifice B from which black body radiation is emitted. The inside walls of the enclosure must be at a uniform temperature. To show that the orifice B behaves as a black body consider the ray of radiation C entering the chamber through B. The ray will suffer many reflections on the inside walls of the enclosure before it emerges at B. Provided the walls of the chamber are not perfectly reflecting the total energy of the radiation will have been absorbed by the many reflections before the ray can emerge. The orifice is then totally absorbing all radiation that enters it. It is a black body.

To show that the orifice must also radiate as a black body first consider a body in a radiant flux at any single wavelength. If that body did not radiate energy at that wavelength as fast as it absorbed it, it would rapidly get warmer than its environment. In practice a body will be at thermal equilibrium with its surroundings so it must be radiating energy as it receives it.

Therefore the emissivity ε of a body must equal its absorbance α. The orifice B which is a black body absorber must also be a black body radiator.

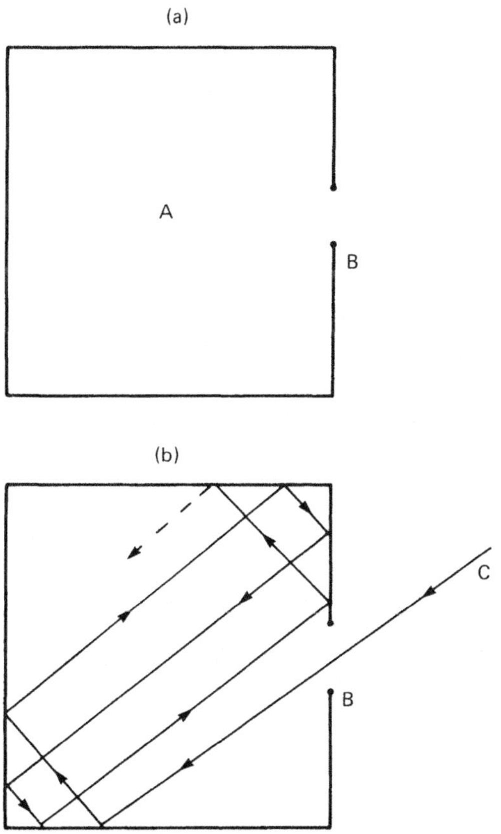

Figure 1.44 (a) Black body radiator, (b) absorption of ray of radiation by black body radiator.

In practice a sighting hole in a furnace will radiate as a black body if the furnace and its contents are in thermal equilibrium and provided it does not contain a gas or flame which absorbs or radiates preferentially in any wavelength band. However, the radiation from the sighting hole will only be black body radiation provided everything in the furnace is at the same temperature. When all objects in the furnace are at the same temperature all lines of demarcation between them will disappear. If a cold object is introduced to the furnace it will be absorbing more energy than it is radiating, the rest of the furnace will be losing more radiation than it receives. Under these conditions the radiation will no longer be black body radiation but will be dependent upon the emissivity of the furnace walls.

Prevost's theory of exchanges Two bodies A and B in a perfectly heat-insulated space will both be radiating and both be absorbing radiation. If A is hotter than B it will radiate more energy than B. Therefore B will receive more energy than it radiates and consequently its temperature will rise. By contrast body A will lose more energy by radiation than it receives so its temperature will fall. This process will continue until both bodies reach the same temperature. At that stage the heat exchanged from A to B will be equal to that exchanged from B to A.

A thermometer placed in a vessel to measure gas temperature in that vessel will, if the vessel walls are cooler than the gas, indicate a temperature lower than the gas temperature because it will radiate more heat to the vessel walls than it receives from them.

Black body radiation: Stefan–Boltzmann law The total power of radiant flux of all wavelengths R emitted into the frontal hemisphere by a unit area of a perfectly black body is proportional to the fourth power of the temperature Kelvin:

$$R = \sigma T^4 \tag{1.30}$$

where σ is the Stefan–Boltzmann constant, having an accepted value of $5.670\,32 \times 10^{-8}$ W.m^{-2}.K^{-4}, and T is the temperature Kelvin.

This law is very important, as most total radiation thermometers are based upon it. If a receiving element at a temperature T_1 is arranged so that radiation from a source at a temperature T_2 falls upon it, then it will receive heat at the rate of σT_2^4, and emit it at a rate of σT_1^4. It will, therefore, gain heat at the rate of $\sigma(T_2^4 - T_1^4)$. If the temperature of the receiver is small in comparison with that of the source, then T_1^4 may be neglected in comparison with T_2^4, and the radiant energy gained will be proportional to the fourth power of the temperature Kelvin of the radiator.

1.6.1.2 The distribution of energy in the spectrum: Wien's laws

When a body is heated it appears to change colour. This is because the total energy and distribution of radiant energy between the different wavelengths, is changing as the temperature rises. When the temperature is about 500 °C the body is just visibly red. As the temperature rises, the body becomes dull red at 700 °C, cherry red at 900 °C, orange at 1100 °C, and finally white hot at temperatures above 1400 °C. The body appears white hot because it radiates all colours in the visible spectrum.

It is found that the wavelength of the radiation of the maximum intensity gets shorter as the temperature rises. This is expressed in Wien's displacement law:

$$\lambda_m T = \text{constant} = 2898\ \mu\text{m}.\text{K} \tag{1.31}$$

where λ_m is the wavelength corresponding to the radiation of maximum intensity, and T is the temperature Kelvin. The actual value of the spectral radiance at the wavelength λ_m is given by Wien's second law:

$$L_{\lambda_m} = \text{constant} \times T^5 \tag{1.32}$$

where L_{λ_m} is the maximum value of the spectral radiance at any wavelength, i.e. the value of the radiance at λ_m, and T is the temperature Kelvin. The constant does not have the same value as the constant in equation (1.31). It is important to realize that it is only the maximum radiance at one particular wavelength which is proportional to T^5, the total radiance for all wavelengths is given by the Stefan–Boltzmann law, i.e. it is proportional to T^4.

Wien deduced that the spectral concentration of radiance, that is, the radiation emitted per unit solid angle per unit area of a small aperture in a uniform temperature enclosure in a direction normal to the area in the range of wavelengths between λ and $\lambda + \delta\lambda$ is $L_\lambda . \delta\lambda$ where

$$L_\lambda = \frac{C_1}{\lambda^5 . e^{C_2/\lambda T}} \tag{1.33}$$

where T is the temperature Kelvin and C_1 and C_2 are constants. This formula is more convenient to use and applies with less than 1 per cent deviation from the more refined Planck's radiation law used to define IPTS-68 provided $\lambda T < 3 \times 10^3$ m.K.

In 1900 Planck obtained from theoretical considerations based on his quantum theory, the expression

$$L_\lambda = \frac{C_1}{\lambda^5 (e^{C_2/\lambda T} - 1)} \tag{1.34}$$

where the symbols have the same meaning, and $C_2 = 0.014\,388$ m.K.

These laws also enable the correction to be calculated for the presence of an absorbing medium

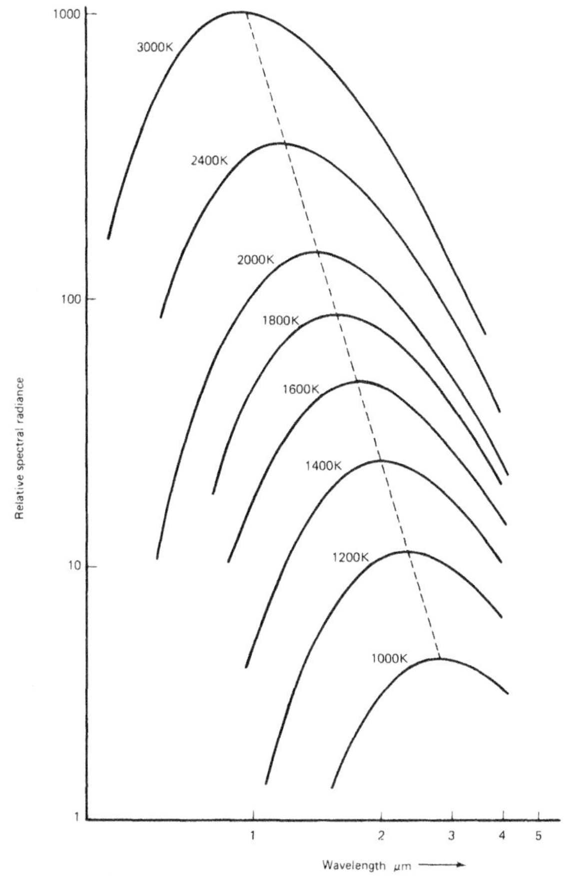

Figure 1.45 Spectral energy distribution with temperature.

such as glass in the optical pyrometer, and also the correction required for changes in the spectral emissive power of the radiating surface.

The variation of spectral radiance with wavelength and temperature of a black body source is given by Figure 1.45.

1.6.2 Radiation thermometer types

Since the energy radiated by an object is a function of its absolute temperature this is a suitable property for the non-contact and non-intrusive measurement of temperature. Instruments for temperature measurement by radiation are called radiation thermometers. The terms pyrometer or radiation pyrometer were formerly used.

There are four principal techniques for the measurement of temperature by the radiation from a hot body: total radiation, pyroelectric, photoelectric, and optical.

Instruments using the first three of these techniques are normally constructed in the same general physical form. Figure 1.46 shows the general format of one of these instruments. It consists of a cylindrical metal body made of aluminium alloy, brass or plastic. One end of the body carries a lens, which depending on the wavelength range required, consists of germanium, zinc sulphide, quartz, glass or sapphire. The opposite end carries the electrical terminations for connecting the sensing head to its signal conditioning module. A typical size of such a sensing head is 250 mm long by 60 mm diameter. A diagrammatic sketch of the construction of the instrument is shown in Figure 1.47. Infrared energy from a target area on the object whose temperature is to be measured is focused by the lens onto the surface of the detector. This energy is converted to an electrical signal which may be amplified by a head amplifier on the circuit board. Power is supplied to the instrument and the output transmitted down a cable which is connected to terminals in the termination box. In instruments working in the near-infrared region where the lens is transparent to visible light a telescope can be provided, built into the instrument, so that it can be focused and aligned by looking through the lens.

A primary advantage of radiation thermometers, especially when used to measure high temperatures, is that the instrument measuring head can be mounted remote from the hot zone in an area cool enough not to exceed the working temperature of the semiconductor electronics, typically about 50–75 °C. However, where the instrument has to be near the hot region, such as attached to the wall of a furnace, or where it needs to be of rugged construction, it can be housed in an air- or water-cooled housing. Such a housing is shown in Figure 1.48.

The function of the lens as indicated above is to concentrate the radiation from the source onto the surface of the sensor. This also has the great advantage that the instrument reading is substantially independent of the distance from the source, provided

Figure 1.46 General-purpose radiation thermometer (courtesy Land Infrared Ltd.)

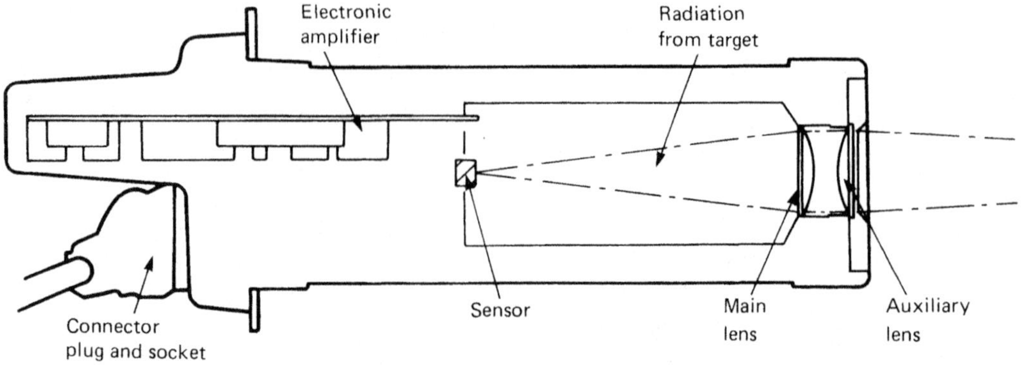

Figure 1.47 Diagram of radiation thermometer.

the source is large enough for its image to fully fill the area of the sensor. The lens material depends on the wavelength to be passed. This will normally be a function of the temperature range for which the instrument is specified. For lower temperatures the lens material will be chosen to give a wide wavelength passband. For higher temperatures a narrower passband may be acceptable. Of course the higher the temperature to be measured the shorter the wavelength that needs to be passed by the lens. Table 1.15 shows the wavelength passband of some lens materials.

Table 1.15 Wavelengths transmitted by lens materials

Lens material	*Passband* (μm)
Pyrex	0.3–2.7
Fused silica	0.3–3.8
Calcium fluoride	0.1–10
Arsenic trisulphide	0.7–12
Germanium	2–12
Zinc selenide	0.5–15

To achieve a wider wavelength range the focusing can be achieved with a concave mirror. Figure 1.49 shows diagrammatically the general arrangement of a reflection instrument.

A special application of mirror focusing for radiation thermometry is in the temperature measurement of stars and other astronomic bodies. The thermopile, or more usually a semiconductor detector, is cooled with liquid nitrogen or helium to increase its sensitivity to very small amounts of radiation. It is located at the focus of a reflecting astronomical telescope. The telescope is directed to the body whose temperature is to be measured so that its image is focused on the detector. The whole assembly forms a very sensitive radiation thermometer with the ability to detect temperatures down to a few tens of kelvins.

1.6.2.1 Total radiation thermometer

In this type of instrument, the radiation emitted by the body whose temperature is required, is focused on a suitable thermal-type receiving element. This receiving element may have a variety of forms. It may be a resistance element, which is usually in the form of a very thin strip of blackened platinum, or a thermocouple or thermopile. The change in temperature of the receiving element is then measured as has already been described.

In a typical radiation thermopile a number of thermocouples made of very fine strips are connected

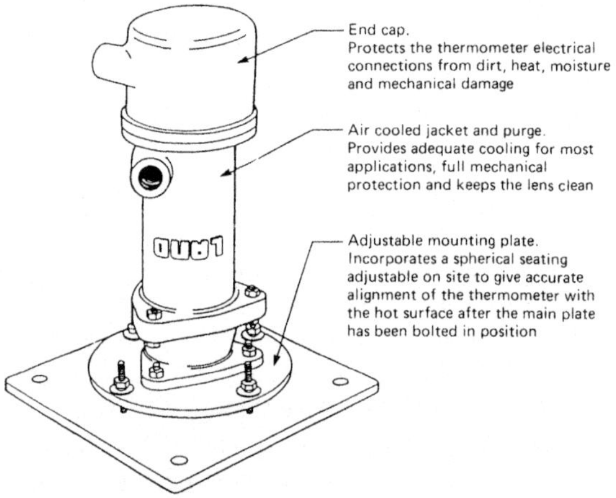

Figure 1.48 Air-cooled housing for radiation thermometer (courtesy Land Infrared Ltd.).

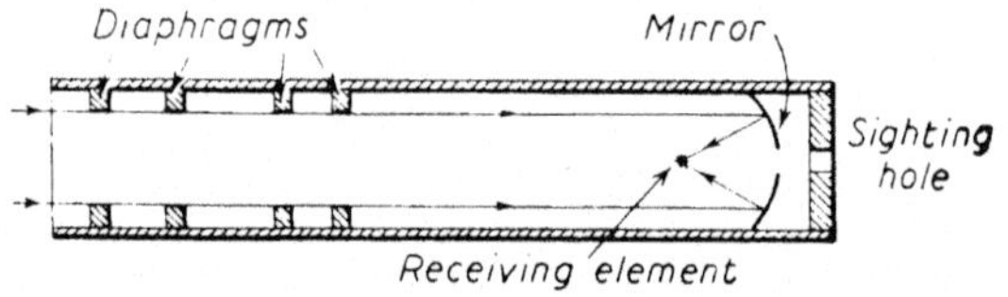

Figure 1.49 Mirror-focused radiation thermometer (courtesy Land Infrared Ltd.).

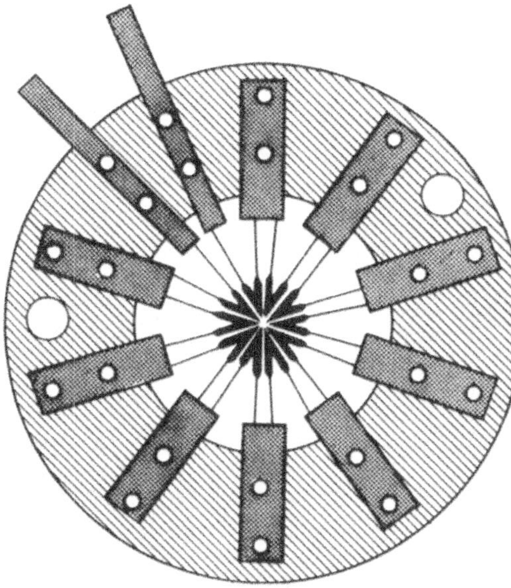

Figure 1.50 Thermopile for use in total radiation pyrometer.

in series and arranged side by side, or radially as in the spokes of a wheel, so that all the hot junctions, which are blackened to increase the energy-absorbing ability, fall within a very small target area. The thermoelectric characteristics of the thermopiles are very stable because the hot junctions are rarely above a few hundred degrees Celsius, and the thermocouples are not exposed to the contaminating atmosphere of the furnace. Stability and the fact that it produces a measurable e.m.f. are the main advantages of the thermopile as a detector. In addition, thermopiles have the same response to incoming radiant energy regardless of wavelength within the range 0.3–20 μm. The main disadvantage of the thermopile is its comparatively slow speed of response which depends upon the mass of the thermocouple elements, and the rate at which heat is transferred from the hot to the cold junctions. Increase in this rate of response can only be attained by sacrificing temperature difference with a resultant loss of output. A typical industrial thermopile of the form shown in Figure 1.50 responds to 98 per cent of a step change in incoming radiation in 2 seconds. Special thermopiles which respond within half a second are obtainable but they have a reduced e.m.f. otuput.

In order to compensate for the change in the thermopile output resulting from changes in the cold junction temperature an ambient temperature sensor is mounted by the cold junctions. Alternative thermal detectors to thermopiles are also used. Thermistors and pyroelectric detectors are currently in use. The advantage of thermistors is that they can be very small and so have a quick speed of response. Their main disadvantage is their non-linearity, though this is not so great a disadvantage as with a direct measurement of temperature because provision has to be made to linearize the radiated energy signal anyway.

Correction for emissivity When the temperature of a hot object in the open is being measured, due regard must be given to the correction required for the difference between the emissivity of the surface of the object and that of a perfectly black body.

The total radiant flux emitted by the source will be given by

$$R=\sigma A T_a^4 \qquad (1.35)$$

where ε is the total emissivity of the body, A is the area from which radiation is received, σ is the Stefan–Boltzmann constant, and T the actual temperature of the body.

This flux will be equal to that emitted by a perfectly black body at a temperature T_a, the apparent temperature of the body:

$$R=\sigma A T_a^4 \qquad (1.36)$$

Equating the value of R in equations (1.35) and (1.36):

$$\varepsilon\sigma A T^4=\sigma A T_a^4$$

$$T^4=\frac{T_a^4}{\varepsilon}$$

$$T=\frac{T_a}{\sqrt[4]{\varepsilon}} \qquad (1.37)$$

The actual correction to be applied to the apparent temperature is given in Figure 1.51. Table 1.16 shows the emissivity of some metals at different temperatures.

The radiation from a hot object can be made to approximate much more closely to black body radiation by placing a concave reflector on the surface. If the reflectivity of the reflecting surface is r, then it can be shown that the intensity of the radiation which would pass out through a small hole in the reflector is given by

$$R=\frac{\varepsilon}{1-r(1-\varepsilon)}\sigma T^4 \qquad (1.38)$$

where R is the radiation intensity through the hole, ε is the emissivity of the surface, σ is the Stefan–Boltzmann constant, and T the temperature in kelvin. With a gold-plated hemisphere, the effective emissivity of a surface of emissivity 0.6 is increased by this method to a value of 0.97.

Surface radiation thermometer A surface radiation thermometer manufactured by Land Infrared Ltd. uses the above principle, see Figure 1.52. This instrument uses a thermopile sited on a small hole in a gold-

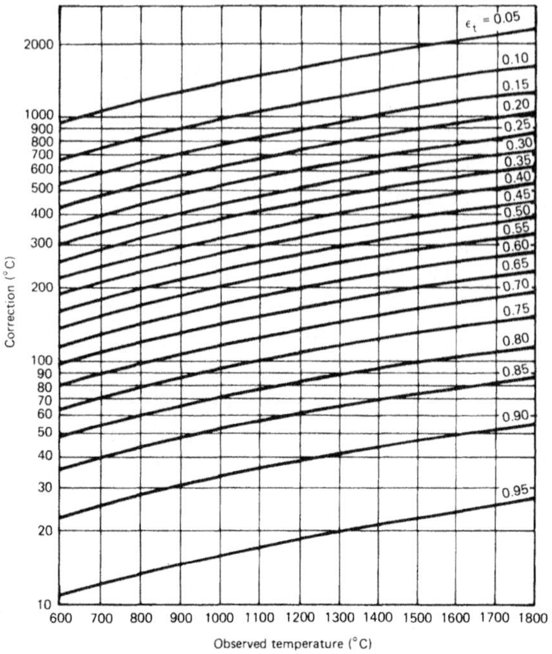

Figure 1.51 Emissivity corrections to the readings of a total radiation thermometer.

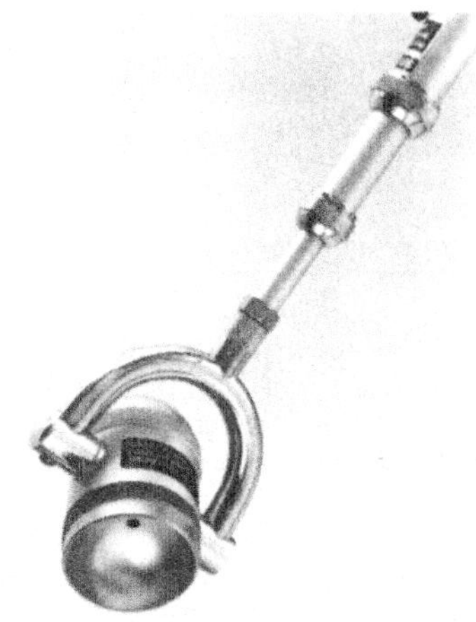

Figure 1.52(a)

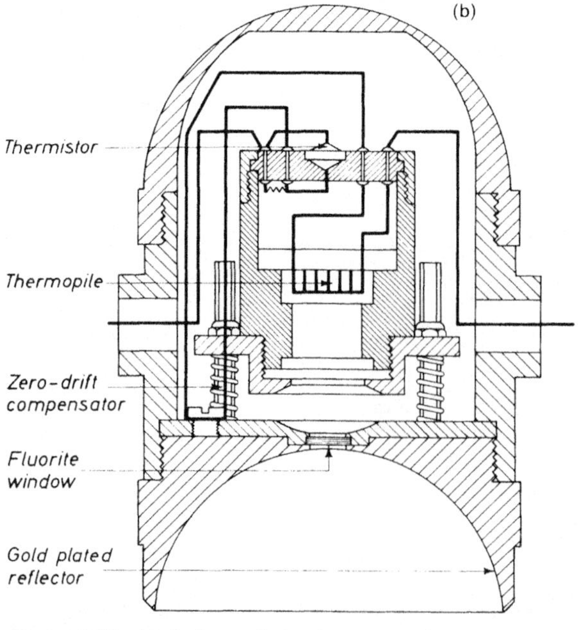

Figure 1.52 (a) Surface radiation thermometer (courtesy Land Infrared Ltd.). (b) cross-section diagram of Land surface radiation thermometer.

plated hemisphere mounted on the end of a telescopic arm.

Gold is chosen for the reflecting surface because it is the best reflector of infrared radiation known, and is not easily tarnished. The hole in the reflector is closed by a fluorite window which admits a wide range of radiation to the thermopile but excludes dirt and draughts. This pyrometer will give accurate surface temperature readings for most surfaces, other than bright or lightly oxidized metals, without any significant error due to surface emissivity changes. The standard instrument covers a temperature range of from 100 to 1300 °C on three scales. A special low temperature version is available for the range 0 to 200 °C. The indicator gives a reading in 5 to 6 seconds, and the pyrometer should not be left on the hot surface for more than this length of time, particularly at high temperatures. The thermistor bridge provides compensation for changes in the sensitivity of the thermopile at high temperatures, but if the head is too hot to touch it is in danger of damage to soldered joints, insulation etc.

Table 1.16 Total emissivity of miscellaneous materials

Total emissivity of unoxidized metals

Material	*25 °C*	*100 °C*	*500 °C*	*1000 °C*	*1500 °C*	*2000 °C*
Aluminium	0.022	0.028	0.060	—	—	—
Bismuth	0.048	0.061	—	—	—	—
Carbon	0.081	0.081	0.079	—	—	—
Chromium	—	0.08	—	—	—	—
Cobalt	—	—	0.13	0.23	—	—
Columbium	—	—	—	—	0.19	0.24
Copper	—	0.02	—	(Liquid 0.15)	—	—
Gold	—	0.02	0.03	—	—	—
Iron	—	0.05	—	—	—	—
Lead	—	0.05	—	—	—	—
Mercury	0.10	0.12	—	—	—	—
Molybdenum	—	—	—	0.13	0.19	0.24
Nickel	0.045	0.06	0.12	0.19	—	—
Platinum	0.037	0.047	0.096	0.152	0.191	—
Silver	—	0.02	0.035	—	—	—
Tantalum	—	—	—	—	0.21	0.26
Tin	0.043	0.05	—	—	—	—
Tungsten	0.024	0.032	0.071	0.15	0.23	0.28
Zinc	(0.05 at 300 °C)					
Brass	0.035	0.035	—	—	—	—
Cast Iron	—	0.21	—	(Liquid 0.29)		—
Steel	—	0.08	—	(Liquid 0.28)		—

Total emissivity ε_t of miscellaneous materials

Material	*Temp.* (°C)	ε_t	*Material*	*Temp.* (°C)	ε_t
Aluminium (oxidized)	200	0.11	Lead (oxidized)	200	0.63
	600	0.19	Monel (oxidized)	200	0.43
Brass (oxidized)	200	0.61		600	0.43
	600	0.59	Nickel (oxidized)	200	0.37
Calorized copper	100	0.26		1200	0.85
	500	0.26	Silica brick	1000	0.80
Calorized copper (oxidized)	200	0.18		1100	0.85
	600	0.19	Steel (oxidized)	25	0.80
Calorized steel (oxidized)	200	0.52		200	0.79
	600	0.57		600	0.79
Cast iron (strongly oxidized)	40	0.95	Steel plate (rough)	40	0.94
	250	0.95		400	0.97
Cast iron (oxidized)	200	0.64	Wrought iron (dull	25	0.94
	600	0.78	oxidized)	350	0.94
Copper (oxidized)	200	0.60	20Ni—25Cr—55Fe	200	0.90
	1000	0.60	(oxidized)	500	0.97
Fire Brick	1000	0.75	60Ni—12Cr—28Fe	270	0.89
Gold enamel	100	0.37	(oxidized)	560	0.82
	100	0.74		100	0.87
Iron (oxidized)	500	0.84	80Ni—20Cr (oxidized)	600	0.87
	1200	0.89		1300	0.89
Iron (rusted)	25	0.65			

Source: 'Temperature, its measurement & control' in *Science & Industry*, American Institute of Physics, Reinhold Publishing Co. (1941).

The instrument may be used to measure the mean emissivity of a surface for all wavelengths up to about 10 μm. This value can be used for the correction of total radiation thermometer readings. A black hemispherical insert is provided with the instrument which can be clipped into the hemispherical reflector to cover the gold. If two measurements are made, one with the gold covered and the other with the gold exposed, the emissivity can readily be deduced from the two measurements. A graph provided with the instrument enables the emissivity to be derived easily from the two readings, while a second graph gives an indication of the error involved in the temperature measurement of the hot body.

Calibration of total radiation thermometers A total radiation thermometer may be calibrated by sighting it through a hole into a black body enclosure of known temperature. A special spherical furnace was developed by the British Iron and Steel Research Association for this purpose. The furnace consisted of a sphere 0.3 m in diameter consisting of a diffusely reflecting material. For temperatures up to 1300 °C stainless steel, 80Ni 20Cr alloy, or nickel may be used. For temperatures up to 1600 °C silicon carbide is necessary, and for temperatures up to 3000 °C graphite may be used provided it is filled with argon to prevent oxidation. The spherical core is uniformly wound with a suitable electrical heating element, completely enclosed in a box containing thermal insulation. For calibration of radiation thermometers up to 1150 °C a hole of 65 mm diameter is required in the cavity, but above this temperature a 45 mm hole is sufficient.

Where the larger hole is used a correction for the emissivity of the cavity may be required for very accurate work. Two sheathed thermocouples are usually placed in the furnace, one near the back and the other just above the sighting hole. Comparison of the two measured temperatures indicates when the cavity is at a uniform temperature.

Calibration may be carried out by comparing the thermometer and thermocouple temperature, or the test thermometer may be compared with a standard radiation thermometer when both are sighted on to the radiating source which may or may not be a true black body.

Cylindrical furnaces may also be used with a thermocouple fitted in the sealed end of the cylinder, which is cut on the inside to form a series of 45° pyramids.

A choice of three aperture sizes is available at the open end. For temperatures up to 1100 °C the furnace is made of stainless steel but for higher temperatures refractory materials are used. For further details see *The Calibration of Thermometers* (HMSO 1971). Figure 1.53 shows typical black body furnaces.

Furnace temperature by radiation thermometer Conditions in a furnace which might otherwise be considered as perfectly black body conditions may be upset by the presence of flame, smoke or furnace gases. In these conditions, a total radiation thermometer generally indicates a temperature between that of the furnace atmosphere and the temperature which would be indicated if such an atmosphere were not present.

A thick luminous flame may shield the object almost completely. Non-luminous flames radiate and absorb energy only in certain wavelength bands, principally because of the presence of carbon dioxide and water vapour. The error due to the presence of these gases can be reduced by using a lens of Pyrex which does not transmit some of these wavelengths, so that the instrument is less affected by variations in quantity of these gases. Where appreciable flame, smoke and gas are present it is advisable to use a closed-ended sighting tube, or provide a purged sighting path by means of a blast of clean dry air.

Errors in temperature measurement can also occur owing to absorption of radiation in the cold atmosphere between a furnace and the thermometer. To ensure that the error from this source does not exceed 1 per cent of the measured temperature, even on hot damp days, the distance between thermometer lens and furnace should not exceed 1.5 m if a glass lens is used, 1 m if the lens is silica, and 0.6 m if it is of fluorite.

1.6.2.2 *Pyroelectric techniques*

Pyroelectric detectors for thermal radiation are a comparatively recent introduction. Pyroelectric materials, mainly ceramics, are materials whose

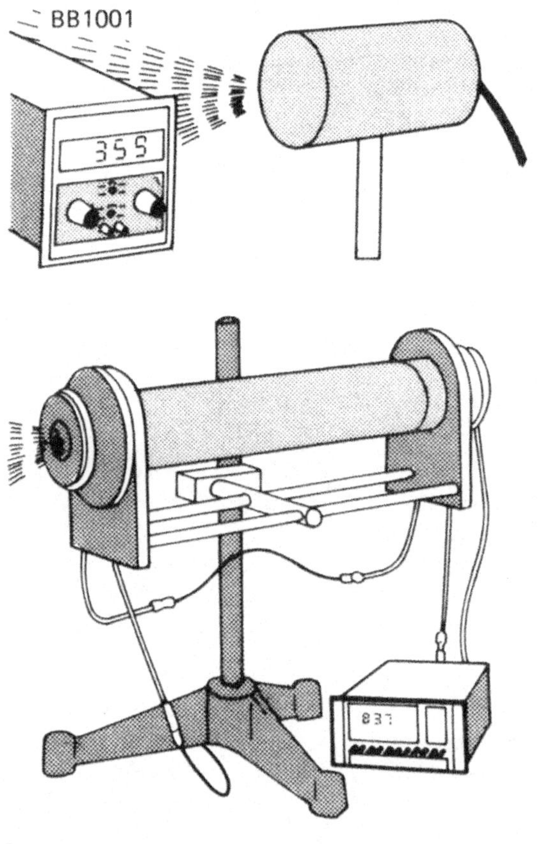

Figure 1.53 Black body radiators (courtesy Polarisers Technical Products).

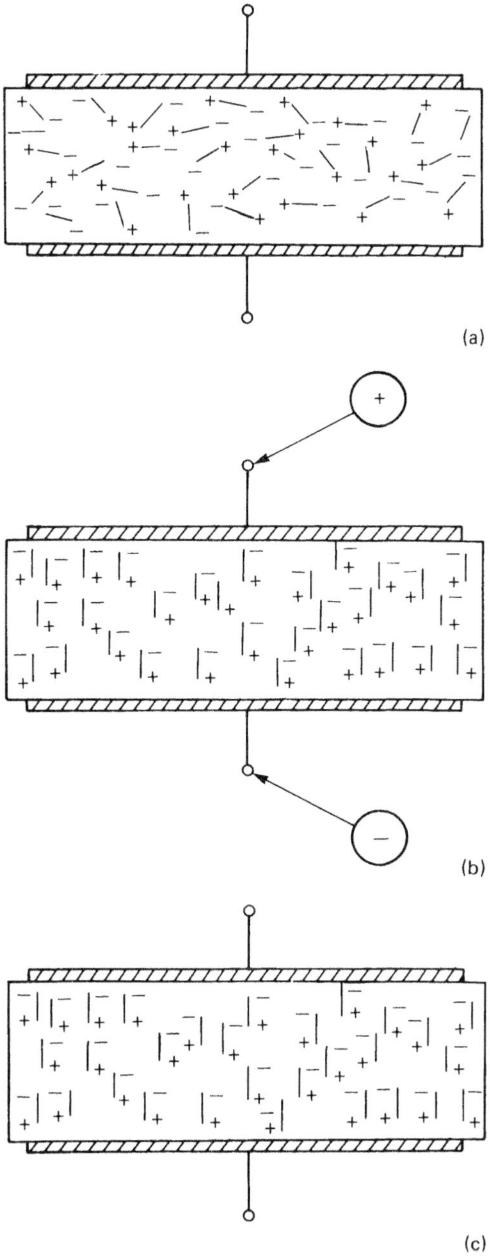

Figure 1.54 Pyroelectric effect.

molecules have a permanent electric dipole due to the location of the electrons in the molecules. Normally these molecules lie in a random orientation throughout the bulk of the material so that there is no net electrification. Also at ambient temperatures the orientations of the molecules are essentially fixed. If the temperature is raised above some level characteristic to the particular material, the molecules are free to rotate. This temperature is called the Curie temperature by analogy with the magnetic Curie temperature.

If a piece of pyroelectric ceramic is placed between two electrodes at ambient temperature the molecular dipoles are fixed in a random orientation, Figure 1.54(a). If it is then heated above its Curie temperature and an electrical potential applied to the electrodes, thus generating an electric field in the ceramic, the molecules will all align themselves parallel to the field, Figure 1.54(b). On cooling the ceramic back to ambient temperature and then removing the applied potential the molecules remain aligned. Figure 1.54(c). The amount of the polarization of the ceramic and therefore the magnitude of the resulting external electric field is a constant Σ which is a function of the material. If the field due to the applied voltage was E and the polarization P then

$$P = \Sigma E \tag{1.39}$$

If the temperature of the polarized pyroelectric ceramic is raised the molecular dipoles, which are anyway oscillating about their parallel orientation, will oscillate through a greater angle. Figure 1.55 shows one molecular dipole of length x and charge $\pm q$. Its electric moment is qx. If then the dipole oscillates through an average angle of $\pm\theta$ the effective length will be z where

$$z = x \cos \theta \tag{1.40}$$

The angle θ will increase with increasing temperature, thus reducing the electric moment of all the molecular dipoles. The electric moment or polarization of the whole piece of pyroelectric ceramic is of course the sum of all the molecular dipoles. Thus as the temperature rises the polarization of the whole piece of material gets less.

The Curie point is the temperature at which the oscillatory energy of the molecular dipoles is such that they can rotate freely into any position allowing them to return to their random orientation.

As stated above the electric moment M of the whole slice of ceramic is the sum of all the molecular dipole

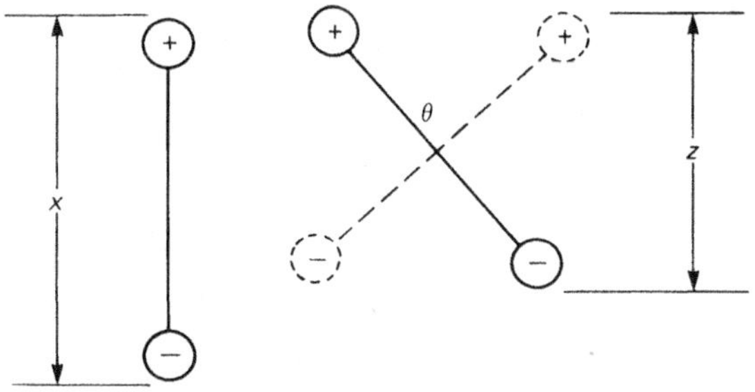

Figure 1.55 Mechanism of pyroelectric effect.

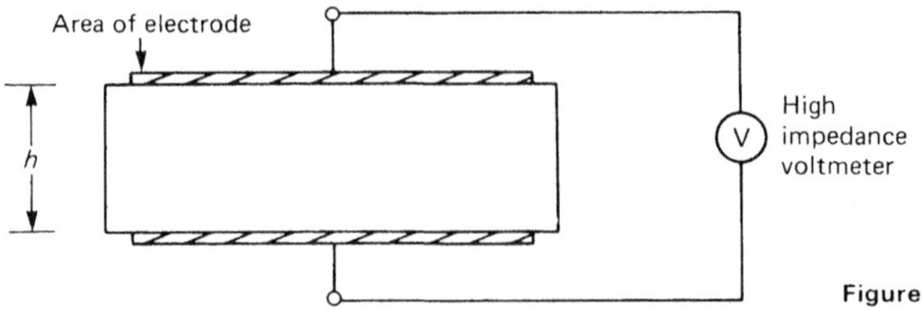

Figure 1.56 Pyroelectric detector.

moments:

$$M = PAh \tag{1.41}$$

where P is the dipole moment per unit volume, h is the thickness of the slice and A is the electrode area; see Figure 1.56.

If the electric charge at the two surfaces of the slice of pyroelectric ceramic is Q_s this has a dipole moment of $Q_s . h$, so that

$$Q_s = PA \tag{1.42}$$

If the temperature of the material rises the polarization is reduced and therefore Q_s becomes less. But if the electrodes are connected by an external circuit to an electrometer or other high impedance detector Q_s is normally neutralized by a charge Q on the electrodes. A reduction of Q_s therefore results in an excess charge on the electrodes and therefore a voltage V is detected.

$$V = Q/C \tag{1.43}$$

where C is the electrical capacitance of the device, for a temperature change of δT the change of charge δQ is given by

$$\delta Q = \Omega . A . \delta T \tag{1.44}$$

where Ω is the pyroelectric coefficient of the material. Therefore the voltage change will be

$$\delta V = \delta Q/C = \Omega A \, \delta T/C \tag{1.45}$$

where C is the electrical capacitance between the electrodes. The pyroelectric coefficient Ω is a function of temperature reducing with a non-linear characteristic to zero at the Curie temperature.

When used as a detector in a radiation thermometer, radiation absorbed at the surface of the pyroelectric slice causes the temperature of the detector to rise to a new higher level. At the start the charge on the electrodes will have leaked away through the external electrical circuit so there will have been zero voltage between the electrodes. As the slice heats up a voltage is detected between the two electrodes. When the device reaches its new temperature, losing heat to its environment at the same rate as it is receiving heat by radiation, the generation of excess charge on the electrodes ceases, the charge slowly leaks away through the electrical circuit and the detected voltage returns to zero. The device detects the change of incident radiation. To detect a constant flux of radiation, i.e. to measure a constant temperature, it is necessary to 'chop' the incident radiation with a rotating or oscillating shutter.

The physical construction of a pyroelectric radiation thermometer is essentially identical to a total radiation instrument except for the location of the radiation-chopping shutter just in front of the detector. Figure 1.57(a) shows the location and Figure 1.57(b) a typical profile of the optical chopper in a pyroelectric radiation thermometer. Figure 1.57(c) shows the graph against time of the chopped radiation together with the resulting electrical signal.

1.6.2.3 Optical (disappearing filament) thermometer

Optical radiation thermometers provide a simple and accurate means for measuring temperatures in the range 600 °C to 3000 °C. Since their operation requires the eye and judgement of an operator they are not suitable for recording or control purposes. However, they provide an effective way of making spot measurements and for calibration of total radiation thermometers.

In construction an optical radiation thermometer is similar to a telescope. However, a tungsten filament lamp is placed at the focus of the objective lens. Figure 1.58 shows the optical arrangement of an optical radiation thermometer. To use the instrument the point where the temperature is required to be known is viewed through the instrument. The current through the lamp filament is adjusted so that the filament disappears in the image. Figure 1.59 shows how the filament looks in the eyepiece against the background of the object, furnace or whatever is to have its temperature measured. At (a) the current through the filament is too high and it looks bright against the light from the furnace, at (c) the current is too low while at (b) the filament is at the same temperature as the background. The temperature of the filament is known from its electrical resistance. Temperature readout is achieved either by a meter measuring the current through the filament or by temperature calibrations on the control resistor regulating the current through the lamp. The filter in the eyepiece shown in Figure 1.58

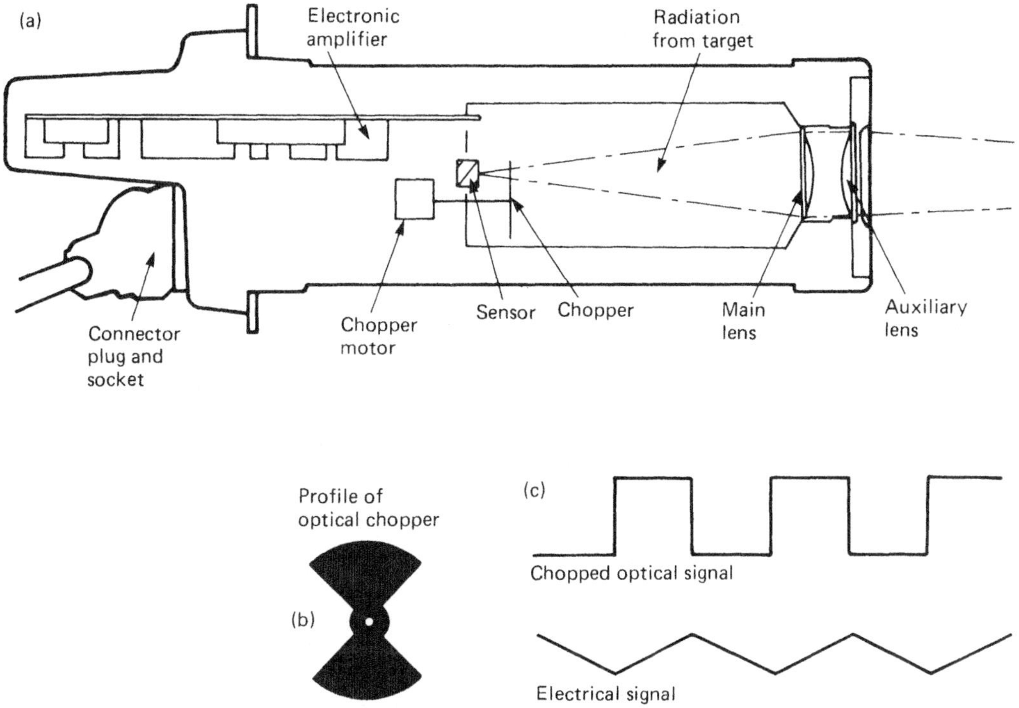

Figure 1.57 Diagram of pyroelectric radiation thermometer.

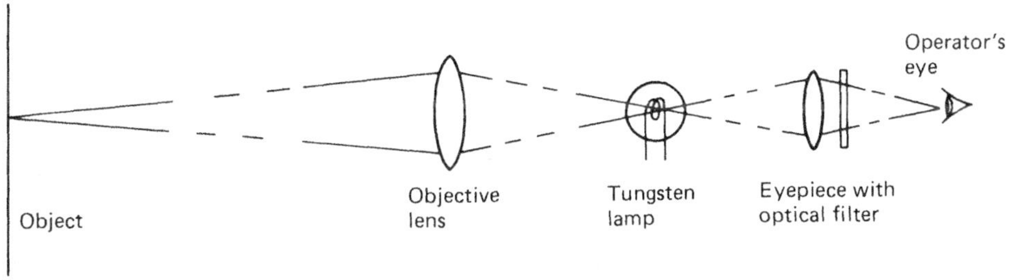

Figure 1.58 Optical system of disappearing filament thermometer.

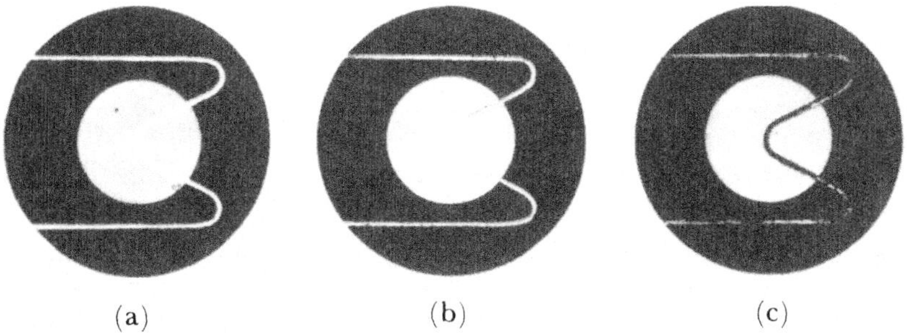

Figure 1.59 Appearance of image in optical thermometer.

passes light at a wavelength around 0.65 μm.

Lamps for optical thermometers are not normally operated at temperatures much in excess of 1500 °C. To extend the range of the instrument beyond this temperature a neutral filter of known transmission factor can be placed in the light path before the lamp. The measurement accuracy of an optical thermometer is typically ±5 K between 800 °C and 1300 °C and ±10 K between 1300 °C and 2000 °C.

Corrections for non-black-body conditions Like the total radiation thermometer, the optical thermometer is affected by the emissivity of the radiation source and by any absorption of radiation which may occur between the radiation source and the instrument.

The spectral emissivity of bright metal surfaces at 0.65 μm is greater than the total emissivity e, representing the average emissivity over all wavelengths. The correction required for the departure from black body conditions is therefore less than in the case of total radiation thermometers.

Due to the fact that a given change of temperature produces a much larger change in radiant energy at 0.65 μm than produced in the average of radiant energy over all wavelengths, the readings of an optical radiation thermometer require smaller corrections than for a total radiation instrument.

The relationship between the apparent temperature T_a and the true temperature T is given by equation (1.46) which is based on Wien's law

$$\frac{1}{T}-\frac{1}{T_a}=\frac{\lambda \log_{10}\varepsilon_\lambda}{6245} \tag{1.46}$$

where λ is the wavelength in micrometers (usually 0.65 μm) and ε_λ is the spectral emissivity at wavelength λ.

1.6.2.4 Photoelectric radiation thermometers

The reading obtained with an optical thermometer shows a lower temperature error than a total radiation thermometer. This is because the emissivity error for a given temperature and a known emissivity is proportional to the wavelength of the radiation used to make the measurement. For instance in the case of oxidized steel at 1000 °C with an emissivity of 0.8 a total radiation thermometer will have an error in excess of 50 degrees while the optical thermometer reading will be within 20 degrees. However the optical thermometer has two major drawbacks. First it is only suitable for spot measurements and requires a skilled operator to use it. Secondly it is not capable of a quick response and is totally unsuitable for control purposes.

Photoelectric radiation thermometers are ideally suited to the short wavelength application. Structurally they are essentially identical to a total radiation thermometer except that the thermal sensor is replaced by a photodiode.

A photodiode is a semiconductor diode, which may be either a silicon or germanium junction diode constructed so that the incident radiation can reach the junction region of the semiconductor. In the case of germanium the diode will be a plain P–N junction, in the case of silicon it may be either a P–N or P–I–N junction. In service the diodes are operated with a voltage applied in the reverse, i.e. non-conduction, direction. Under these conditions the current carriers, i.e. electrons, in the semiconductor do not have sufficient energy to cross the energy gap of the junction. However, under conditions of incident radiation some electrons will gain enough energy to cross the junction. They will acquire this energy by collision with photons. The energy of photons is inversely proportional to the wavelength. The longest wavelength of photons that will, on impact, give an electron enough energy to cross the junction dictates the long wave end of the spectral response of the device. The short wavelength end of the response band is limited by the transparency of the semiconductor material. The choice of germanium or silicon photodiodes is dictated by the temperature and therefore the wavelength to be measured. Silicon has a response of about 1.1 μm to 0.4 μm. The useful passband of germanium lies berween 2.5 μm and 1.0 μm. The exact passband of photodiodes varies somewhat from type to type depending on the manufacturing process used, but the above figures are typical. Normally the range of wavelengths used is reduced to a narrower passband than that detected by the semiconductor sensor. For instance, for general applications above 600 °C a narrow passband centred on 0.9 μm is usually used. Wherever possible silicon is to be preferred as it will tolerate higher ambient temperatures than germanium and in general it has the higher speed of response. Small P–I–N photodiodes can have a frequency response up to several hundred megahertz while P–N devices more usually have a response of several kilohertz.

Like all other semiconductor devices the electrical output of photodiodes is temperature-dependent. It is therefore necessary to construct these radiation thermometers with thermistors or resistance thermometers in close proximity to the photodiode to provide ambient temperature compensation.

1.6.2.5 Choice of spectral wavelength for specific applications

It might seem at first sight that apart from optical radiation thermometers the obvious choice should be to use a total radiation thermometer so as to capture as much as possible of the radiant emission from the target to achieve the maximum output signal. However as already mentioned above, except at the

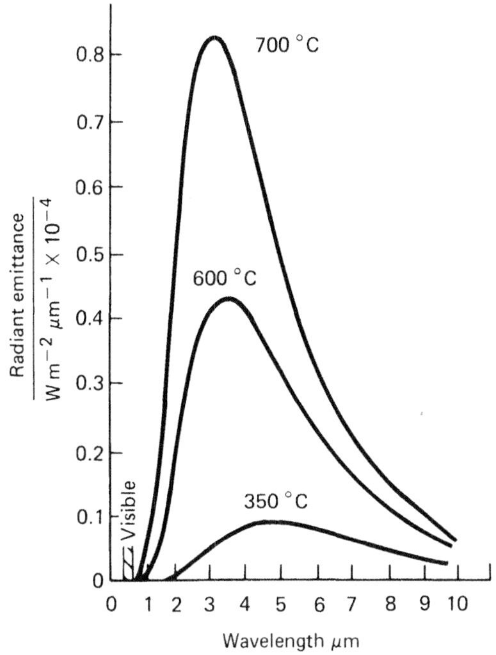

Figure 1.60 Black body radiation characteristics.

lowest temperature ranges, there are several reasons for using narrower wavelength bands for measurement.

Effect of radiant emission against wavelength One reason relates to the rate at which the radiant emission increases with temperature. An inspection of Figure 1.60 will show that the radiant emission at 2 μm increases far more rapidly with temperature than it does at say 6 μm. The rate of change of radiant emission with temperature is always greater at shorter wavelengths. It is clear that the greater this rate of change the more precise the temperature measurement and the tighter the temperature control. On the other hand this cannot be carried to extremes because at a given short wavelength there is a lower limit to the temperature that can be measured. For example, the eye becomes useless below about 600 °C. For these reasons alone we can understand the general rule that the spectral range of the appropriate infrared thermometer shifts to longer wavelengths as the process temperature decreases.

Emittance, reflectance and transmittance Another important reason for the use of different spectral regions relates to the specific emission characteristics of particular target materials. The curves of Figure 1.60 show the emission characteristics of the ideal emitter or black body. No material can emit more strongly than a black body at a given temperature. As discussed previously, however (p. 43), many materials can and do emit less than a black body at the same temperature in various portions of the spectrum. The ratio of the radiant emittance at wavelength λ of a material to that of a black body at the same temperature is called spectral emittance (ε_λ). The value of ε_λ for the substance can range between 0 and 1, and may vary with wavelength. The emittance of a substance depends on its detailed interaction with radiation. A stream of radiation incident on the surface of a substance can suffer one of three fates. A portion may be reflected. Another portion may be transmitted through the substance. The remainder will be absorbed and degraded to heat. The sum of the fraction reflected r, the fraction transmitted t and the fraction absorbed a will be equal to the total amount incident on the substance. Furthermore, the emittance ε of a substance is identical to the absorptance a and we can write

$$\varepsilon = a = 1 - t - r \qquad (1.47)$$

For the black body the transmittance and relectance are zero and the emittance is unity. For any opaque substance the transmittance is zero and

$$\varepsilon = 1 - r \qquad (1.48)$$

An example of this case is oxidized steel in the visible and near-infrared where the transmittance is 0, the reflectance is 0.20 and the emittance is 0.80. A good example of a material whose emittance characteristics change radically with wavelength is glass. Figure 1.61 shows the overall transmission of soda-lime glass. The reflectance of the glass is about 0.03 or less through most of the spectral region shown. At wavelengths below about 2.6 μm the glass is very highly transparent and the emittance is essentially zero. Beyond 2.6 μm the glass becomes increasingly opaque. From this it is seen that beyond 4 μm glass is completely opaque and the emittance is above 0.98.

This example of glass clearly illustrates how the detailed characteristics of the material can dictate the choice of the spectral region of measurement. For example, consider the problem of measuring and controlling the temperature of a glass sheet during manufacture at a point where its temperature is 900 °C. The rule that suggests a short wavelength infrared thermometer, because of the high temperature, obviously fails. To use the region around 1 μm would be useless because the emittance is close to 0. Furthermore, since the glass is highly transparent the radiation thermometer will 'see through' the glass and can give false indications because of a hot wall behind the glass. One can recognize that glass can be used as an effective 'window' with a short wavelength radiation thermometer. By employing the spectral region between 3 and 4 μm the internal temperature of the glass can be effectively measured and controlled. By operating at 5 μm or more the surface temperature of the glass is

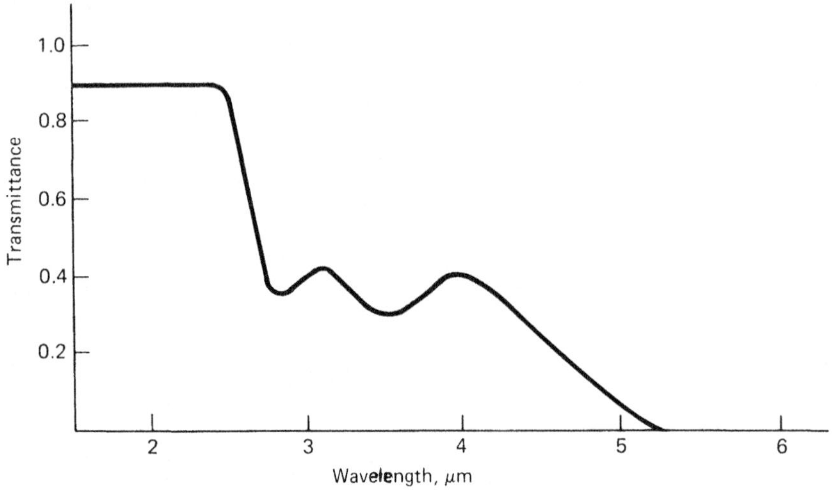

Figure 1.61 Transmittance of one millimetre of soda-lime glass.

measured. Each of these cases represents a practical application of infrared thermometry.

Atmospheric transmission A third important consideration affecting the choice of spectral region is that of the transmission of the atmosphere between the target substance and the radiation thermometer. The normal atmosphere always contains a small but definite amount of carbon dioxide and a variable amount of water vapour. Carbon dioxide strongly absorbs radiation between 4.2 and 4.4 μm and the water vapour absorbs strongly between 5.6 and 8.0 μm and also somewhat in the region 2.6 to 2.9 μm; see Figure 1.62. It is obvious that these spectral regions should be avoided, particularly in the region of the water bands. If this is not done the temperature calibration will vary with path length and also humidity. If the air temperature is comparable to or higher than the target temperature the improperly designed infrared thermometer could provide temperature measurements strongly influenced by air temperatures.

1.6.2.6 Signal conditioning for radiation thermometers

Although the output of a radiation thermometer can be used directly in a voltage or current measuring instrument this is unsatisfactory for two prime reasons. First the energy radiated by a hot body is a function of the fourth power of absolute temperature resulting in a very non-linear scale. Secondly the radiation detectors are themselves sensitive to ambient temperature. This requires either that the radiation thermometer be maintained at a constant temperature or alternatively an ambient temperature sensor is mounted beside the radiation sensor to provide a signal for temperature correction.

To compensate for these two deficiencies in the signal suitable electronic circuits must be used to provide linearization of the signal and to provide automatic temperature correction. It is also necessary to provide correction for the emissivity of the target. Typically the instrument itself carries a small 'head amplifier' to bring the signal up to a suitable level for transmission to the readout instrument. This head amplifier also provides the required ambient tempera-

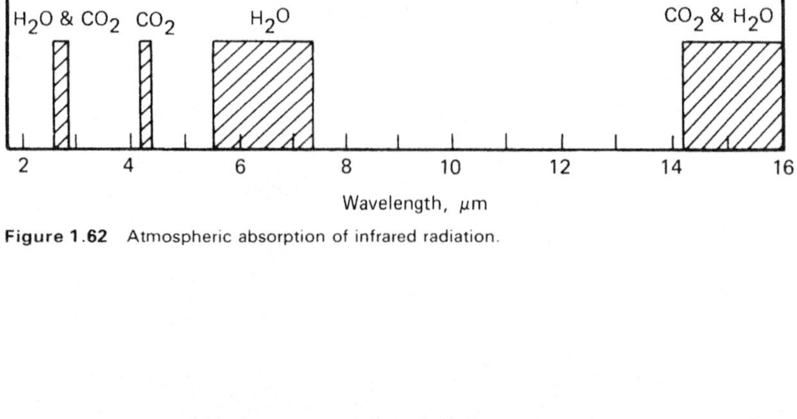

Figure 1.62 Atmospheric absorption of infrared radiation.

Figure 1.63 Radiation thermometer transmitter (courtesy Sirius Instruments Ltd.).

ture compensation circuits. The linearization and compensation for emissivity are provided at the readout module.

Some modern instruments provide the whole signal conditioning circuitry in the main instrument itself. Figure 1.63 shows such an instrument. In this equipment the output is a 4 to 20 milliamp signal linear with temperature and compensated for ambient temperature.

With the growing use of microprocessors in instrumentation several manufacturers are introducing instruments where the linearization and compensation are performed by a microcomputer.

1.6.2.7 Radiation thermometer applications

Infrared thermometers are currently used in a wide range of laboratory and industrial temperature control applications. A few low temperature examples include extrusion, lamination and drying of plastics, paper and rubber, curing of resins, adhesives and paints, and cold rolling and forming of metals.

Some high temperature examples include forming, tempering and annealing of glass, smelting, casting, rolling, forging and heat treating of metals, and calcining and firing of ceramics and cement.

In short, the infrared thermometer can be used in almost any application in the range 0 to 3600 °C where its unique capabilities can turn a seemingly impossible measurement and control problem into a practical working process. Many processes now controlled manually can be converted into continuous, automated systems.

1.7 Temperature measurement considerations

1.7.1 Readout

1.7.1.1 Local readout

If temperature requires to be measured at a particular point on say a chemical plant, what considerations govern the choice of instrument? The obvious first choice to most people is a liquid-in-glass thermometer. However this requires that one must be able to get close enough to read the instrument accurately. A better solution is a dial thermometer. The type of instrument chosen will of course depend upon the accuracy and repeatability required. In general and especially on bigger plants local temperature measurement is for general surveillance purposes only, the measurement is probably not essential but is provided as a cross-check on the control instruments to provide operator confidence. An inexpensive bimetal thermometer is probably adequate. If greater accuracy is required then a capillary-type thermometer (see Sections 1.3.2–1.3.4) with short capillary can be used, or where high accuracy is necessary an electrical technique may be specified. In the case of furnaces a portable radiation instrument may be the best choice.

Of course, on small plant not controlled from a separate control room all measurements will probably be local measurements. It is mainly in this situation that the higher accuracy local readout is required.

1.7.1.2 Remote reading thermometers

The first question to ask in the selection of remote reading instruments is: what is the distance between the measurement point and the readout location? If that distance is less than say 100 metres, capillary instruments may well be the best solution. However if the distance is near the top limit vapour pressure instruments will probably be ruled out. They may also not be usable if there is likely to be big ambient temperature variation at the readout point or along the length of the capillary.

The next question is: what is the height difference between the thermometer bulb and the readout position? Long vertical runs using liquid-in-metal thermometers can cause measurement offsets due to the liquid head in the vertical capillary adding to (or subtracting from) the pressure at the instrument Bourdon tube. In the case of height differences greater than say 10 metres liquid thermometers are likely to be unsuitable. This then reduces the choice to gas-filled instruments. A further consideration when specifying instrumentation on a new plant is that it is convenient from itinerary considerations to use as many instruments of the same type as possible. The choice of instrument is then dictated by the most stringent requirement.

On large installations where many different types of instrument are being installed and especially where pneumatic instrumentation is used, capillary instruments can run into an unexpected psychological hazard. Not infrequently a hard-pressed instrument technician has, on finding he has a too long capillary, been known to cut a length out of the capillary and rejoint the ends with a compression coupling. The result is of course disaster to the thermometer. Where on installation the capillary tube is found to be significantly too long it must be coiled neatly in some suitable place. The choice of that place may depend on the type of instrument. In gas-filled instruments the location of the spare coil is irrelevant but especially with vapour pressure instruments it wants to be in a position where it will receive the minimum of ambient temperature excursions to avoid introduction of measurement errors.

For installations with long distances between the point of measurement and the control room it is almost essential to use an electrical measurement technique. For long runs resistance thermometers are to be preferred to thermocouples for two principal reasons. First the copper cables used for connecting resistance bulbs to their readout equipment are very much less expensive than thermocouple wire or compensating cable. Secondly the resistance thermometer signal is at a higher level and lower impedance than most thermocouple signals and is therefore less liable to electrical interference.

An added advantage of electrical measurements is that, whether the readout is local or remote, the control engineer is given wider options as to the kinds of readout available to him. Not only does he have a choice of analogue or digital readout but he can also have a wider range of analogue readouts, since they are not limited to a rotary dial.

1.7.1.3 Temperature transmitters

On big installations or where a wide variety of different measurements are being made with a wide range of instrumentation it is more usual to transfer the signal from the measurement point to the control area by means of temperature transmitters. This has the great advantage of allowing standardization of the readout equipment. Also in the case of electrical transmission, by say a 4–20 milliamp signal, the measurement is much less liable to degradation from electrical interference. Also the use of temperature transmitters allows the choice of measurement technique to be unencumbered by considerations of length of run to the readout location.

The choice of electrical or pneumatic transmission is usually dictated by overall plant policy rather than the needs of the particular measurement, in this case temperature. However, where the requirement is for electrical temperature measurement for accuracy or other considerations the transmission will also need to be electrical. See Volume 4 (Telemetry).

1.7.1.4 Computer-compatible measurements

With the increasing use of computer control of plants there is a requirement for measurements to be compatible. The tendency here is to use thermocouples, resistance thermometer, or where the accuracy does not need to be so high, thermistors as the measuring techniques. The analogue signal is either transmitted to an interface unit at the control room or to interface units local to the measurement. The latter usually provides for less degradation of the signal.

As most industrial temperature measurements do not require an accuracy much in excess of 0.5 per cent it is usually adequate for the interface unit to work at eight-bit precision. Higher precision would normally only be required in very special circumstances.

1.7.1.5 Temperature controllers

While thermometers, in their widest sense of temperature measurement equipment are used for readout purposes, probably the majority of temperature measurements in industrial applications are for control purposes. There are therefore many forms of dedicated temperature controllers on the market. As briefly described in Section 1.3.5.1, the simplest of these is a thermostat.

Thermostats A thermostat is a device in which the control function, usually electrical contacts but sometimes some other control function such as a valve, is directly controlled by the measurement action. The instrument described in Section 1.3.5.1 uses solid expansion to operate electrical contacts, but any of the other expansion techniques may be used. In automotive applications the thermostat in an engine cooling system is a simple valve directly operated either by vapour pressure or change of state, e.g. the change of volume of wax when it melts.

Thermostats, however, are very imprecise controllers. In the first place their switching differential (the difference in temperature between switch-off and switch-on) is usually several kelvin. Secondly the only adjustment is setpoint.

Contact dial thermometers A first improvement on a thermostat is the use of a contact dial thermometer. The dial of this instrument carries a second pointer, the position of which can be set by the operator. When the indicating pointer reaches the setpoint pointer they make electrical contact with one another. The current that then flows between the pointers operates an electrical relay which controls the load. In this case the switching differential can be very small, typically a fraction of a kelvin.

Proportional temperature controllers Dedicated one-, two- or three-term temperature controllers are available in either pneumatic or electronic options. The use of such controllers is mainly confined to small plants where there is a cost advantage in avoiding the use of transmitters.

In the case of pneumatic controllers the input measurement will be liquid, vapour pressure or gas expansion. The Bourdon tube or bellows used to measure the pressure in the capillary system operates directly on the controller mechanism.

However in recent years there has been an enormous increase in the number of electronic temperature controllers. The input to these instruments is from either a thermocouple or a resistance thermometer. The functions available in these controllers vary from on/off control to full three-term proportional, integral and derivative operation. Some of the more sophisticated electronic controllers use an internal microprocessor to provide the control functions. Some units are available with the facility to control several temperature control loops. Of course the use of an internal microprocessor can make direct computer compatibility a simple matter.

1.7.2 Sensor location considerations

To obtain accurate temperature measurement careful consideration must be given to the siting of temperature sensing probes. Frequently in industrial applications temperature measuring equipment does not live up to the expectations of the plant design engineer. The measurement error is not infrequently ten or even twenty times the error tolerance quoted by the instrument manufacturer.

Large measurement errors in service may be due to the wrong choice of instrument but more frequently the error is due to incorrect location of the measurement points. Unfortunately the location of temperature sensors is dictated by the mechanical design of the plant rather than by measurement criteria.

1.7.2.1 Immersion probes

To minimize errors in the measurement of the temperature of process fluids, whether liquid or gas, it is preferable to insert the sensor so that it is directly immersed in the fluid. The probe may be directly dipped into liquid in an open vessel, inserted through the wall of the vessel, or inserted into a pipe.

Measurement of liquid in vessels Temperature measurement of liquid in a plant vessel may illustrate the dilemma of the control engineer when faced with mechanical problems. Consider Figure 1.64 which represents a vessel filled with liquid and stirred by a double anchor agitator. The ideal place to measure the temperature would be somewhere near the centre of

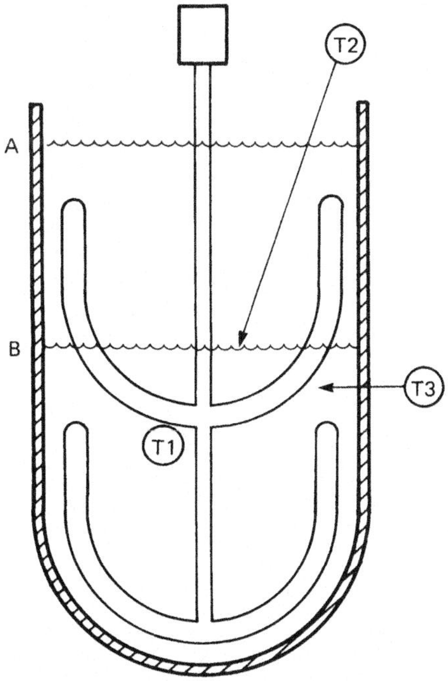

Figure 1.64 Problems associated with temperature measurement in a stirred vessel.

the mass at say T1. The best arrangement would seem to be a dip probe T2. But even though the design level of the liquid is at A in operation the liquid level may fall as low as B leaving probe T2 dry. The only remaining possibility is T3. This is not a very good approach to T1 and is subject to error due to conduction of heat from or to the vessel wall.

An approach that can be used if the temperature measurement is critical is to mount a complete temperature measuring package onto the shaft of the agitator. Wires are then brought up the shaft out of the vessel from whence the temperature signal can be taken off with slip rings, inductively coupled or radio telemetered to a suitable receiver. This is of course only possible where the temperature of the process is within the operating range of the electronics in the measurement package. The use of slip rings is not very satisfactory as they add unreliability, but in the absence of slip rings the package must also carry its own power supply in the form of batteries.

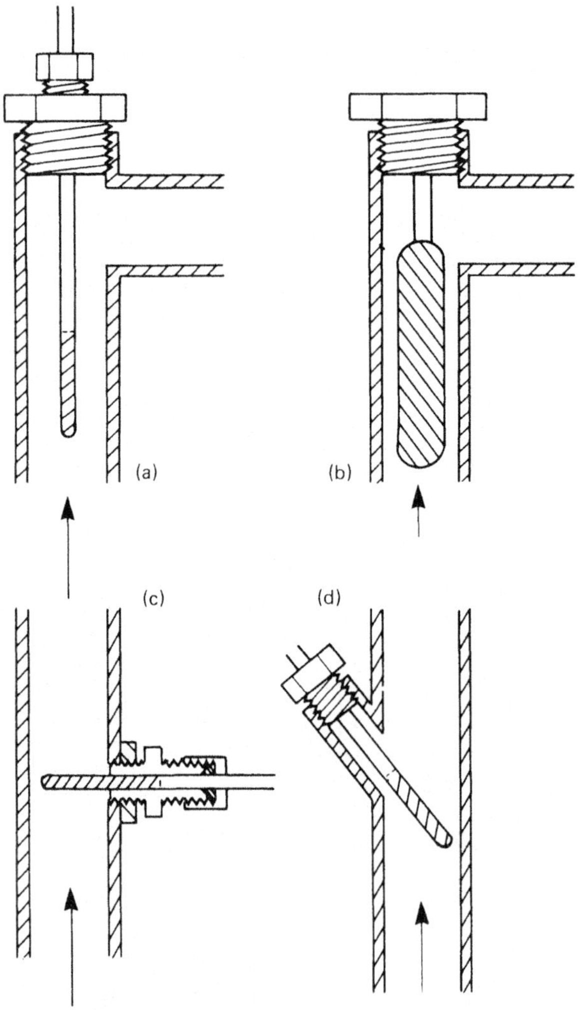

Figure 1.65 Problems associated with location of thermometer probe in pipe: (a) preferred arrangement, (b) probe obstructing pipe, (c) sensitive area of probe not fully immersed, (d) alternative preferred arrangement, sensitive portion of probe shaded.

Probes in pipes or ducts There is frequently a requirement to measure the temperature of a fluid flowing in a pipe. This is usually straightforward but there are still points to watch out for. Figure 1.65 shows three possible configurations for insertion into a pipe. The most satisfactory arrangement is to insert the thermometer probe into the pipe at a bend or elbow. Figure 1.65(a) shows this arrangement. Points to note are:

(a) To ensure that the probe is inserted far enough for the sensitive length to be wholly immersed and far enough into the fluid to minimize thermal conduction from the sealing coupling to the sensor.
(b) To insert the probe into the direction of flow as indicated. The reasons for this are to keep the sensor ahead of the turbulence at the bend which could cause an error due to local heating and to remove the effects of cavitation that could occur at the tip of a trailing probe. Figure 1.65(b) shows the problem that can arise in small pipes where the probe can cause serious obstruction to the flow.

Where it is not possible to put the thermometer at a bend in the pipe we can insert it radially provided the pipe is big enough. Great care should be taken to ensure complete immersion of the sensitive portion of the probe. Figure 1.65(c) illustrates this problem. A better solution is diagonal insertion as shown at (d). Again the probe should point into the direction of flow.

When measuring temperature in large pipes or ducts it must be remembered that the temperature profile across the pipe may not be constant. This is especially true for large flue stacks and air-conditioning ducts. The centre liquid or gas is usually hotter (or colder in refrigerated systems) than that at the duct wall. In horizontal ducts carrying slow-moving air or gas the gas at the top of the duct will be significantly hotter than that at the bottom of the duct. In these circumstances careful consideration must be given as to how a representative measurement can be obtained; it may well be necessary to make several measurements across the duct and average the readings.

1.7.2.2 Radiation errors

Gas temperature measurements present extra problems compared with temperature measurements in liquids. The difficulties arise from two sources. First the relatively low thermal conductivity and specific

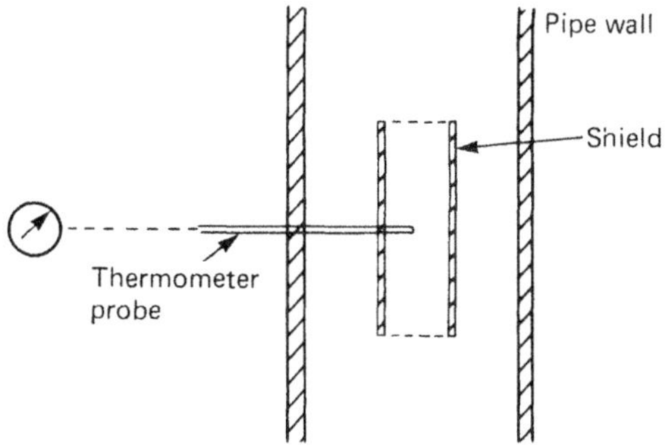

Figure 1.66 Radiation shield for gas temperature measurement.

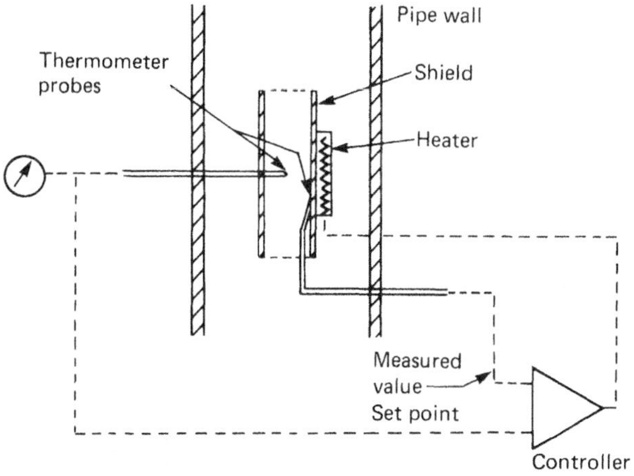

Figure 1.67 Active radiation shield.

heat of gases result in a poor heat transfer from the gas to the sensing element. This results in a slow response to temperature changes. Secondly since most gases are transparent at least to a substantial part of the thermal radiation spectrum significant measurement errors are lilely to occur, as mentioned in Section 1.6. Consider a thermometer bulb inserted into a pipe containing a gas stream. The walls of the pipe or duct are likely to be at a different temperature to the gas, probably but not necessarily cooler. This means that while the thermometer is being warmed by receiving heat by contact with the gas it is also losing heat by radiation to the pipe wall and if the wall is cooler than the gas the thermometer will lose more heat than it receives and will therefore register a lower temperature than the true gas temperature. Likewise if the pipe wall is hotter than the gas then the thermometer reading will be too high. This error can be reduced by surrounding the sensitive part of the thermometer probe with a cylindrical shield with its axis parallel to the pipe axis. This shield will reach a temperature intermediate between that of the pipe wall and that of the gas, Figure 1.66. Where more precise measurements are required an active shield may be employed. In this case a second thermometer is attached to the shield which is also provided with a small heater. This heater's output is controlled via a controller so that the two thermometers, the one in the gas and the one on the shield, always indicate identical temperatures. In this state the thermometer will be receiving exactly the same amount of radiation from the shield as it radiates back to the shield. Figure 1.67 shows this arrangement.

1.7.2.3 Thermometer pockets, thermowells

The direct immersion of temperature sensing probes into process fluid, while being the optimum way to get an accurate measurement, has its disadvantages. First it has disadvantages from the maintenance point of view: normally the sensing probe cannot be removed while the plant is on stream. Secondly in the case of corrosive process streams special corrosion-resistant materials may need to be used. Standard temperature gauges are normally only available in a limited range of materials, typically brass, steel, stainless steel or ceramic, so a sheath or thermometer pocket or thermowell can be used to protect the temperature sensing probe.

The use of a thermometer pocket does degrade the measurement accuracy of the instrumentation.

Figure 1.68 shows a thermometer pocket mounted in the wall of a steam-jacketed process vessel. The thermometer probe receives heat from the wall of the pocket by conduction where it touches it and by radiation at other places. The inner wall of the pocket receives heat from the process fluid and by conduction in this case from the steam jacket of the vessel. In the case of a short pocket the heat conducted along the

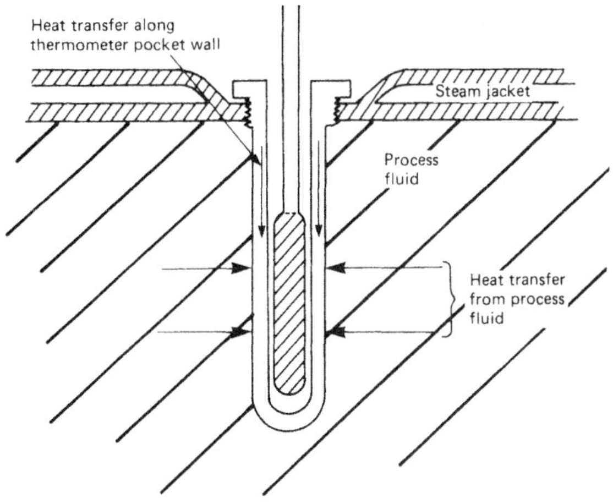

Figure 1.68 Thermometer pocket or thermowell.

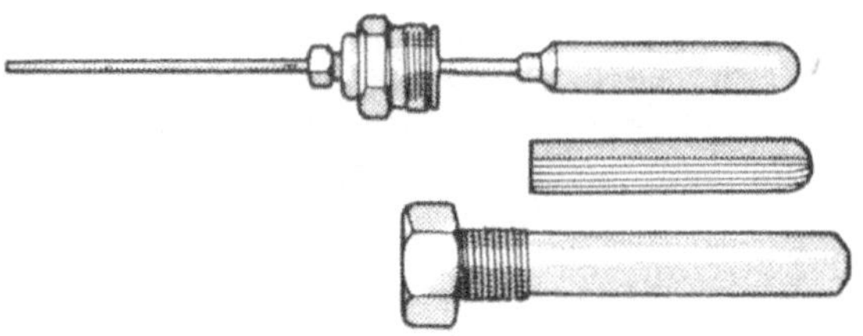

Figure 1.69 Taylor thermospeed separable well system (courtesy Taylor Instruments).

pocket can cause a significant measurement error, causing too high a reading. In the situation where the outer jacket of the vessel is used for cooling the vessel, for example, a cooling water jacket, the heat flow will be away from the sensing probe and consequently the error will be a low measurement. This conduction error is only significant where the thermometer pocket is short or where the pocket is inserted into a gas stream. To minimize the error the length of the pocket should be at least three times the length of the sensitive area of the probe.

The use of a thermowell or pocket will also slow down the speed of response of an instrument to temperature changes. A directly immersed thermometer probe will typically reach thermal equilibrium within 30 to 90 seconds. However the same probe in a thermometer pocket may take several minutes to reach equilibrium. This delay to the instrument response can be improved in those cases where the pocket is mounted vertically pointed downwards or in any position where the closed end is generally lower than the mouth, by filling it with a heat-transfer liquid. This liquid is usually a silicone oil.

An alternative method for improving the rate of heat transfer between the pocket and the bulb is illustrated in Figure 1.69. A very thin corrugated aluminium or bronze sleeve is inserted between the bulb and pocket on one side. This forces the bulb over to the other side, ensuring metal-to-metal contact on this side, while on the other side, the sleeve itself, being made of aluminium which has a high thermal conductivity, provides a reasonable path for the heat. In addition the bulb should be placed well down in the pocket to reduce the possibility of errors due to heat conducted by the pocket to the outside with consequent reduction of the temperature at the bulb.

The errors associated with thermal conduction along the thermometer pocket are of course more critical in the case of gas temperature measurement, as the thermal transfer from gas to thermometer is not nearly as good as it is from liquid.

1.7.2.4 Effect of process fluid flow rate

Two sources of error in temperature measurement are clearly identified.

Frictional heating Where the process fluid flows past a probe at high velocity there is, especially in the case of gases, a frictional heating effect. The magnitude of the effect is not easily evaluated but it is advisable if possible to site the probe at a location where the fluid velocity is low.

Conductive cooling Resistance thermometers and thermistors depend for their operation on an electric current flowing through them. This current causes a small heating effect in the sensor. When such a sensor is used for liquid temperature measurement the relatively high specific heat of most liquids ensures that this heat is removed and the sensor temperature is that of the liquid. However, in gas measurement the amount of heat removed is a function of the gas velocity and thus a variable source of error can arise dependent on flow rate. In a well designed instrument this error should be very small but it is a potential source of error to be borne in mind.

Cavitation Liquid flowing past a thermometer probe at high speed is liable to cause cavitation at the downstream side of the probe. Apart from any heating effect of the high flow rate the cavitation will generate noise and cause vibration of the probe. This vibration is likely in due course to cause deterioration or premature catastrophic failure of the probe.

1.7.2.5 Surface temperature measurement

Where the temperature of a surface is to be measured this can be done either with a temperature probe cemented or clamped to the surface or where a spot measurement is to be made a sensor can be pressed against the surface. In the former arrangement, which is likely to be a permanent installation, the surface in the region of the sensor itself can be protected from heat loss by lagging with thermally insulating material. Provided heat losses are minimized the measurement error can be kept small. Errors can be further reduced where the sensor is clamped to the surface by coating the surface and the sensor with heat-conducting grease. This grease is normally a silicone grease heavily loaded with finely ground alumina. A grease loaded with beryllium oxide has better heat transfer properties. However, since beryllium oxide is very toxic this grease must be handled with the greatest of care.

Where spot measurements are to be made, using for instance a hand-held probe, it is difficult to get accurate readings. The normal practice is to use a probe mounted on a spring so that it can take up any reasonable angle to press flat against the surface to be measured. The mass of the probe tip is kept as small as possible, usually by using a thermocouple or thermistor, to keep the thermal mass of the probe to a minimum. Again accuracy can be improved somewhat by using thermally conducting grease. Figure 1.40 shows a typical hand-held probe.

1.7.3 Miscellaneous measurement techniques

Temperature measurement may be the primary measurement required for the control of a plant. There are, however, many cases where temperature measurement is a tool to get an indication of the conditions in a plant. For instance in distillation columns it is more convenient and quicker to judge the compositions of the offtake by temperature measurement than to install on-line analysers and as a further bonus the cost of temperature measurement is very significantly less than the cost of analysers.

The reverse situation also exists where it is not possible to gain access for a thermometer to the region where the temperature requires to be known. In this instance some indirect measurement technique must be resorted to. One case of indirect measurement that has already been dealt with at some length is the case of radiation thermometers.

1.7.3.1 Pyrometric cones

At certain definite conditions of purity and pressure, substances change their state at fixed temperatures. This fact forms a useful basis for fixing temperatures, and is the basis of the scales of temperature.

For example, the melting points of metals give a useful method of determining the electromotive force of a thermocouple at certain fixed points on the International Practical Temperature Scale as has been described.

In a similar way, the melting points of mixtures of certain minerals are used extensively in the ceramic industry to determine the temperature of kilns. These minerals, being similar in nature to the ceramic ware, behave in a manner which indicates what the behaviour of the pottery under similar conditions is likely to be. The mixtures, which consist of silicate minerals such as kaolin or china clay (aluminium silicate), talc (magnesium silicate), felspar (sodium aluminium silicate), quartz (silica), together with other minerals such as calcium carbonate, are made up in the form of cones known as Seger cones. By varying the composition of the cones, a range of temperature between 600 °C and 2000 °C may be covered in convenient steps.

A series of cones is placed in the kiln. Those of lower melting point will melt, but eventually a cone is found which will just bend over. This cone indicates the temperature of the kiln. This can be confirmed by the fact that the cone of next higher melting point does not melt.

Since the material of the cone is not a very good conductor of heat, a definite time is required for the cone to become fluid, so that the actual temperature at which the cone will bend will depend to a certain extent upon the rate of heating. In order to obtain the maximum accuracy, which is of the order of ± 10 °C, the cones must, therefore, be heated at a controlled rate.

1.7.3.2 Temperature-sensitive pigments

In many equipment applications it is necessary to ensure that certain components do not exceed a specified temperature range. A typical case is the electronics industry where it is essential that semiconductor components remain within their rather limited operating range, typically -5 °C to 85 °C or for equipment to military specification -40 °C to 125 °C. These components are too small to fix all but the finest thermocouples to them. To deal with this situation temperature-sensitive paints can be used. These paints contain pigments which change colour at known temperatures with an accuracy of ± 1 °C. The pigments are available either having a reversible or a non-reversible colour change, the latter being the more usually used. In the case above, a semiconductor component in an electronic machine can have two spots of paint put on its case having colour changes at say 0 °C and 110 °C. On subsequent inspection, perhaps after equipment failure, it can be seen at once whether that component has been outwith its temperature tolerance.

As an alternative to paint these pigments are available on small self-adhesive labels. In either case they are available for temperatures within the range of 0 °C to about 350 °C in steps of about 5 degrees.

1.7.3.3 Liquid crystals

A number of liquids, mainly organic, when not flowing tend to form an ordered structure with for instance all the molecules lying parallel to one another. This structure is maintained against the thermal agitation by weak intermolecular bonding such as hydrogen bonding. These bonds hold the structure until the weak bonds between the molecules get broken as will occur when the liquid begins to flow. The structure can also be changed by electric fields, magnetic fields or temperature. Different compounds respond to different stimuli. Most poeple will be familiar with the liquid crystal displays on digital watches and pocket calculators. These displays use compounds sensitive to electric fields.

However, in this section we are interested in those liquid crystalline compounds that respond primarily to temperature. The compounds involved are a group of compounds derived from or with molecular structures similar to cholesterol. They are therefore

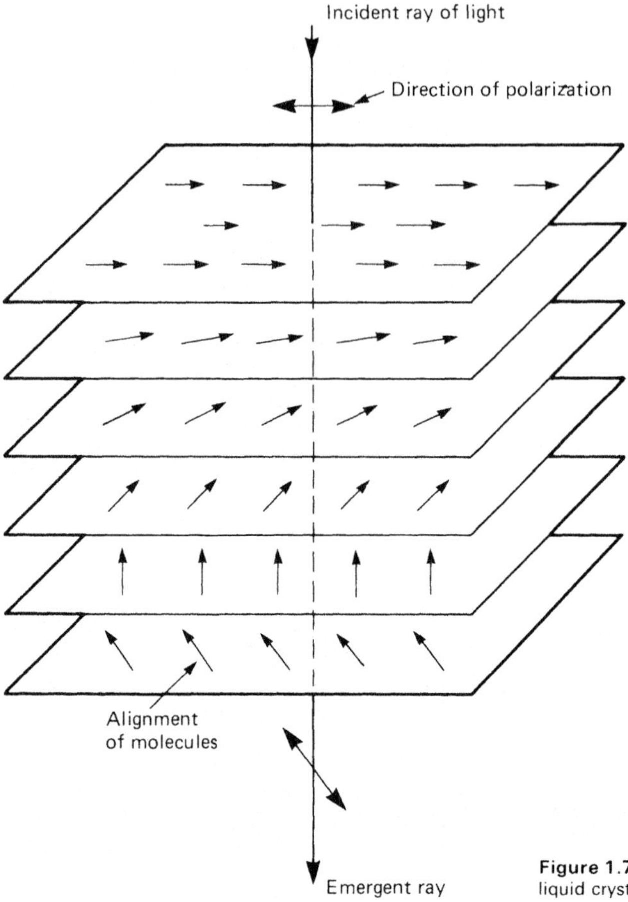

Figure 1.70 Rotation of the plane of polarization of light by liquid crystal.

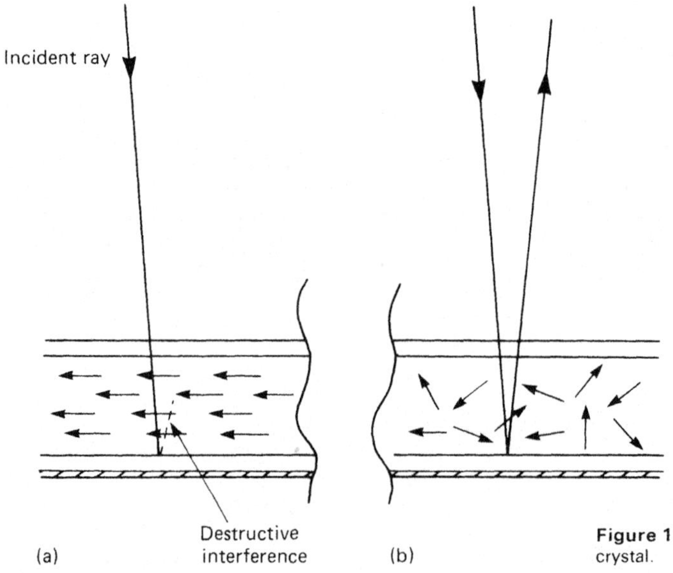

Figure 1.71 Destructive interference of reflected ray in liquid crystal.

called cholesteric compounds. Cholesteric liquids are extremely optically active as a consequence of their forming a helical structure. The molecules have a largely flat form and as a result lie in a laminar arrangement. However, the molecules have side groups which prevent them lying on top of one another in perfect register. The orientation of one layer of molecules lies twisted by a small angle compared to the layer below. This helical structure rotates the plane of polarization of light passing through the liquid in a direction perpendicular to the layers of molecules. Figure 1.70 illustrates this effect diagrammatically. The optical effect is very pronounced, the rotation of polarization being of the order of 1000° per millimetre of path length. The laminar structure can be enhanced by confining the cholesteric liquid between two parallel sheets of suitable plastic. The choice of polymer for this plastic is based on two prime requirements, first it is required to be transparent to light and secondly it should be slightly chemically active so that the liquid crystal molecules adjacent to the surface of the polymer are chemically bonded to it with their axes having the required orientation.

When used for temperature measurement the liquid crystal is confined between two sheets of transparent plastic a few tens of micrometres apart. The outer surface of one plastic layer is coated with a reflective layer, see Figure 1.71. In (a) a light ray enters the sandwich and travels to the bottom face where it is reflected back. Since the liquid crystal is in its ordered form it is optically active. The reflected ray interferes destructively with the incident ray and the sandwich looks opaque. In (b), however, the liquid crystal is above the temperature at which the ordered structure breaks up. The material is no longer optically active and the light ray is reflected back in the normal way – the material looks transparent.

The temperature at which the ordered structure breaks up is a function of the exact molecular structure. Using polarized light a noticeable change in reflected light occurs for a temperature change of 0.001 K. In white light the effect occurs within a temperature range of 0.1 K. Both the appearance of the effect and the exact temperature at which it occurs can be affected by addition of dyes or other materials.

1.7.3.4 Thermal imaging

In Section 1.6 the measurement of temperature by infrared and visual radiation was discussed in some detail. This technique can be extended to measure surface temperature profiles of objects. This is known as thermal imaging. The object to be examined is scanned as for television but at a slower rate and in the infrared region instead of the optical part of the spectrum. The signal so obtained is displayed on a visual display unit. This then builds up an image of the object as 'seen' by the infrared radiation from its surface. As well as producing a 'picture' of the object, the temperature of the surface is indicated by the colour of the image, producing a temperature map of the surface. Surface temperatures can be so imaged to cover a wide range from sub-ambient to very high temperatures. The technique has a very high resolution of temperature of the order of a small fraction of a Kelvin. Applications are to be found in such diverse fields as medicine and geological survey from space.

The technique is dealt with in very much greater detail in Volume 3, Chapter 2.

1.7.3.5 Turbine blade temperatures

In the development and design of gas turbines there is a requirement to measure the temperature and

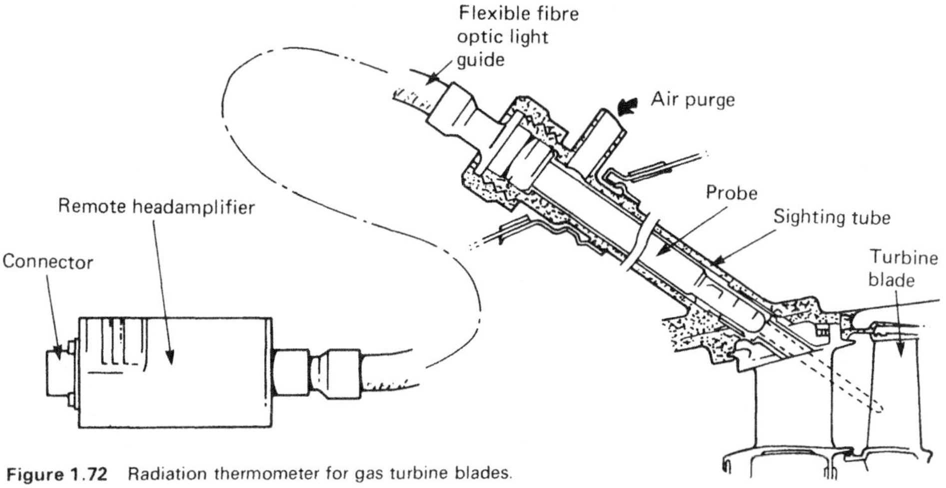

Figure 1.72 Radiation thermometer for gas turbine blades.

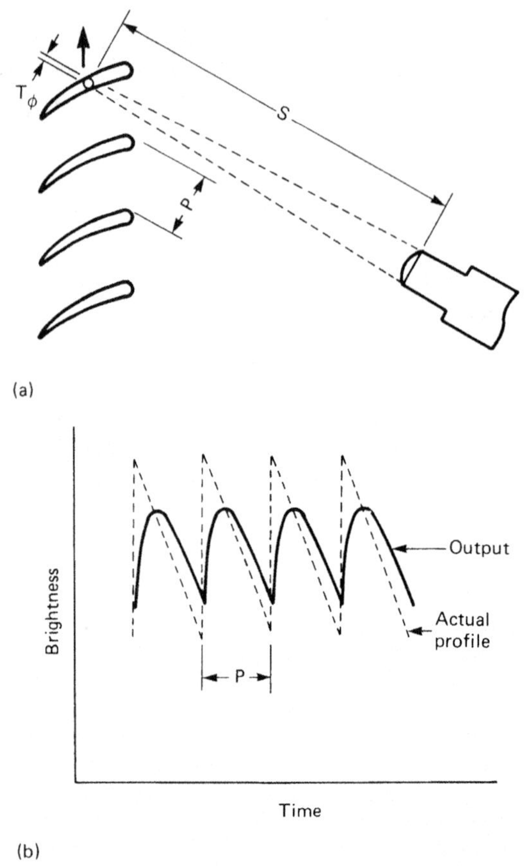

Figure 1.73 Measurement of the temperature profile of gas turbine blades: (a) geometry of focusing of thermometer, (b) temperature profile as 'seen' by radiation thermometer and electrical output.

temperature profile of the turbine rotor blades. This presents some problems as the turbine may be running at speeds of the order of 25 000 revolutions per minute. The rotor may consist of say 50 blades so that the time available to measure each blade temperature profile as it passes a point will be about a microsecond. A technique has been developed by Land Infrared Ltd. to carry out this measurement using fibre optic radiation thermometers. In this arrangement a small optical probe is inserted through the turbine wall and focused onto the rotor blades. The probe is connected by a fibre optic cable to a detector head amplifier unit nearby. Figure 1.72 shows a schematic diagram of focusing a measurement head. By designing the probe so that it focuses on a very small target area it is possible to 'read' a turbine blade temperature profile as it passes the target spot. Figure 1.73 shows the installation arrangement schematically at (A) and at (B) shows the theoretical and actual signal from the radiation thermometer. The degradation between the theoretical and actual signal is a function of the speed of response of the detector and the frequency bandwidth of the electronics. The theoretical signal consists of a sawtooth waveform. The peak represents the moment when the next blade enters the target area. The hottest part of the blade is its leading edge, the temperature falling towards the trailing edge. The signal falls until the next blade enters the field. The output from the thermometer can be displayed, after signal conditioning, on an oscilloscope or can be analysed by computer.

1.8 References

ASTM, *Manual on Use of Thermo-couples in Temperature Measurement*, ASTM Special Technical Publication 470B, (1981)

Billing, B. F. and Quinn, T. J. (eds), *Temperature Measurement 1975*, Adam Hilger, (1975)

Eckert, E. R. G. and Goldstein, R. J. (eds), *Measurements in Heat Transfer*, McGraw-Hill, (1976)

HMSO, *The Calibration of Thermometers*, (1971)

Kinzie, P. A. *Thermo-couple Temperature Measurement*, Wiley, (1973)

Quinn, T. J., *Temperature*, Academic Press, (1983)

2 Chemical analysis – introduction

W. G. CUMMINGS

2.1 Introduction to chemical analysis

Fifty years ago analytical chemistry depended almost entirely on measurements made gravimetrically and by titrimetry and students were taught that the essential steps in the process were sampling, elimination of interfering substances, the actual measurement of the species of concern and finally the interpretation of results. Each step required care, and, often, substances were analysed completely so that the components could be checked to total to within an acceptable reach of 100 per cent.

Classical analytical methods are still used from time to time, generally for calibrating instruments, but during the last thirty years the analytical chemistry scene has changed considerably. Spectroscopy and other physical methods of analysis are now widely used and a comprehensive range of chemical measuring instruments has been developed for specific techniques of analysis. This has meant that chemical analysis is now carried out as a cooperative effort by a team of experts, each having extensive knowledge of his own specialist technique, e.g. infrared absorption, emission spectrography, electrochemistry, gas chromatography, while also having considerable knowledge of the capabilities of the methods used by other members of the team.

Thus the analytical chemist has become more than just a chemist measuring the chemical composition of a substance; he is now a problem solver with two more steps in the analytical process – one at the beginning, 'definition of the problem', and another at the end, 'solution to the problem'. This means that the analytical chemist may measure things other than narrowly defined chemical composition – he may decide, for example, that pH measurements are better than analysis of the final product for controlling a process or that information on the valency states of compounds on the surface of a metal is more important than determining its composition.

Many elegant techniques have now become available for the analytical chemist's armoury with beautifully constructed electronic instruments, many complete with microprocessors or in-built computers. However, the analytical chemist should beware of becoming obsessed solely with the instruments that have revolutionized analytical chemistry and remember that the purpose of his work is to solve problems. He must have an open and critical mind so as to be able to evaluate the analytical instruments available – it is not unknown for instrument manufacturers in their enthusiasm for a new idea to emphasize every advantage of a technique without mentioning major disadvantages. It should also be remembered that, although modern analytical instrumentation can provide essential information quickly, misleading information can equally easily be obtained by inexperienced or careless operators and chemical measuring instruments must be checked and recalibrated at regular intervals.

Choosing the correct analytical technique or instrument can be difficult because several considerations have to be taken into account. First of all one must ensure that the required range of concentrations can be covered with an accuracy and precision that is acceptable for the required purpose. Then one must assess the frequency with which a determination must be made in order to set the time required for an analysis to be made or the speed of response of an instrument. This is particularly important if control of an on-going process depends on results of an analysis but is of less importance when the quality of finished products is being determined where ease of handling large numbers of samples may be paramount. Many requirements are conflicting and decisions have to be made on speed versus accuracy, cost versus speed, cost versus accuracy, and correct decisions can only be made with a wide knowledge of analytical chemistry and of the advantages and limitations of the many available analytical techniques. An important consideration is the application of the analytical instrument. This can be in a laboratory, in a rudimentary laboratory or room in a chemical plant area or working automatically on-stream. It is obvious that automatic on-stream instrumentation will be much more complex and expensive than simple laboratory instruments because the former must withstand the hostile environment of chemical plant and be capable of coping with temperature changes and plant variables without loss of accuracy. Such instruments have to be constructed to work for long continuous periods without exhibiting untoward drift or being adversely affected by the materials in the plant stream being monitored. Laboratory instruments on the other hand can be much more simple. Here the essential is a robust, easy-to-use instrument for a unique determination. Temperature compensation can be made by manual adjustment of controls at the time of making a determination and the instrument span can be set by use of standards each time the instrument is used. Thus there is no problem with drift. Laboratory instruments in general-purpose laboratories, however, can be as complex and costly as on-stream instruments but with different requirements. Here flexibility to carry out several determinations on a wide variety of samples is of prime importance but again temperature compensation and span adjustment can be carried out manually each time a determination is made. More expensive instruments use microprocessors to do such things automatically and these are becoming common in modern laboratories. Finally, although the cost of an analytical instrument depends on its complexity and degree of automation, there are other costs which should not be forgotten. Instrument maintenance charges can be appreciable and there is also the cost of running an instrument. The latter can range from almost nothing in the case of visible and ultraviolet spectrometers to several thousand pounds a year for argon supplies to inductively coupled plasma spectrometers. Many automatic analytical instruments require the preparation of reagent solutions and this, too, can involve an appreciable man-power requirement – also something which should be costed.

More detailed analysis of the factors affecting the costing of analytical chemistry techniques and instrumentation is beyond the scope of this chapter, but this whole volume of *Instrument Technology* gives details and comparisons of analytical instrumentation for many applications. It is arranged with large chapters on electrochemical and spectrochemical techniques and further chapters on the applications of gas analysis and humidity measuring instruments. For completeness, the remainder of this chapter contains brief descriptions of chromatography, thermal analysis and polarography.

2.2 Chromatography

2.2.1 General chromatography

Around 1900 M.S. Tswett used the adsorbing power of solids to separate plant pigments and coined the term chromatography for the method. It was then not used for twenty years; then the method was rediscovered and used for the separation of carotenes, highly unsaturated hydrocarbons to which various animal and plant substances (e.g. butter and carrots) owe their colour.

Chromatography is thus a separating procedure with the actual measurement of the separated substance made by another method, such as ultraviolet absorption or thermal conductivity, but as it is such a powerful analytical tool it will be dealt with here as an analytical method.

All chromatographic techniques depend on the differing distributions of individual compounds in a mixture between two immiscible phases as one phase (the mobile phase) passes through or over the other (the stationary phase). In practice the mixture of compounds is added to one end of a discrete amount of stationary phase (a tubeful) and the mobile phase is then introduced at the same end and allowed to pass along the stationary phase. The mixture of compounds is eluted, the compound appearing first at the other end of the stationary phase being that which has the smallest distribution into the stationary phase. As the separated compounds appear at the end of the stationary phase they are detected either by means of unique detectors or by general-purpose detectors which sense the compound only as an impurity in the mobile phase.

The apparatus used varies according to the nature of the two phases. In gas chromatography the mobile phase is a gas with the stationary phase either a solid or a liquid. This is described in detail in Chapter 5. Liquid chromatography covers all techniques using liquid as a mobile phase – these are column chromatography (liquid/liquid or liquid/solid), paper chromatography and thin layer chromatography.

2.2.2 Paper chromatography and thin layer chromatography

In paper chromatography the separation is carried out on paper, formerly on ordinary filter papers but more recently on papers specially manufactured for the purpose. These are made free from metallic impurities and have reproducible thickness, porosity and arrangement of cellulose fibres.

The paper used (which must not have been dried) contains adsorbed water and so paper chromatography can be regarded as an absorption process. However, the characteristics of the paper can be changed by applying specific liquids to it. Silicone oils, paraffin oil, petroleum jelly and rubber latex can be used to give a paper with non-polar liquid phases. Specially treated papers are also available, such as those containing ion exchange resins. Papers for paper chromatography can also be made of glass fibres or nylon as well as cellulose.

In thin layer chromatography, instead of using paper, a thin layer of an adsorbing substance such as

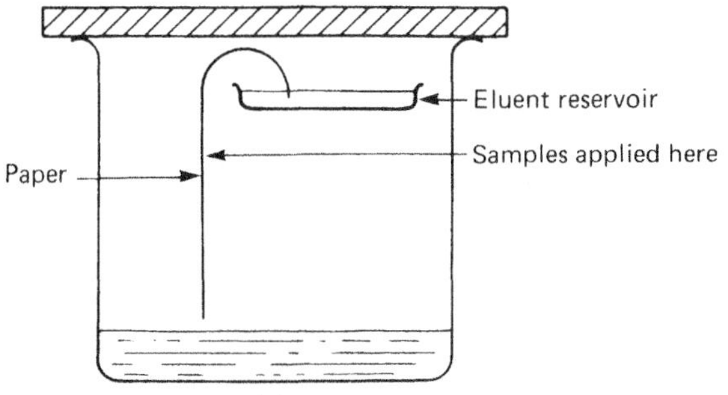

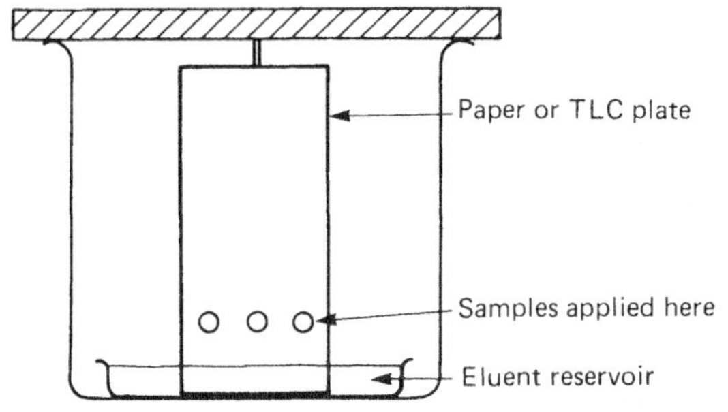

Figure 2.1 Apparatus for paper or thin-layer chromatography: (a) descending eluent used with paper chromatography, (b) ascending eluent used with paper chromatography or TLC.

silica gel is coated onto a glass or plastic plate. A very small volume of sample ($\sim 30\,\mu l$) is transferred onto one end of the plate which is then placed in a closed tank dipping into a solvent, the mobile phase. As the mobile phase moves along the plate the components of the sample are separated into a series of spots at different distances from the sample starting position. Figure 2.1 shows alternative arrangements. The location of the spots can be identified by their colour, or if colourless by spraying the plate with a reagent that produces a visible colour (or UV-detectable absorbance) with the compounds of interest. The position of the spots identifies the compound, the intensity of the colour, the concentration.

To establish a method for a particular mixture of compounds one has to select suitable adsorbents, solvents or mixtures of solvents, and a sensitive and selective reagent for detecting the separated compounds. There are many textbooks which discuss this in detail and give applications of the technique.

The apparatus used for measuring the separated substances in both paper and thin layer chromatography is quite straightforward laboratory-type equipment, for example, visible/ultraviolet spectrometers to determine the colour density or the UV absorbance of the spots.

Thin layer chromatography is generally found to be more sensitive than paper chromatography, development of the chromatogram is faster and it is possible to use a wider range of mobile phases and reagents to detect the position of the spots. Uses include the determination of phenols, carcinogenic polynuclear aromatic hydrocarbons, non-ionic detergents, oils, pesticides, amino acids and chlorophylls.

2.2.2.1 High performance liquid chromatography

Although liquid chromatography in columns was used by Tswett at the beginning of this century, an improved, quantitative version of the technique, high performance liquid chromatography (HPLC), has been fully developed and used only recently. By using precision instruments determination of trace organic and inorganic materials at concentrations of 10^{-6} to 10^{-12} g are possible. There are also several advantages of HPLC over other chromatographic techniques. HPLC is more rapid and gives better separations than classical liquid chromatography. It also gives better reproducibility, resolution and accuracy than thin layer chromatography, although the latter is generally the more sensitive technique. A large variety of separation methods is available with HPLC: liquid/liquid; liquid/solid; ion exchange and exclusion chromatography; but, again, the sensitivity obtainable is less than with gas chromatography.

Classical column liquid chromatography, in which the mobile liquid passed by gravity through the column of stationary phase, was used up to about 1946–50. In these methods a glass column was packed with a stationary phase such as silica gel and the sample added at the top of the column. Solvent, the mobile phase, was then added at the top of the column and this flowed through under the force of gravity until the sample components were either separated in the column or were sequentially eluted from it. In the latter case components were identified by refractive index or absorption spectroscopy. This type of elution procedure is slow (taking several hours) and the identification of the components of the sample is difficult and time-consuming.

Modern high performance liquid chromatography equipment has considerably better performance and is available from many chemical measuring instrument manufacturers. The main parts of a general-purpose HPLC apparatus are as shown in Figure 2.2.

The system consists of a reservoir and degassing system, a gradient device, a pump, a pulse dampener, a pre-column, a separating column and a detector.

Reservoir and degassing system The capacity of the reservoir is determined by the analysis being carried out, generally 1 litre is suitable. If oxygen is soluble in the solvent being used, it may need to be degassed. This can be done by distilling the solvent, heating it with stirring, or by applying a reduced pressure.

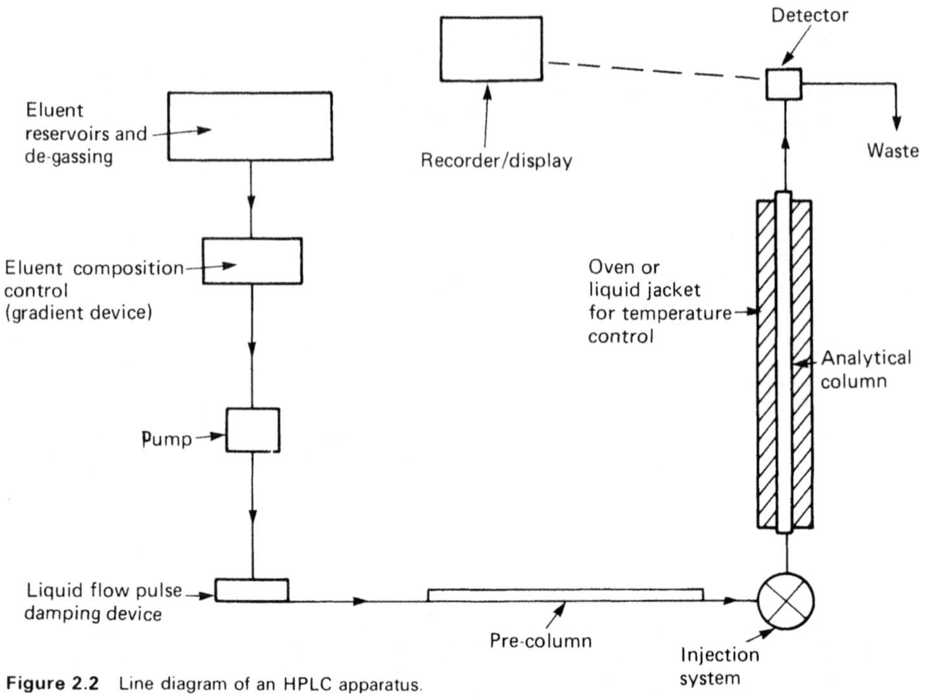

Figure 2.2 Line diagram of an HPLC apparatus.

Gradient devices If one wishes to change the composition of the mobile phase during the separation this can be done by allowing another solvent to flow by gravity into a stirred mixing vessel that contains the initial solvent and feeds the pump. This change of solvent mix is known as generating a solvent gradient.

A better way is to pump the solvents separately into a mixing tube; the desired gradient (composition) can be obtained by programming the pumps. This is elegant but expensive.

Pumps Suitable pumps deliver about 10 ml of solvent per minute at pressures up to 70 bar. These can be pressurized reservoirs, reciprocating pumps, motor-driven syringes or pneumatically operated syringes. It is essential to arrange for pulseless liquid flow and pulse damping may be required. This can be done by using small-bore tubes of small volume or by using sophisticated constant pressure control equipment.

Pre-column The solvent (the mobile phase) must be presaturated with the stationary liquid phase in the pre-column so that the stationary phase is not stripped off the analytical column.

Sample introduction Samples can be injected onto the analytical column by injection by syringe through a septum or by means of a sample loop. Injection via a septum can be difficult because of the very high pressures in the column – an alternative is stop-flow injection, where the solvent flow is stopped, the sample injected and then solvent flow and pressure restored. However, this can cause problems from the packing in the column shifting its position.

Analytical columns Very smooth internal walls are necessary for efficient analytical columns and very thick-walled glass tubing or stainless steel are the preferred materials. Connections between injection ports, columns and detectors should be of very low volume and inside diameters of components should be of similar size. Tubing of 2–3 mm internal diameter is most often used and temperature control is sometimes necessary. This can be done by water-jacketing or by containing the columns within air-ovens.

Stationary phases A very wide variety of materials can be used as solid stationary phases for HPLC – a summary of materials to use has been compiled (R. E. Majors, *Am. Lab.*, **4**(5), 27, May 1972). Particle sizes must be small: e.g. 35–50 μm and 25–35 μm.

There are various methods of packing the stationary phase into the column. Materials such as ion exchange resins, which swell when they come into contact with a solvent, must be packed wet as a slurry. Other materials are packed dry with the column being vibrated to achieve close packing. Packed columns should be evaluated before use for efficiency (a theoretical plate height of about 0.1 mm), for permeability (pressure required), and speed. (Theoretical plate height is a measure of the separating efficiency of a column analogous to the number of separating plates in a liquid distillation column.)

Guidance on column packing materials can be obtained from manufacturers such as Pechiney-St Gobain, Waters Associates, E.M. Laboratories, Reeve Angel, Dupont and Separations Group.

Mobile phase The mobile phase must have the correct 'polarity' for the desired separation, low viscosity, high purity and stability, and compatibility with the detection system. It must also dissolve the sample and wet the stationary phase.

Detectors Commercially available detectors used in HPLC are fluorimetric, conductiometric, heat of absorption detector, Christiansen effect detector, moving wire detector, ultraviolet absorption detector and the refractive index detector. The last two are the most popular.

Ultraviolet detection requires a UV-absorbing sample and a non-UV-absorbing mobile phase. Temperature regulation is not usually required.

Differential refractometers are available for HPLC but refractive index measurements are temperature-sensitive and good temperature control is essential if high sensitivity is required. The main advantage of the refractive index detector is wide applicability.

HPLC has been applied successfully to analysis of petroleum and oil products, steroids, pesticides, analgesics, alkaloids, inorganic substances, nucleotides, flavours, pharmaceuticals and environmental pollutants.

2.3 Polarography and anodic stripping voltammetry

2.3.1 Polarography

Polarography is an electrochemical technique and a specific polarographic sensor for the on-stream determination of oxygen in gas streams is described in Chapter 5. However, there are also many laboratory polarographic instruments; these are described briefly here together with the related technique of anodic stripping voltammetry.

2.3.1.1 Direct current polarography

In polarography an electrical cell is formed with two electrodes immersed in the solution to be analysed. In the most simple version of the technique (d.c. polarography) the anode is a pool of mercury in the bottom of the cell (although it is often preferable to use a large capacity calomel electrode in its place) and the cathode consists of a reservoir of mercury connected to a fine glass capillary with its tip below the surface of the solution. This arrangement allows successive fine drops of mercury to fall through the solution to the anode at the rate of one drop of mercury every 3 or 4 seconds. Figure 2.3 shows the arrangement in practice. The voltage applied across the two electrodes is slowly increased at a constant rate and the current flowing is measured and recorded. Figure 2.4 shows the step type of record obtained, the oscillations in the magnitude of the current are due to the changing surface area of the mercury drop during the drop life.

The solutions to be analysed must contain an 'inert' electrolyte to reduce the electrical resistance of the solution and allow diffusion to be the major transport mechanism. These electrolytes can be acids, alkalis, or citrate, tartrate and acetate buffers, as appropriate. The cells are designed so that oxygen can be removed from the solution by means of a stream of nitrogen, for otherwise the step given by oxygen would interfere with other determinations. The voltage range can run from +0.2 to −2.2 volts with respect to the calomel electrode. At the positive end the mercury electrode itself oxidizes; at the negative end the 'inert' electrolyte is reduced.

The potential at which reduction occurs in a given base electrolyte, conventionally the half-wave potential, is characteristic of the reducible species under consideration and the polarogram (the record obtained during polarography) thus shows the reducible species present in the solution. The magnitude of the diffusion current is a linear function of the concentration of the ion in solution. Thus, in Figure 2.4, $E_{1/2}$ is characteristic of cadmium in a hydrochloric acid electrolyte and I_d is a measure of the amount of cadmium. The limit of detection for d.c. polarography is about 1 ppm.

2.3.1.2 Sampled d.c. polarography

One disadvantage of the simple polarographic technique is that the magnitude of diffusion current has to be measured on a chart showing current oscillations (Figure 2.4). As these are caused by the changing surface area of the mercury drop during its lifetime an improvement can be made by using sampled d.c. polarography in which the current is measured only during the last milliseconds of the drop life. To do this the mercury drop time must be mechanically controlled. The resulting polarogram has the same shape as the d.c. polarogram, but is a smooth curve without large oscillations.

2.3.1.3 Single-sweep cathode ray polarography

Another modification to d.c. polarography is single-

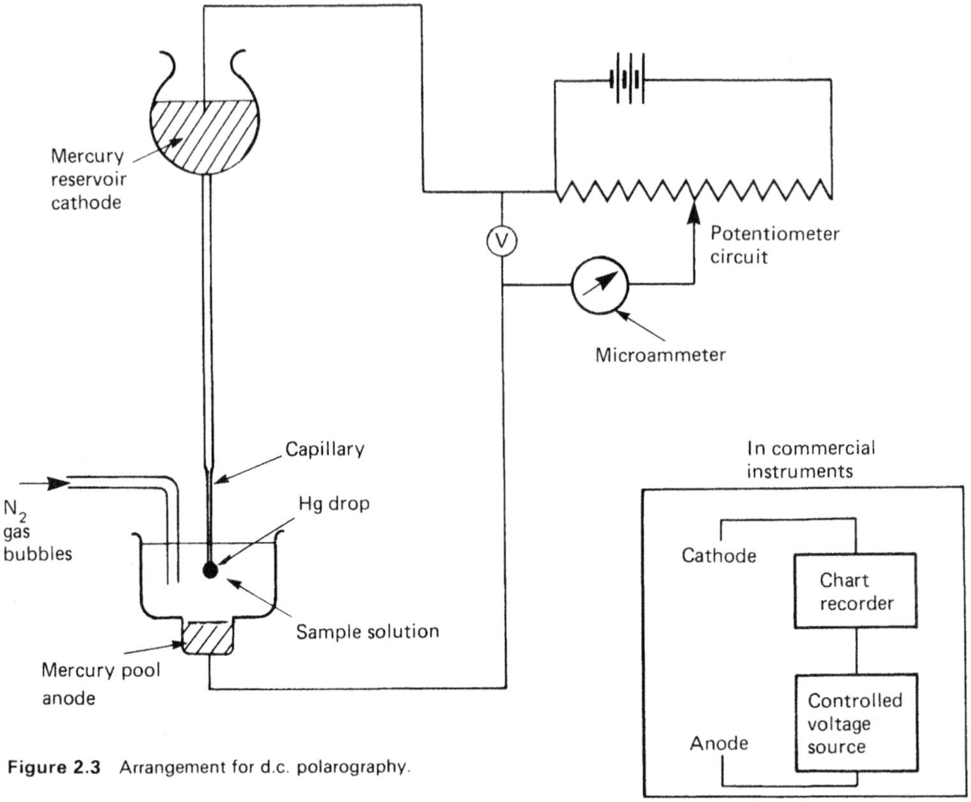

Figure 2.3 Arrangement for d.c. polarography.

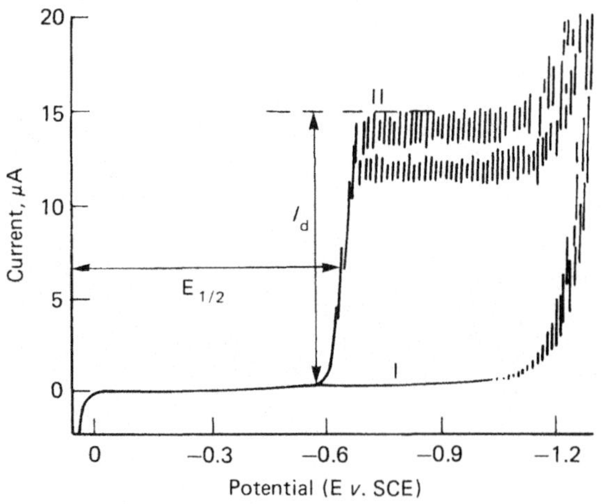

I : DC Polarogram 1*M* HCl
II : DC Polarogram of 5.0 × 10^{-4}*M* Cd (ii) in 1*M* HCl
I_d : Diffusion current
$E_{1/2}$: Half-wave potential

Figure 2.4 Polarograms of cadmium in hydrochloric acid (reprinted by courtesy of EG & G Princeton Applied Research and EG & G Instruments Ltd.).

sweep cathode ray polarography. Here an increasing d.c. potential is applied across the cell but only once in the life of every mercury drop. Drop times of about 7 seconds are used; the drop is allowed to grow undisturbed for 5 seconds at a preselected fixed potential, and a voltage sweep of 0.3 volts per second is applied to the drop during the last 2 seconds of its life. The sharp decrease in current when the drop falls is noted by the instrument and the sweep circuits are then automatically triggered back to zero. After the next 5 seconds drop growing time another voltage sweep is initiated, is terminated by the drop fall, and so on. The use of a long persistence cathode ray tube enables the rapid current changes to be followed easily with the trace remaining visible until the next sweep. Permanent records can be made by photography.

A characteristic of this technique is the peaked wave (Figure 2.5(a)) obtained compared with classical d.c. polarography. This peak is not a polarographic maximum, but is due to the very fast voltage sweep past the deposition potential causing the solution near the drop surface to be completely stripped of its reducible species. The current therefore falls and eventually flattens out at the diffusion current level. The peak height is proportional to concentration in the same way as the diffusion current level but

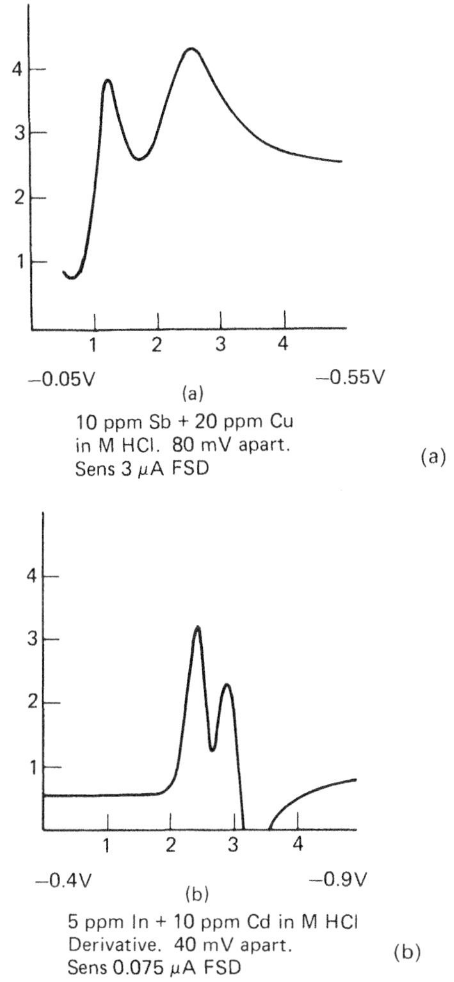

Figure 2.5 Single-sweep cathode ray polarograms (courtesy R. C. Rooney) (a) direct; (b) derivative.

sensitivity is increased. Resolution between species is enhanced by the peaked waveform and even this can be improved by the use of a derivative circuit, see Figure 2.5(b). Also, because of the absence of drop growth oscillations, more electronic amplification can be used. This results in the sensitivity of the method being at least ten times that of conventional d.c. polarography.

2.3.1.4 Pulse polarography

The main disadvantage of conventional d.c. polarography is that the residual current, due mainly to the capacitance effect continually charging and discharging at the mercury drop surface, is large compared with the magnitude of the diffusion current when attempting to determine cations at concentrations of 10^{-5} mol l^{-1} or below. Electronic methods have again been used to overcome this difficulty and the most important techniques are pulse and differential pulse polarography.

In normal pulse polarography the dropping mercury electrode is held at the initial potential to within about 60 milliseconds of the end of the drop life. The potential is then altered in a stepwise manner to a new value and held there for the remainder of the drop life. During the last 20 milliseconds of this the current is measured and plotted against the applied potential. Each new drop has the potential increased to enable the whole range of voltage to be scanned. The change in current that occurs when the voltage is stepped comes from the current passed to charge the double-layer capacitance of the electrode to the new potential. This decays very rapidly to zero. There is also a Faradaic current which is observed if the potential is stepped to a value at which an oxidation or reduction reaction occurs. This decays more slowly and is the current that is measured. This technique gives detection limits from 2 to 10 times better than d.c. polarography, Figure 2.6, but it is still not as sensitive as differential pulse polarography.

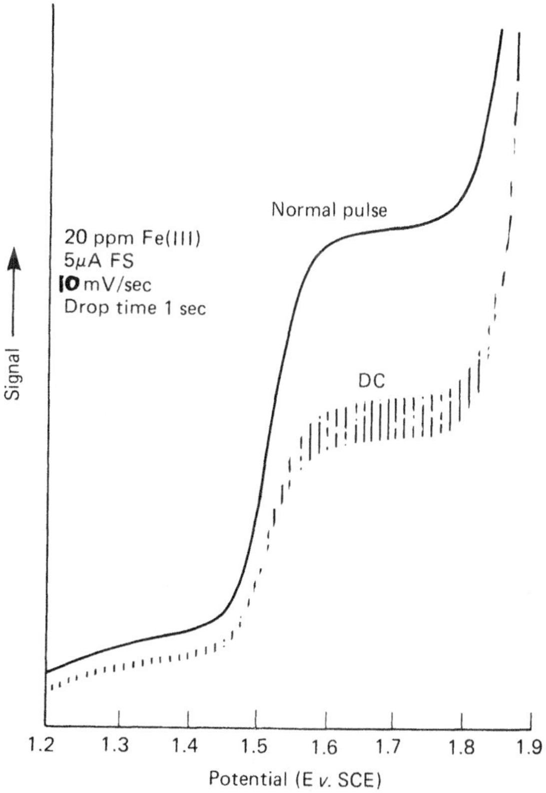

Figure 2.6 Normal pulse and d.c. polarograms for iron in ammonium tartrate buffer, pH 9 (reprinted by courtesy of EG & G Princeton Applied Research and EG & G Instruments Ltd.).

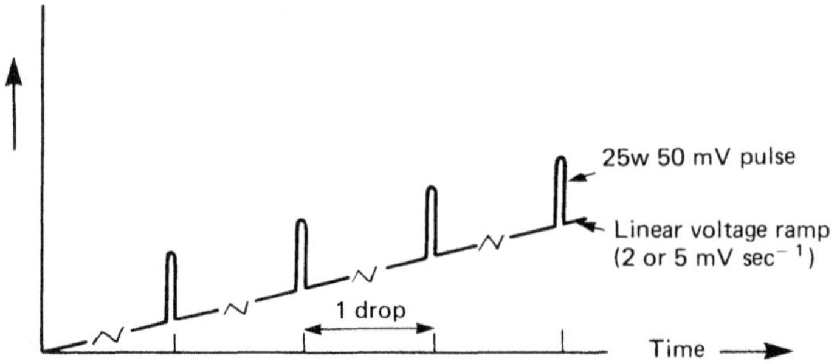

Figure 2.7 Voltage waveform for differential pulse polarography.

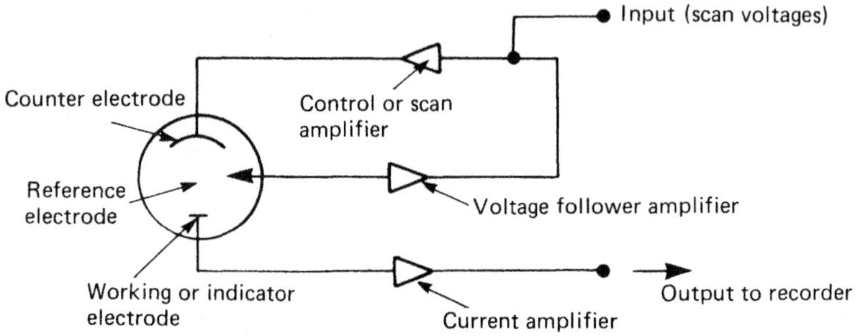

Figure 2.8 Practical arrangement for differential pulse polarography.

2.3.1.5 Differential pulse polarography

The most important of modern polarographic techniques is that of differential pulse polarography. Here a 25 or 50 mV amplitude pulse is superimposed at fixed time intervals on the normal linear increasing voltage of 2 or 5 $mV\,s^{-1}$ with the mercury drop being dislodged mechanically and so arranged that the pulse occurs once during the lifetime of each drop, Figure 2.7. The current is measured over a period of about 0.02 seconds just before the pulse is applied and during 0.02 seconds towards the end of the drop life. The difference between the two measurements is recorded as a function of the applied d.c. potential. In practice, a three-electrode potentiostatic arrangement is used, Figure 2.8. The polarograms obtained in this way are peak shaped (Figure 2.9), there is increased resolution between any two species undergoing reduction, and a great increase in sensitivity which is mainly a function of the reduction in measured capacitance current. There is a linear relationship between peak height and the concentration of the species being determined and limits of detection can be as low as 10^{-8} mol l^{-1}. The sensitivity of the technique can be varied by varying the pulse height; the peak height increases with increased pulse height but the resolution between peaks suffers (Figure 2.10). A comparison of the sensitivities of d.c., sampled d.c., normal pulse and differential pulse polarography is shown in Figure 2.11.

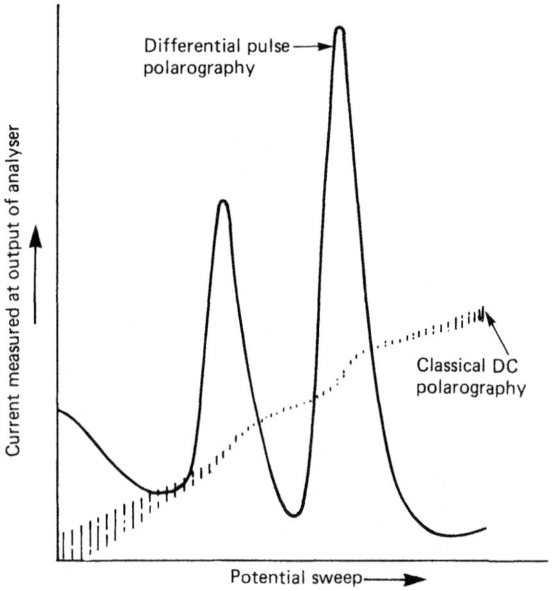

Figure 2.9 Differential pulse polarogram.

2.3.1.6 Applications of polarography

Polarographic methods can be used for analysing a wide range of materials. In metallurgy Cu, Sn, Pb, Fe, Ni, Zn, Co, Sb and Bi can be determined in light and zinc-based alloys, copper alloys and aluminium bronze; the control of effluents is often carried out

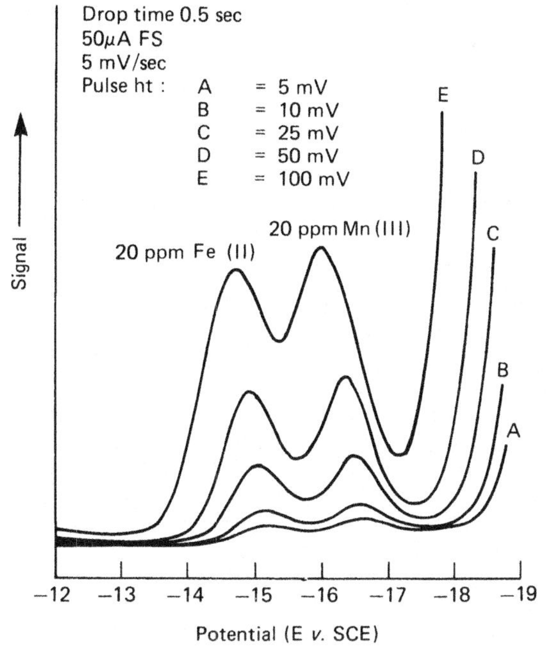

Figure 2.10 Effect of pulse height on peak height and resolution (reprinted by courtesy of EG & G Princeton Applied Research and EG & G Instruments Ltd.).

using polarographic methods. Cyanide concentrations down to ~0.1 ppm can be determined and sludges and sewage samples as well as fresh and sea waters can be analysed. Trace and toxic elements can be determined polarographically in foodstuffs and animal feed, in soils and in pharmaceutical products. In the latter, some compounds are themselves polarographically reducible or oxidizable, for example, ascorbic acid, riboflavin, drugs such as phenobarbitone and ephedrine and substances such as saccharine. Body fluids, plastics and explosives can also be analysed by polarographic techniques.

2.3.2 Anodic stripping voltammetry

Anodic stripping voltammetry is really a reversed polarographic method. Metals that are able to form amalgams with mercury, e.g. Pb, Cu, Cd and Zn can be cathodically plated onto a mercury drop using essentially the same instrumentation as for polarography and then the amalgamated metal is stripped off again by changing the potential on the mercury drop linearly with time in an anodic direction. By recording the current as a function of potential, peaks are observed corresponding to the specific species present in the test solution; the heights of the peaks are proportional to concentration.

In practice, it is not very convenient to use a mercury drop as cathode and several other types of electrode have been used including a rotating ring-disc electrode. The most often used, especially for water and environmental analysis is a wax-treated mercury-coated graphite rod. This, together with a silver/silver chloride reference electrode and a platinum counter electrode are immersed in the test solution (Figure 2.12) and the plating out and metal stripping carried out. Figure 2.13 illustrates the plating and stripping steps and Figure 2.14 shows a typical recording of the peak heights of Cd, In, Pb, Cu and Bi. As with polarography, various electronic modifications have been made to the basic technique and the stripping step has also been carried out with a.c. or pulsed voltages superimposed on the linear variation of d.c. voltage. Details of these systems can be found in reviews of the subject. Equipment for this technique is available at reasonable cost and units can be obtained for simultaneous plating of up to 12 samples with sequential recording of the stripping stages.

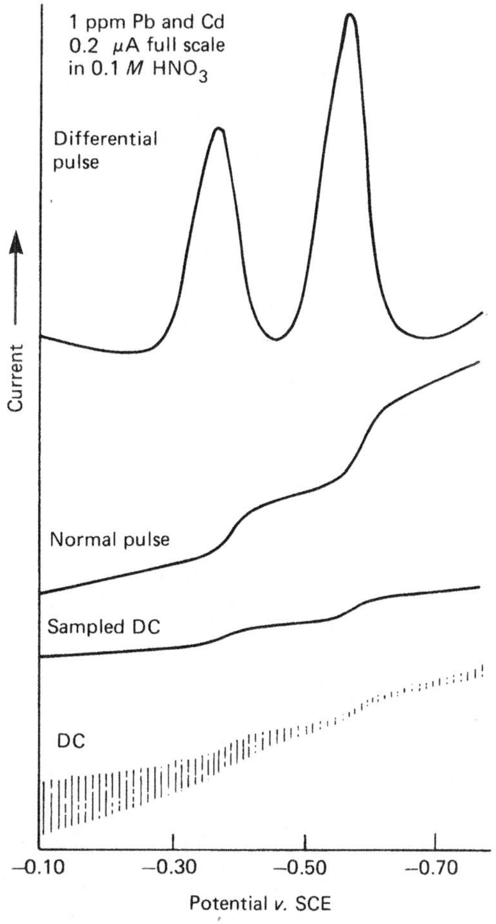

Figure 2.11 Comparison of polarographic modes (reprinted by courtesy of EG & G Princeton Applied Research and EG & G Instruments Ltd.).

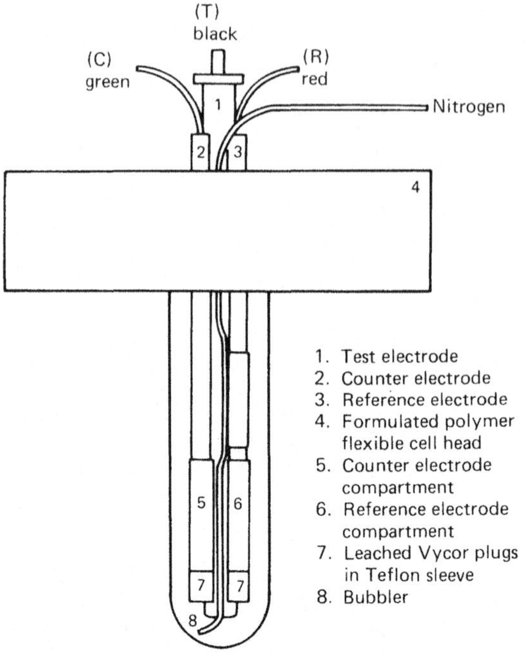

Figure 2.12 Cell arrangement for anodic stripping voltammetry (courtesy International Laboratory).

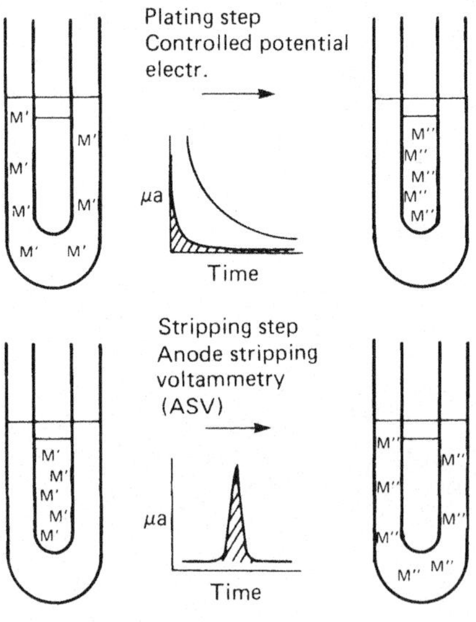

Figure 2.13 Plating and stripping steps (courtesy International Laboratory).

With anodic stripping voltammetry small samples (mg) can be used or very low concentrations of species determined because the plating step can be used as a concentration step. Plating times from 5 to 30 minutes are common depending on the required speed and accuracy of the analysis. Figure 2.14 was obtained using a 30-minute plating time. Good precision and accuracy can be obtained in concentration ranges as low as 0.1 to 10 μg per litre and this, combined with the fact that small samples can be used, means that the technique is most attractive for trace-metal characterization in the analysis of air, water, food, soil and biological samples.

2.4 Thermal analysis

No work on instrumental methods of determining chemical composition would be complete without mention of thermal analysis. This is the name applied to techniques where a sample is heated or cooled while some physical property of the sample is recorded as a function of temperature. The main purpose in making such measurements is most often not to evaluate the variation of the physical property itself but to use the thermal analysis record to study both the physical and chemical changes occurring in the sample on heating.

There are three main divisions of the technique depending on the type of parameter recorded on the thermal analysis curve. This can be (a) the absolute value of the measured property such as sample weight, (b) the difference between some property of the sample and that of a standard material, e.g. their temperature difference (these are differential measurements), and (c) the rate at which the property is changing with temperature or time, e.g. the weight loss, these are derivative measurements.

A convention has grown up for thermal analysis nomenclature and recommendations of the International Confederation for Thermal Analysis are that the term 'thermogravimetry' (TG) be used for measuring sample weight, 'derivative thermogravimetry' (DTG) for rate of weight loss, and 'differential thermal analysis' (DTA) for measuring the temperature difference between sample and standard. There are also many other terms relating to specific

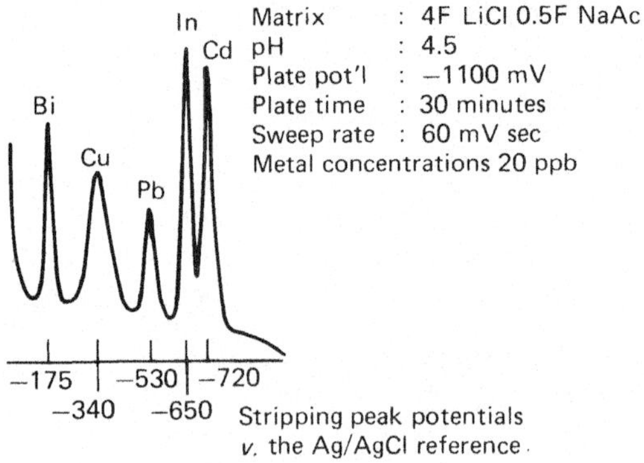

Figure 2.14 Stripping peak potentials (courtesy International Laboratory).

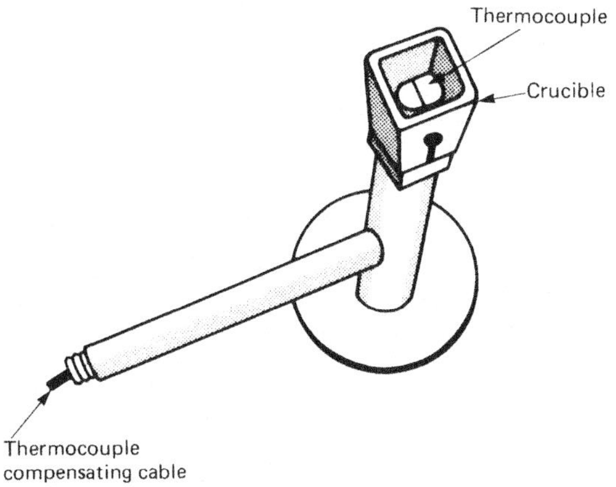

Figure 2.15 Cup for thermal analysis of cast iron.

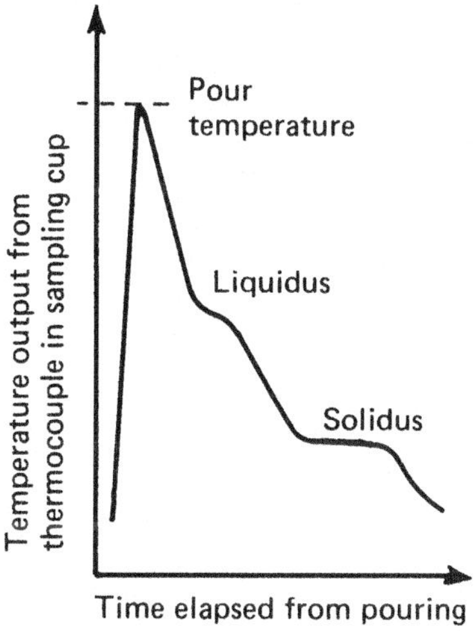

Figure 2.16 Cooling profile during cooling of liquid cast iron.

heat measurement, magnetic susceptibility, evolved gases, etc.

During the past 20 years a wide choice of commercially available equipment has become available and thermal analysis is now widely used as a tool in research and product control.

One particular application is to the composition of cast iron in terms of its carbon, silicon and phosphorus content, which can be calculated from the temperatures at which it freezes. As it is an alloy the freezing occurs at two temperatures, the liquidus and the solidus temperatures. At both temperatures the change of state of the metal releases latent heat. The temperatures at which the liquidus and solidus occur can be measured by the use of equipment made by Kent Industrial Measurements Ltd. To make the measurement a sample of liquid iron is poured into a special cup made from resin-bonded sand into which a small type K thermocouple is mounted, Figure 2.15. As the iron cools and passes through its two changes of state its temperature is monitored by the thermocouple. The graph showing the cooling against time, Figure 2.16, has two plateaux, one at the liquidus and one at the solidus. To complete the analysis the signal from the thermocouple is processed by a microcomputer which calculates and prints out the required analysis.

Figures 2.17–2.22 show other applications of thermogravimetry and derivative thermogravimetry to commercial samples and are largely self-explanatory.

In commercial thermal analysis instruments the sample is heated at a uniform rate while its temperature and one or more of its physical properties are measured and recorded. A typical arrangement is shown in Figure 2.22(a). The measuring unit has a holder to fix the position of the sample in the furnace, a means of controlling the atmosphere around the sample, a thermocouple for measuring the sample

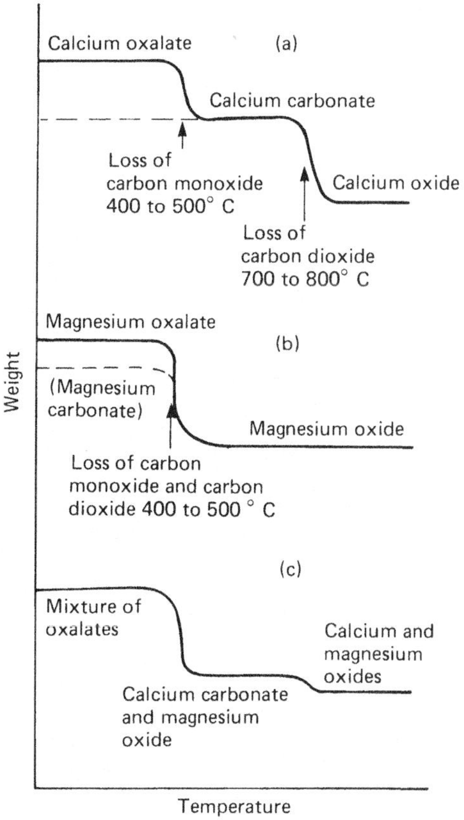

Figure 2.17 Weight-loss curves for calcium and magnesium oxalates and a precipitated mixture. (Reproduced by permission from *Thermal Analysis* by T. Daniels, published Kogan Page, Ltd.)

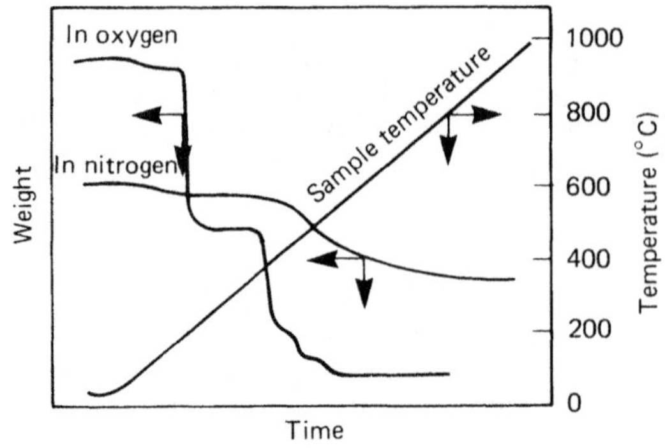

(a) TG CURVES FOR A COAL SAMPLE IN OXYGEN AND NITROGEN (FISHER TG SYSTEM)

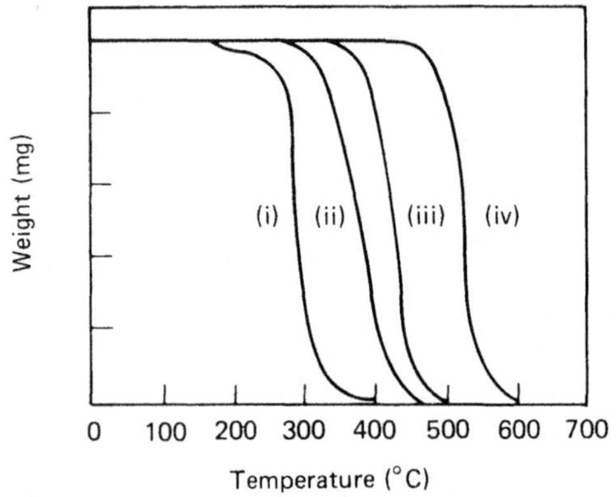

(b) TG CURVES FOR (i) POLYHEXAFLUOROPROPYLENE, (ii) POLYPROPYLENE, (iii) POLYETHYLENE, AND (iv) POLYTETRAFLUOROETHYLENE (Du Pont TG SYSTEM)

Figure 2.18 Thermal and thermo-oxidative stability of organic materials. (Reproduced by permission from *Thermal Analysis* by T. Daniels, published Kogan Page, Ltd.)

temperature and the sensor for the property to be measured, e.g. a balance for measuring weight. The design of the property sensor has to be such that it will function accurately over a wide temperature range and it is most important to ensure that the atmosphere around the sample remains fixed, be it an inert gas, a reactive gas or a vacuum.

The temperature control unit consists of a furnace and a programming unit, the function of which is to alter the sample temperature (not the furnace temperature) in a predetermined manner. The recording unit receives signals from the property sensor and the sample thermocouple, amplifies them and displays them as a thermal analysis curve. Figure 2.22(b) shows arrangements for differential instruments where the sample material and a reference material are placed in identical environments with sensors to measure the difference in one of their properties. The differential signal is amplified and recorded as in the basic system. In derivative instruments (Figure 2.22(c)) a derivative generator, such as an electro-optical device or an electronic unit is incorporated to compute the derivative of an input signal. Generally, both the derivative signal and the signal from the property being measured are recorded on the thermal analysis curve. It is, of course, possible to combine both modifications, thereby recording the derivative of a differential signal.

Most measuring units are designed specifically for a particular thermal analysis technique but furnaces, programmers, amplifiers and recorders are common to all types of instrument. Instrument manufacturers therefore generally construct a basic control unit containing programming and recording facilities to which can be connected modules designed for specific thermal analysis techniques.

Detailed description of the design of thermal

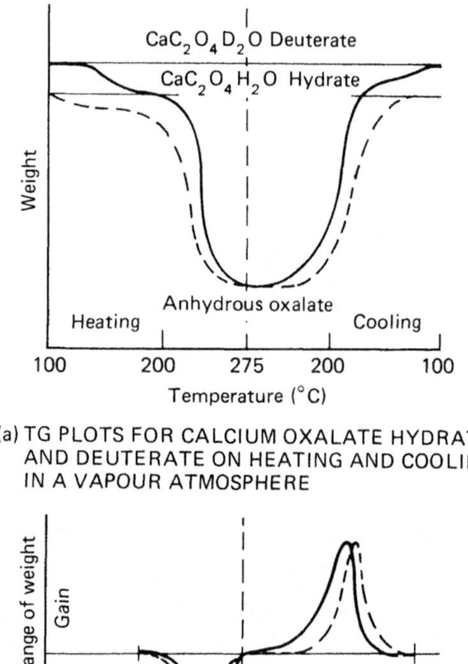

(a) TG PLOTS FOR CALCIUM OXALATE HYDRATE AND DEUTERATE ON HEATING AND COOLING IN A VAPOUR ATMOSPHERE

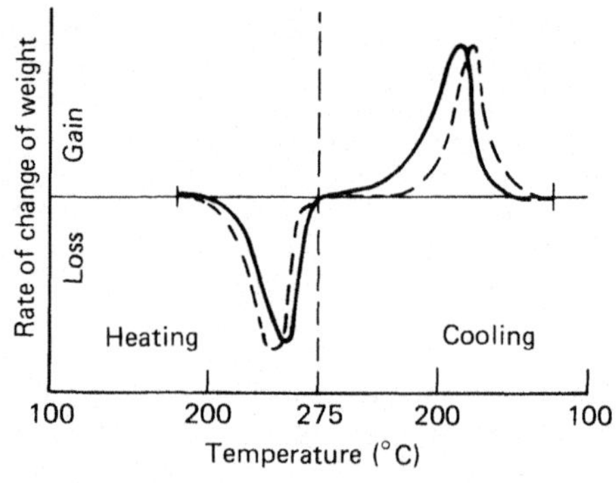

(b) DIG CURVES

Figure 2.19 The use of vapour atmospheres in TG. (Reproduced by permission from *Thermal Analysis* by T. Daniels, published Kogan Page, Ltd.)

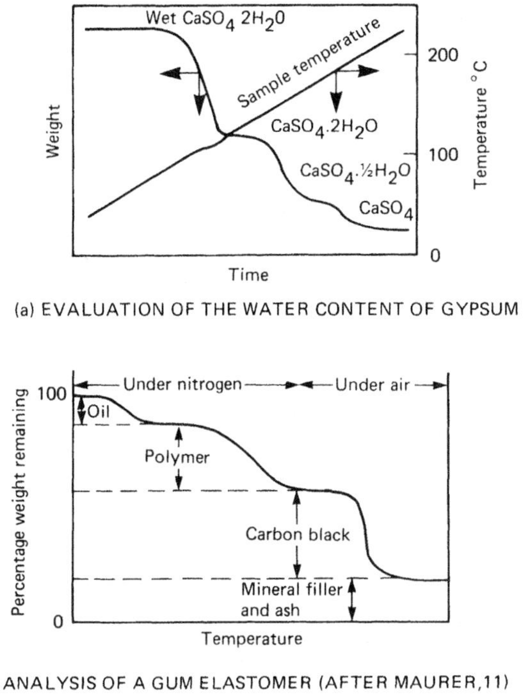

Figure 2.20 Analysis of commercial materials by TG. (Reproduced by permission from *Thermal Analysis* by T. Daniels, published Kogan Page, Ltd.)

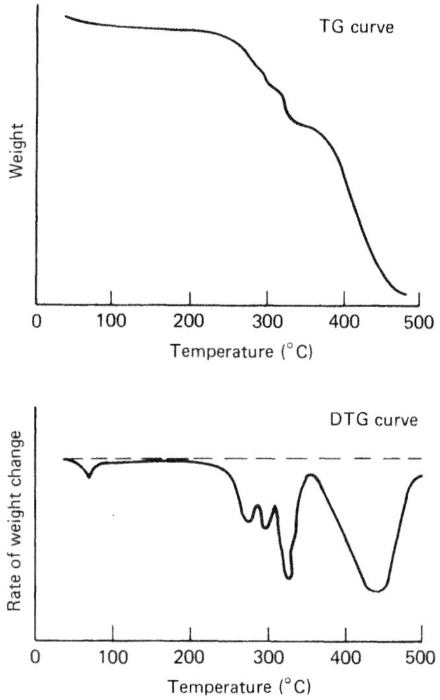

Figure 2.21 Dehydration and reduction of $\alpha Fe_2O_3 . H_2O$ on heating in hydrogen. (Reproduced by permission from *Thermal Analysis* by T. Daniels, published Kogan Page, Ltd.)

analysis instruments, their applications and the precautions necessary to ensure good results are beyond the scope of this volume, but there are several well written books on the topic.

2.5 Further reading

Bristow, P. A., *Liquid Chromatography in Practice*, Lab. Data Florida, (ISBN 0 9504933 1 1)

Daniels, T., *Thermal Analysis*, Kogan Page, (1973)

Fried, B. and Sherma, J., *Thin Layer Chromatography: Techniques and Applications*, Marcel Dekker, New York, (1982)

Heyrovsky, J. and Zuman, P., *Practical Polarography*, Academic Press, (1968)

Kirkland, J. J. (ed.), *Modern Practice of Liquid Chromatography*, Wiley Interscience, New York, (1971)

Meites, L., *Polarographic Techniques* (2nd edn), Interscience, (1965)

Perry, S. G., Amos, R. and Brewer, P. I., *Practical Liquid Chromatography*, Plenum, New York, (1972)

Snyder, L. R. and Kirkland, J. J., *Introduction to Modern Liquid Chromatography*, Wiley Interscience, New York, (1974)

Touchstone, J. C. and Rogers, D. (eds), *Thin Layer Chromatography Quantitative, Environmental and Clinical Applications*, Wiley, New York, (1980)

Wendland, W. W., *Thermal Methods of Analysis*, Interscience, (1964)

Wiedemann, H. G. (ed.), *Thermal Analysis* Vols 1–3, Birkhäuser Verlag, Basle and Stuttgart, (1972)

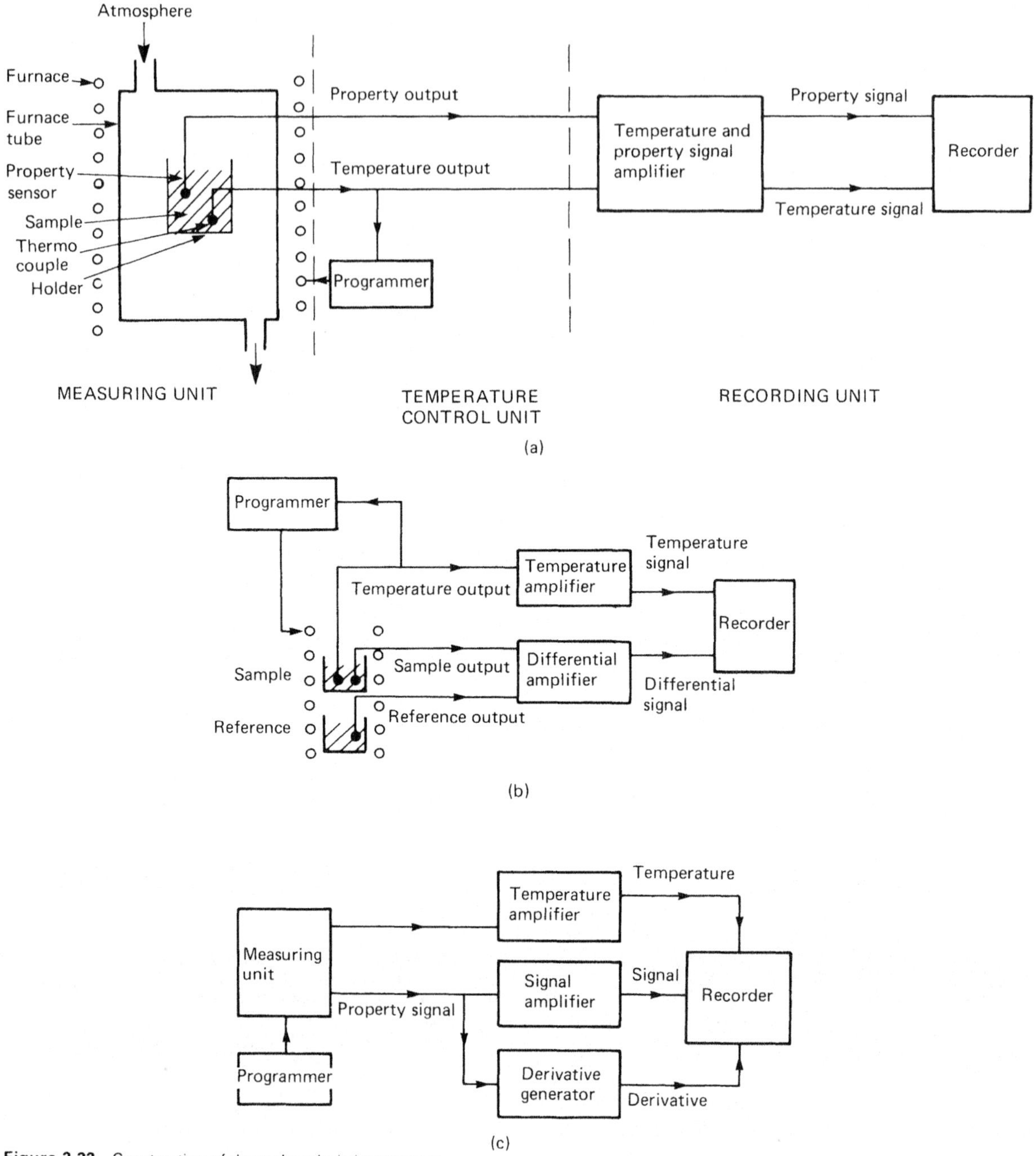

Figure 2.22 Construction of thermal analysis instruments. (Reproduced by permission from *Thermal Analysis* by T. Daniels, published Kogan Page, Ltd.): (a) basic thermal analysis system, (b) differential instrument, (c) derivative instrument.

3 Chemical analysis – spectroscopy

A. C. SMITH

The analysis of substances by spectroscopic techniques is a rather specialized field and cannot be covered in full depth in a book such as this. However, some fifteen techniques will be covered – giving the basic principles for each, descriptions of commercial instruments and where possible, their use as on-line analysers.

Details of other techniques may be found in modern physics textbooks and greater detail of those techniques which are described may be found in literature provided by instrument manufacturers such as Pye Unicam, Perkin-Elmer, Hilgers, Applied Research Laboratories. There are also many textbooks devoted to single techniques.

Some aspects of measurements across the electromagnetic spectrum are dealt with in Volume 3, Chapter 2.

3.1 Absorption and reflection techniques

3.1.1 Infrared

Measurement of the absorption of infrared radiation enables the quantity of many gases in a complex gas mixture to be measured in an industrial environment. Sometimes this is done without restricting the infrared frequencies used (Dispersive). Sometimes only a narrow frequency band is used (Non-dispersive).

3.1.1.1 Non-dispersive infrared analysers

Carbon monoxide, carbon dioxide, nitrous oxide, sulphur dioxide, methane and other hydrocarbons and vapours of water, acetone, ethyl alcohol, benzene and others may be measured in this way. (Oxygen, hydrogen, nitrogen, chlorine, argon and helium do not absorb infrared radiation and are therefore ignored.) An instrument to do this is illustrated in Figure 3.1(a). Two beams of infrared radiation of equal energy are interrupted by a rotating shutter which allows the beams to pass intermittently but simultaneously through an analysis cell assembly and a parallel reference cell, and hence into a Luft-pattern detector.

The detector consists of two sealed absorption

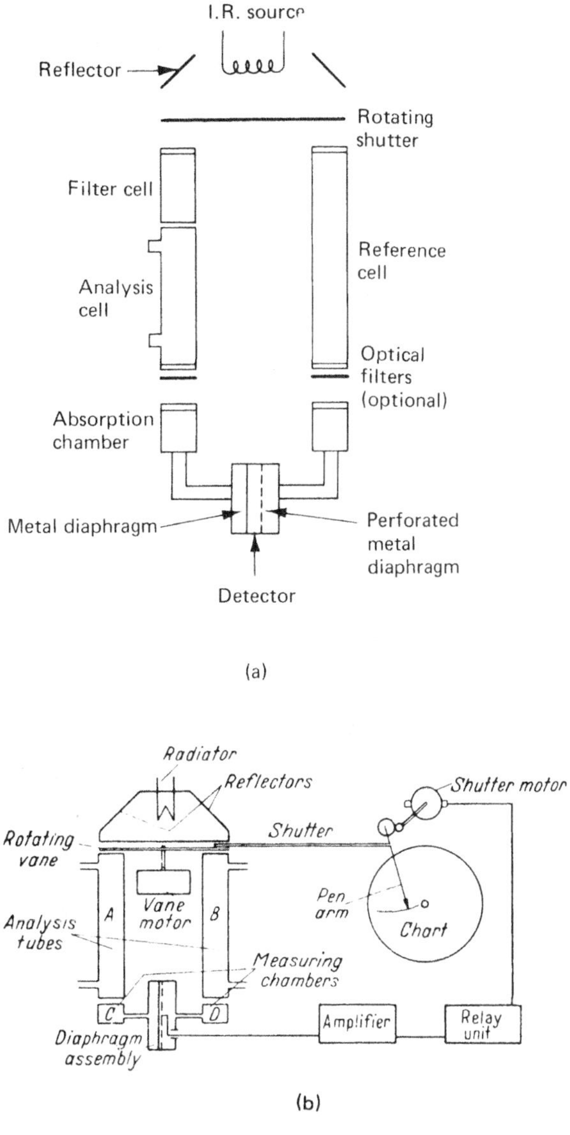

Figure 3.1 (a) Luft-type infrared gas analyser (courtesy Grubb Parsons), (b) infrared gas analyser of the concentration recorder.

chambers separated by a thin metal diaphragm. This diaphragm, with an adjacent perforated metal plate, forms an electrical capacitor. The two chambers are filled with the gas to be detected so that the energy characteristic of the gas to be measured is selectively absorbed.

The reference cell is filled with a non-absorbing gas. If the analysis cell is also filled with a non-absorbing gas equal energy enters both sides of the detector. When the sample is passed through the analysis cell, the component to be measured absorbs some of the energy to which the detector is sensitized, resulting in an imbalance of energy causing the detector diaphragm to be deflected and thus changing the capacitance. This change is measured electrically and a corresponding reading is obtained on the meter.

Any other gas also present in the sample will not affect the result unless it has absorption bands which overlap those of the gas being determined. In this event, filter tubes containing the interfering gas or gases can be included in one or both optical paths, so that the radiations emerging from these tubes will contain wavelengths which can be absorbed by the gas to be detected but will contain very little radiation capable of being absorbed by the interfering gases in the sample, since such radiations have already been removed.

The length of absorption tube to be used depends upon the gas being estimated and the concentration range to be covered. The energy absorbed by a column of gas l cm long and containing a concentration c of absorbing component is approximately $Elkc$, where E is the incident energy and k is an absorption constant, provided that kcl is small compared with unity. Thus at low concentrations it is advantageous to use long absorption paths provided kcl remains small and the relationship between energy absorbed and the measured concentration remains reasonably linear. At higher concentrations the energy absorbed is $E[1-\exp(-kcl)]$, and the relationship between energy absorbed and concentration departs greatly from linearity when absorption exceeds 25 per cent. When the absorption reaches this value it is, therefore, necessary to reduce the length of the absorption cell, and the product $c \times l$ should be kept approximately constant.

The most convenient method of calibrating the instrument is to pass mixtures of the pure gas and air of known composition through the measuring cell and note the output for each concentration of measured gas. For day-to-day checking a simple internal calibrating device is fitted, and it is only necessary to adjust the sensitivity control until a standard deflection is obtained.

The instrument is usually run from a.c. mains through a constant voltage transformer. Where utmost stability is required an a.c. voltage stabilizer may be used, as the constant voltage transformer converts frequency variations to voltage changes. Generally, the instrument is insensitive to temperature changes, although the gas sensitivity depends on the temperature and pressure of the sample gas in the absorption tube, since it is the number of absorbing molecules in the optical path which determines the meter deflection. For instruments sensitive to water vapour the detecting condenser has a temperature coefficient of sensitivity of 3 per cent per kelvin and it is therefore necessary to maintain the detector at a constant temperature.

The approximate maximum sensitivity to certain gases is given in Table 3.1.

Table 3.1 Sensitivity of non-dispersive infrared analyser

Gas	*Minimum concentration for full-scale deflection,* (Vol. %)	*Gas*	*Minimum concentration for full-scale deflection,* (Vol. %)
CO	0.05	NO_2	0.1
CO_2	0.01	SO_2	0.02
H_2O	0.1	HCN	0.1
CH_4	0.05	Acetone	0.25
C_2H_4	0.1	Benzene	0.25
N_2O	0.01		

Errors due to zero changes may be avoided by the use of a null method of measurement illustrated in Figure 3.1(b). The out-of-balance signal from the detector is amplified, rectified by a phase-sensitive rectifier, and applied to a servo system which moves a shutter to cut off as much energy from the radiation on the reference side as has been absorbed from the analysis side, and so restore balance. The shutter is linked to the pen arm which indicates the gas concentration.

On-line infrared absorption meter using two wavelengths In order to overcome the limitations of other infrared analysers and provide a rugged reliable drift-free analyser for continuous operation on a chemical plant, ICI Mond Division developed an analyser based on the comparison of the radiation absorbed at an absorption band with that at a nearby wavelength. By use of this comparison method many of the sources of error such as the effect of variation in the source intensity, change in the detector sensitivity or fouling of the measurement cell windows are greatly reduced.

The absorption at the measurement wavelength (λ_m) is compared with the nearby reference wavelength (λ_r) at which the measured component does not absorb. The two measurements are made alternately using a single absorption path and the same source and detecting system.

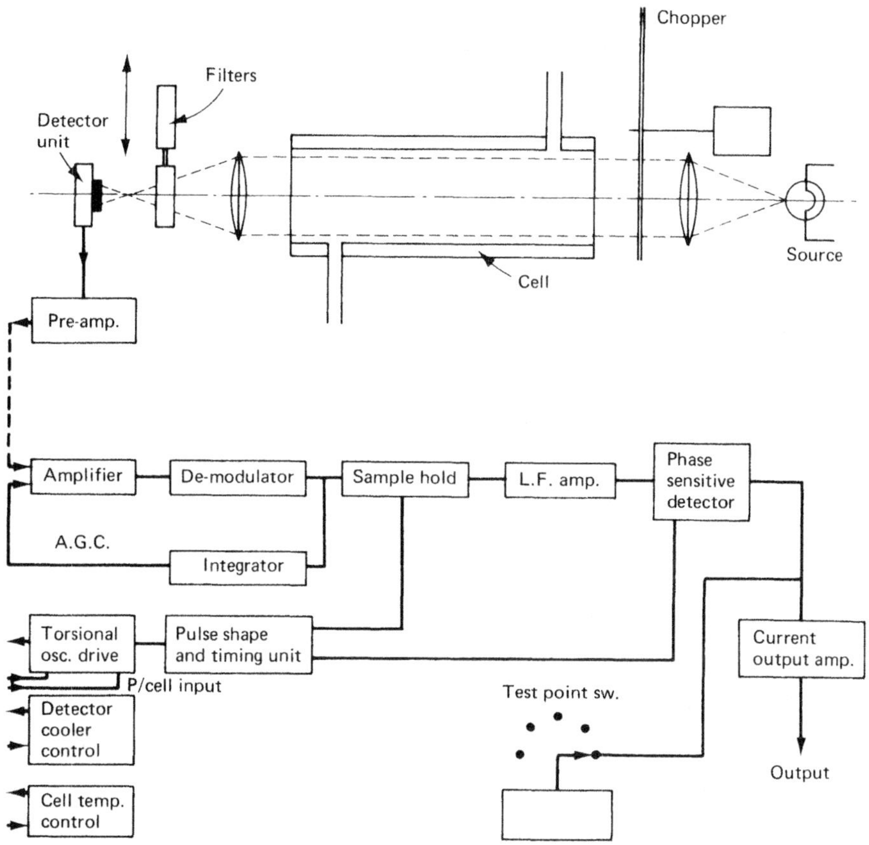

Figure 3.2 Dual wavelength comparison method (courtesy Feedback Instruments Ltd.).

The principle of the system is illustrated in Figure 3.2. The equipment consists of two units, the optical unit and the electronics unit, which are connected by a multicore cable. The source unit contains a sealed infrared source which consists of a coated platinum coil at the focus of a calcium fluoride collimating lens. A chopper motor with sealed bearings rotates a chopper disc which modulates the energy beam at 600 Hz. The source operates at low voltage, and at a temperature well below the melting point of platinum. It is sealed in a nitrogen atmosphere. Energy from the source passes through the absorption cell to the detector unit. A calcium fluoride lens focuses the energy onto an indium antimonide detector. This is mounted on a Peltier cooler in a sealed unit. The temperature is detected by a thermistor inside the sealed module. A pre-amplifier mounted in the detector unit amplifies the signal to a suitable level for transmission to the electronics unit. Between the lens and the detector module two interference filters, selected for the measurement and reference wavelengths, are interposed alternately in the beam, at about 6 Hz, so that the detector receives chopped energy at a level corresponding alternately to the measurement and reference transmission levels. Its output is a 600 Hz carrier modulated at 6 Hz.

The two filters are mounted on a counterbalanced arm, attached to a stainless steel torsion band. An iron shoe at the opposite end of the arm moves in and out of the gap in an electromagnet. It also cuts two light beams which illuminate two silicon phototransistors. The light is provided by two aircraft-type signal lamps which are under-run to ensure very long life. A drive circuit in the electronics unit causes the system to oscillate at its own natural frequency. One of the photocells provides positive feedback to maintain the oscillation, and the other provides negative feedback to control the amplitude. There are no lubricated parts in the detector unit, and the whole can be hermetically sealed if desired.

The absorption cell is a thick-walled tube with heavy flanges. Standard construction is in mild steel, nickel-plated, but type 316 stainless steel construction is available where required. The windows are of calcium fluoride, sealed with Viton O-rings and retaining rings. A heater wire is wound on the cell, and the sample gas passes through a tube in thermal contact along the length of the cell before entering it at the end. Provision is made for rodding out tubes and entries in case of blockage. A thermistor embedded in the cell wall detects the cell temperature which is controlled by a circuit in the electronics unit. The cell is thermally insulated and sealed inside a plastics bellows. The enclosed space is coupled to the purge system. The two end units each have a sealing window so there is a double seal between the cell and the interior of the detector and source units. Since the source is inside a further sealed module, there is minimal danger of the hot source being exposed to leakage from the sample cell. The gaps between the three units are normally sealed with neoprene gaskets and the whole device is sufficiently well sealed to maintain a positive purge pressure of at least 2 cm water gauge with a purge gas consumption of 8.3 cm^3/s. For use with highly flammable sample gases, the sealing gaskets at either end of the absorption cell may be replaced by vented gaskets. In this case a relatively large purge flow may be maintained around the cell, escaping to atmosphere across the windows. Thus, any leak at the windows can be flushed out.

To facilitate servicing on site the source, detector, torsional vibrator, lamps, pre-amplifier and source voltage control are all removable without the use of a soldering iron. Since the single-beam system is tolerant to window obscuration and the internal walls of the absorption cell are not polished, cell cleaning will not be required frequently, and in many cases adequate cleaning may be achieved *in situ* by passing solvent or detergent through the measuring cell. There is no need to switch the instrument off while doing this. If it becomes necessary the cell can be very quickly removed and disassembled.

The electronics unit contains the power supplies together with signal processing circuits, temperature control circuits, output and function check meter operating controls and signal lamps. The housing is of cast-aluminium alloy, designed for flush panel mounting. The circuitry is mostly on five plug-in printed circuit boards. The indicating meter, controls and signal lamps are accessible through a window in the door. The unit is semi-sealed, and a purge flow may be connected if sealed glands are used at the cable entry. The signal processing circuits are contained on two printed circuit boards. Output from the pre-amplifier is applied to a gain-controlled amplifier which produces an output signal of 3 V peak-to-peak mean. Thus the mean value of $I_r + I_m$ is maintained constant. The signal is demodulated and smoothed to obtain the 6 Hz envelope waveform. A sample-and-hold circuit samples the signal level near the end of each half-cycle of the envelope, and this produces a square wave whose amplitude is related to $I_r - I_m$. Since $I_r + I_m$ is held constant, the amplitude is actually proportional to $(I_r - I_m)/(I_r + I_m)$ which is the required function to give a linearized output in terms of sample concentration. This signal is amplified and passed to a phase-sensitive detector, consisting of a pair of gating transistors which select the positive and negative half-cycles and route them to the inverting and non-inverting inputs of a differential amplifier. The output of this amplifier provides the 0–5 V output signal.

The synchronizing signals for the sample/hold and phase-sensitive detector circuits are derived from the torsional oscillator drive circuit via appropriate time delays. The instrument span is governed by selection of feedback resistors in the low frequency amplifier, and a fine trim is achieved by adjusting the signal level at the gain-controlled amplifier. This is a preset adjustment – no operator adjustment of span is considered necessary or desirable. A front panel zero adjustment is provided. This adds an electrical offset signal at the phase-sensitive detector. The system is normally optically balanced (i.e. $I_r = I_m$) at some specified concentration of the measured variable (usually zero).

The current output and alarm circuits are located on a separate printed circuit board. The voltage output is applied to an operational amplifier with selected feedback and offset signals to produce 0–10 mA, 5–20 mA or 10–50 mA output. The required output is obtained by soldered selector links. The output current is unaffected by load resistances up to 1 kΩ at 50 mA, or 5 kΩ at 10 mA.

A front panel alarm-setting potentiometer provides a preset signal which is compared with the analyser output voltage in a differential amplifier. The output of this opens a relay if the analyser output exceeds a preset value, which may be either a low or high analyser output as required. The alarm condition is indicated by two signal lamps on the panel and the system can be arranged to operate external alarms, or shut-down circuits.

The power to the cell heater and the detector cooler is controlled from a bridge circuit containing thermistors which detect the temperatures of the absorption cell and detector.

The indicating meter on the front panel has a calibrated output scale, and is used in conjunction with a selector switch to monitor key points in the circuit, in particular the degree of obscuration in the measuring cell. By choosing the appropriate absorption bands the analyser may be made suitable for a wide range of gases or liquids. For gases, it may be used for CO_2, CO, SO_2, CH_4, C_2H_6, C_2H_4, C_6H_6, C_2H_2, NH_3, N_2O, NO, NO_2, $COCl_2$, H_2O, with ranges of 0–300 ppm and 0–100 per cent.

It may also be used for measuring water in ketones, hydrocarbons, organic acids, alcohols, glucols and oils. The accuracy is ±1 per cent and the response time for 90 per cent change is 3 s.

The instrument is marketed by Anatek Ltd. as the

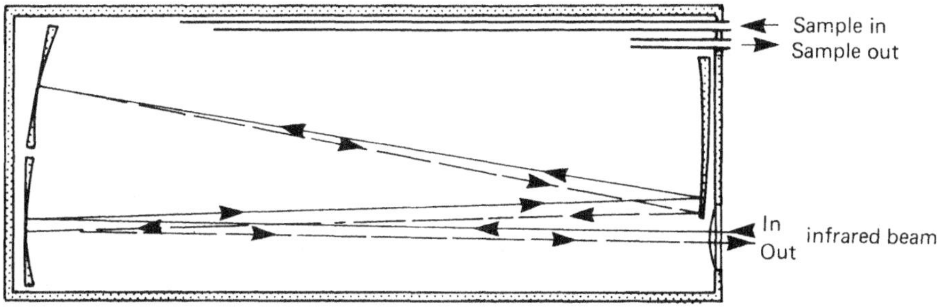

Figure 3.3 Internal view of multiple reflections of variable long path cell (courtesy The Foxboro Company).

PSA 401 process stream analyser.

Another instrument based on the same principle is the Miran II Infra Red process analyser – the chief difference being the sample cell used for gas and liquid streams. These cells are either long path gas cells or multiple internal reflection cells. The gas cells which are normally manufactured in stainless steel have a variable path length (see Figure 3.3). Energy passes through the sample gas and reflects one or more times off the mirrors in the cell before striking the detector. The path length can be adjusted between 0.75 and 20.25 metres by suitable adjustment of the mirrors. These gas cells are used to analyse the presence of low concentrations of components in gases or for those gases requiring a long path length to enhance sensitivity at a weak analytical wavelength.

In a multiple internal reflection cell, the infrared beam is directed along or around an optical crystal through which the beam passes (Figure 3.4). As the beam is reflected on the sample crystal interface, it slightly penetrates the liquid. These penetrations form a path whose length is dependent on the number of reflections. The energy is absorbed at the analytical wavelength proportionally to concentration just as in other types of cells. The crystal used is made of KRS (a composite of thallium bromide and iodide). Ordinary transmission cells have limited applicability for high concentrations, viscous or aqueous streams. In many cases, the infrared beam is grossly attenuated or the sample cannot be pumped through such cells. Multiple internal reflection overcomes these problems.

The applications to which this instrument has been put include (a) for gases: the determination of phosgene in methane and plastic production, methane and carbon dioxide in synthetic and natural gases in the range 1 ppm to 100 per cent, (b) for liquids: water in acetone distillation, petroleum waste treatments, urea in fertilizer production and isocyanates in urethane and plastic production in the range 50 ppm to 50 per cent and (c) for solids: the percentage weight of film coatings such as inks and polymers and film thickness for nylon and polythene (up to 0.025 mm).

3.1.1.2 Dispersive infrared analysis

The previous section was devoted to analysis using only one absorption frequency. However, all organic compounds give rise to a spectrum in the infrared in which there are many absorption frequencies giving a complete fingerprint of that compound. Dispersive infrared can be used, amongst other things, to identify a substance, for the determination of molecular structure for reaction kinetic studies, and for studies of hydrogen bonding.

In Figure 3.5 is shown a simplified layout of a typical double-beam spectrophotometer. A source provides radiation over the whole infrared spectrum, the monochromator disperses the light and then selects a narrow frequency range, the energy of which is measured by a detector – the latter transforms the energy received into an electrical signal which is then amplified and registered by a recorder or stored in a computer for further processing. The light path and ultimate focusing on the detector is determined by precision manufactured mirrors.

Light from the radiation source S is reflected by mirrors M_1 and M_2 to give identical sample and reference beams. Each of these focuses upon vertical entrance slits S_1 and S_2, the sample and reference cells being positioned in the two beams near their foci. Transmitted light is then directed by a mirror M_3 on to a rotating sector mirror (or oscillating plane mirror) M_4. The latter serves first to reflect the sample beam towards the monochromator entrance slit S_3 and then as it rotates (or oscillates), to block the sample beam and allow the reference beam to pass on to the entrance slit. A collimating mirror M_5 reflects parallel light to a prism P, through which it passes only to be reflected

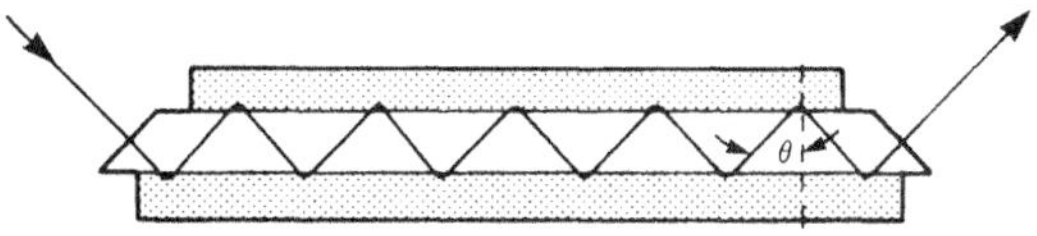

Figure 3.4 Principle of MIR sampling technique (courtesy Foxboro Analytical).

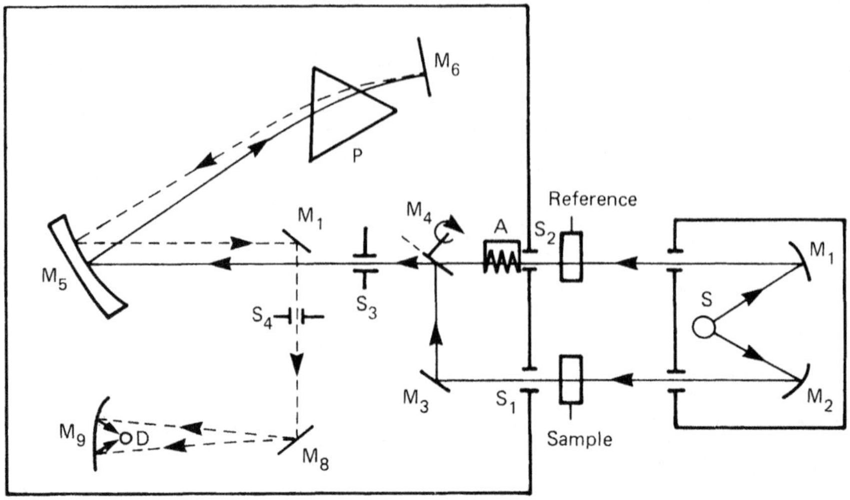

Figure 3.5 Simplified spectrophotometer.

back again through the prism by a rotatable plane mirror M_6. The prism disperses the light beam into its spectrum. A narrow range of this dispersed light becomes focused on a plane mirror M_7 which reflects it out through the exit slit. A further plane mirror M_8, reflects the light to a condenser M_9 which focuses it sharply on the detector D. When the energy of the light transmitted by both sample and reference cells is equal, no signal is produced by the detector. Absorption of radiation by the sample results in an inequality of the two transmitted beams falling on the detector and a pulsating electrical signal is produced. This is amplified and used to move an attenuator A across the reference beam, cutting down the transmitted light until an energy balance between the two beams is restored. The amount of reference beam reduction necessary to balance the beam energies is a direct measure of the absorption by the sample.

The design and function of the major instrument components now described have a significant influence on its versatility and operational accuracy.

Source IR radiation is produced by electrically heating a Nernst filament (a high resistance, brittle element composed chiefly of the powdered sintered oxides of zirconium, thorium and cerium held together by a binding material) or a Globar (SiC) rod. At a temperature in the range 1100–1800 °C depending on the filament material, the incandescent filament emits radiation of the desired intensity over the wavelength range 0.4–40 μm.

Monochromator The slit width, and optical properties of the components are of paramount importance. The wavelength range covered by different prisms is shown in Table 3.2. Gratings allow better resolution than is obtainable with prisms.

Detector This is usually a bolometer or thermocouple. Some manufacturers use a Golay pneumatic detector which is a gas-filled chamber that undergoes a pressure rise when heated by radiant energy. One wall of the chamber functions as a mirror and reflects a light beam directed at it onto a photocell – the output of the photocell bearing a direct relation to the gas chamber expansion.

The infrared spectra of liquids and gases may be obtained by direct study of undiluted specimens. Solids, however, are usually studied after dispersion in one of a number of possible media. These involve reduction of the solid to very small particles which are then diluted in a mill, pressed into an alkali halide disc at 1500–3300 bar or spread as pure solid on a cell plate surface.

Table 3.2 Prism frequency ranges

Prism material	*Glass*	*Quartz*	*CaF_2*	*LiF*	*NaCl*	*KBr(CsBr)*	*CsI*
Useful frequency range (cm^{-1})	above 3500	above 2860	5000–1300	5000–1700	5000–650	1100–285	1000–200
Wavelength range (μm)	below 2.86	below 3.5	2.0–7.7	2.0–5.9	2–15.4	9–35	10–50

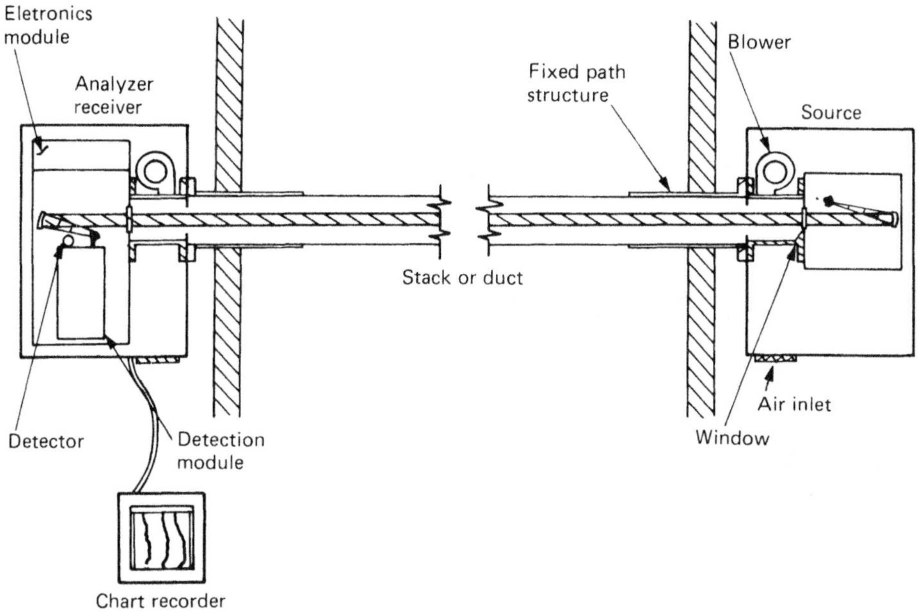

Figure 3.6 EDC flue gas analyser system (courtesy Environmental Data Corp.).

The interpretation of the spectra – particularly of mixtures of compounds – is a complex problem and readers should consult textbooks on infrared analysis.

3.1.2 Absorption in UV, visible and IR

One instrument that uses absorption in the UV, visible and IR is the Environmental Data Corporation stack-gas monitoring system. It is designed to measure from one to five component gases simultaneously. Depending on requirements, the components may include CO_2, NO, CO, SO_2, H_2O, NH_3, hydrocarbons and opacity or any other gases with selected spectral absorption bands in the UV, visible or IR. The basis of the system is shown in Figure 3.6. It consists of a light source, receiver, mounting hardware and recorder. Each gas monitoring channel is similar in basic operation and calibration. The instrumentation can be mounted on a stack, duct or other gas stream. A polychromatic beam of light, from a source in an enclosure on one side, is collimated and then passed through the gas to an analyser on the opposite side. Signals proportional to the gas concentrations are transmitted from analyser to recorder.

Most gases absorb energy in only certain spectral regions. Their spectra are often quite complex with interspersed absorbing and non-absorbing regions. The analyser section of the instrument isolates the wavelengths characteristic of the gases of interest and measures their individual intensities. Both the intensity at a specific wavelength where the gas uniquely absorbs (A) and the intensity at a nearby region where the gas is non-absorbing (B) are alternately measured with a single detector 40 times per second. Any light level change, whether due to source variation, darkening of the window, scattering by particulates, water drops or aerosols in the gas stream affects both A and B leaving the ratio unchanged. This ratio gives a reading that is free of interferences, instrumental drift etc. Most gases obey approximately Beer's law:

$$B = A\,e^{-\alpha c l}$$

or

$$\ln(B/A) = -\alpha c l$$

or

$$c = \frac{\ln(A/B)}{\alpha l}$$

where α is absorption coefficient (known), l is path length (fixed), and c is sample concentration (unknown).

The system response is almost instantaneous and is averaged by damping circuits to typically one second.

The stack gas is separated from the source and analyser enclosures by means of optical surfaces, such as mirrors or windows. These windows are kept clean by an air curtain system. Self-contained blowers continually renew the air curtains, preventing the gases from contacting the windows directly (see Figure 3.7).

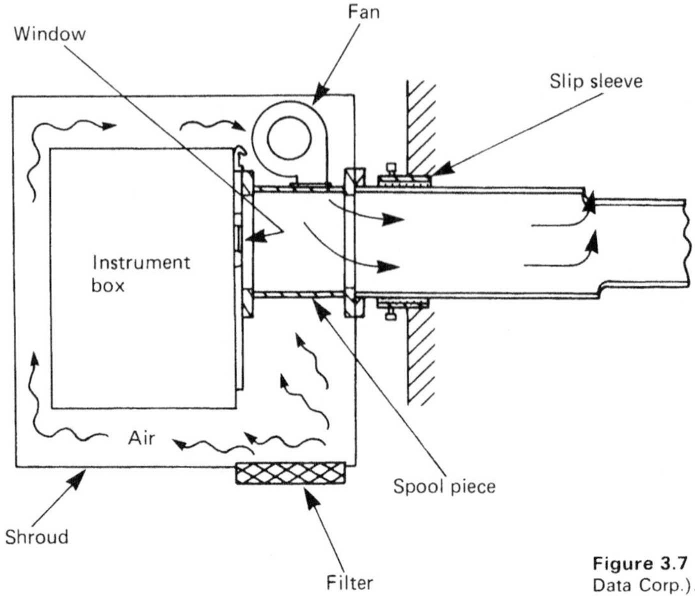

Figure 3.7 EDC flue gas analyser (courtesy Environmental Data Corp.).

The flow volume and pressure of the purge air is designed for each application to allow a well defined shear by the flue gas. Thus a known and fixed path length is provided.

When measuring opacity the instrument measures the reduction in transmission in the visible portion of the spectrum.

Typical ranges covered by the instrument are:

NO	0–25 ppm to 0–5000 ppm
CO	0–500 ppm to 0–3000 ppm
CO_2	0–15%
SO_2	0–25 ppm to 0–10 000 ppm
C–H	0–25 ppm to 0–6000 ppm
H_2O	0–1000 ppm to 0–80%
NH_3	0–100 ppm

3.1.3 Absorption in the visible and ultraviolet

Two instruments are worthy of note here. The first is the Barringer remote sensing correlation spectrometer designed for the quantitative measurement of gases such as nitrogen oxides or sulphur dioxide in an optical path between the instrument and a suitable source of visible and ultraviolet radiant energy. The sensor is designed for maximum versatility in the remote measurement of gas clouds in the atmosphere using the day sky or ground-reflected solar illumination as the light source. It may also be used with artificial sources such as quartz-iodine or high pressure Xe lamps. Very simply the sensor contains two telescopes to collect light from a distant source, a two-grating spectrometer for dispersion of the incoming light, a disc-shaped exit mask or correlator and an electronics system (see Figure 3.8). The slit arrays are designed to correlate sequentially in a positive and negative sense with absorption bands of the target gas by rotation of the disc in the exit plane. The light modulations are detected by photomultiplier tubes and processed in the electronics to produce a voltage output which is proportional to the optical depth (expressed in ppm meters) of the gas under observation. The system automatically compensates for changes in average source light intensity in each channel. The basic principle of this method rests on comparison of energy in selected proportions of the electromagnetic spectrum where absorption by the target gas occurs in accordance with the Beer–Lambert law of absorption.

Typically, this instrument covers the range 1–1000 ppm m or 100–10 000 ppm m, this unit being the product of the length of the optical path through the gas and the average concentration (by volume) over that length.

The second instrument which covers absorption in the visible in liquids, is the Brinkmann Probe Colorimeter. This instrument is basically a standard colorimeter consisting of a tungsten light source, the output from which passes through one of a series of interchangeable filters covering the wavelength range 420–880 nm, then through a light pipe at the end of which is a probe cell. This cell has a reflecting mirror at one end and so the optical path length is twice the length of the cell. The light then returns to the instrument via a second light pipe to a photomultiplier, the output of which is amplified and fed to a recorder in the usual way. This instrument is ideal for

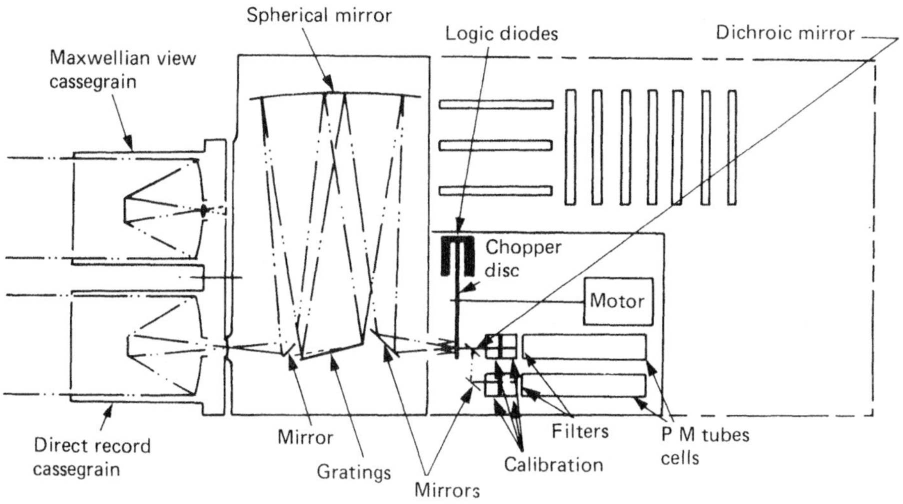

Figure 3.8 Barringer remote sensing correlation spectrometer.

measuring turbidity in liquids and has the advantage that very small volumes of liquid (down to 0.5 ml) may be examined. Its other uses include general quality control, chemical analyses, pollution control and food processing. Most of these applications make use of the fact that different elements will form coloured solutions with reagents. The absorption of these coloured solutions is then proportional to the concentration of that particular element.

3.1.4 Measurements based on reflected radiation

Just as measurements of moisture, or other components, may be made by comparison at two wavelengths of transmitted infrared radiation, the method will work equally well by measuring the attenuation when infrared is reflected or back-scattered. The principle is illustrated in Figure 3.9.

For water measurement of paper or granulated material on a conveyor belt, the intensity of the reflected beam at the moisture absorption wavelength of 1.93 μm may be compared with the intensity at a reference wavelength of 1.7 μm. The beams are produced by interposing appropriate filters contained in a rotating disc in front of a lamp producing appropriate radiation. The radiation is then focused onto the measured material, and the reflected beam focused onto a lead sulphide photoelectric cell. By measuring the ratio of the intensity of radiation at two wavelengths, the effects of source variation, detector sensitivity and drift in the electronic circuitry are minimized. Furthermore, calibration has shown that for a number of materials the results are substantially independent of the packing density.

However, if the measured material is a strong absorber of radiation a powerful source of radiation such as a water-cooled quartz halogen lamp may be necessary.

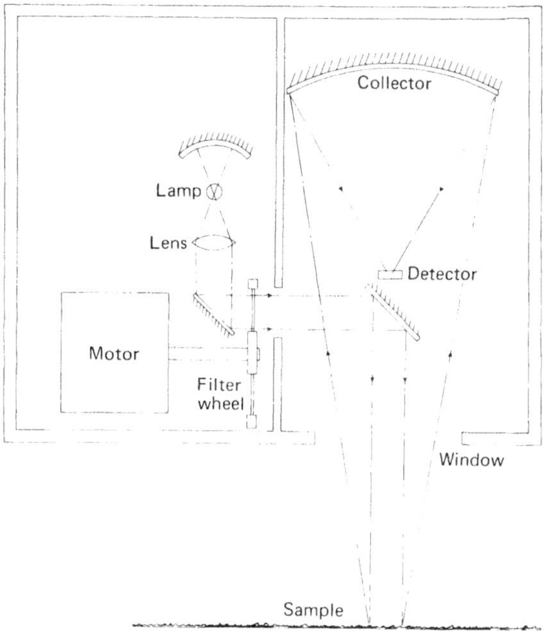

Figure 3.9 Backscatter infrared gauge (courtesy Infra-red Engineering Ltd.).

With this type of instrument on-line measurement of the moisture content of sands, clay, dust or flake, refractory mixtures, paper, textiles, feeding stuffs and a wide range of other materials may be undertaken with an accuracy of ± 1 per cent of instrument full scale.

3.1.5 Chemiluminescence

When some chemical reactions take place, energy may be released as light. This phenomenon is known as chemiluminescence. There are many instruments which make use of this effect for the determination of the concentration of oxides of nitrogen and for ozone. The principles are described in Chapter 5.

3.2 Atomic techniques – emission, absorption and fluorescence

3.2.1 Atomic emission spectroscopy

This is one of the oldest of techniques employed for trace analysis. Because of its relative simplicity, sensitivity and ability to provide qualitative information quickly, it has been widely used in both industrial and academic analytical problems. It can be used for the analysis of metals, powders and liquids and is used extensively in the steel and non-ferrous alloy industries, and the advent of inductively coupled plasma sources for producing spectra has made the technique invaluable for the analysis of some 70 elements in solution – down to concentrations of 1 ppb and less.

The basic principles of the technique are as follows.

Each atom consists of a nucleus around which revolve a set of electrons. Normally these electrons follow orbits immediately adjacent to the nucleus. If energy is imparted to the atom by means of a flame or an electric arc or spark, then it undergoes excitation and its electrons move into orbits further removed from the nucleus. The greater the energy, the further from the nucleus are the orbits into which the electrons are moved. When sufficient energy is imparted to the electron, it may be torn from the atom and the atom becomes a positively charged ion. Atoms will not remain in this excited state, especially when removed from the source of energy and they return to their original states with electrons falling to lower orbits. This electron transition is accompanied by a quantum of light energy. The size of this pulse of light energy and its wavelength depend on the positions of the orbits involved in the transition.

The energy emitted is

$$E = h\nu$$

where h is Planck's constant, and ν is the frequency of the radiation. Or

$$E = hc/\lambda$$

where c is the velocity of light and λ the wavelength. Hence the greater the light energy quantum, the shorter is the wavelength of the light emitted.

Only the outer, valence electrons participate in the emission of spectral lines. The number of valence electrons in an atom differs for chemical elements. Thus the alkali elements, sodium, lithium, potassium etc. contain only one electron in their outer shell and these elements have simple spectra. Such elements as manganese and iron have five or six valence electrons and their spectra are very complex. Generally speaking, the structure of an atom is closely bound up with its optical spectrum. Thus if a mixture of atoms (as found in a sample) are excited by applying energy, then quantities of light are emitted at various wavelengths, depending on the elements present. The intensity of light corresponding to one element bears a relationship to the concentration of that element in the sample.

In order to sort out the light emitted, use is made of a spectroscope. In Figures 3.10–3.12 are shown the layout of a medium quartz spectroscope, a Littrow spectrograph and a spectroscope using a diffraction grating. This last employs the principle, due to Rowland, of having the grating on a concave surface. There are many other configurations. In all cases, each instrument contains three main components, a slit, a dispersive device such as a prism or diffraction grating to separate radiation according to wavelength, and a suitable optical system to produce the spectrum lines which are monochromatic images of the slit. These images may be recorded on a photographic plate, or by suitable positioning of exit slits, mirrors and photomultiplier tubes, the light intensity may be recorded electronically.

3.2.1.1 *Dispersive devices*

Prisms Prisms are usually made of glass or quartz and their dispersive ability is based on the variation of the index of refraction with wavelength. As the incident light beam enters the transparent material, it bends towards the normal according to Snell's law:

$$n_1 \sin i = n_2 \sin r$$

where n_1 is the refractive index of air, n_2 is the refractive index of the prism material, i is angle of incidence, and r is angle of refraction. Shorter wavelengths are deviated

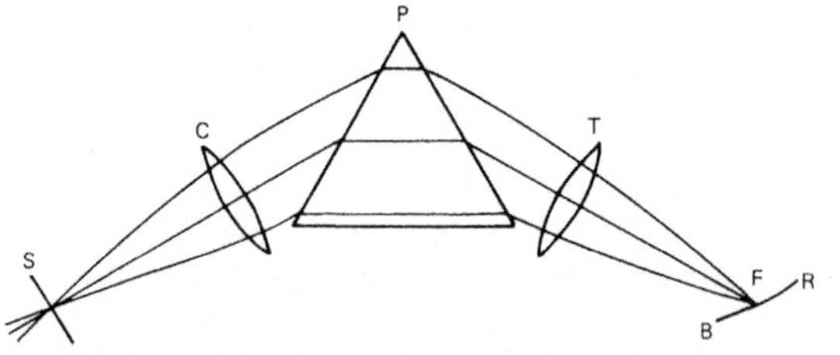

Figure 3.10 Optical system of a simple spectroscope. S, silt; C, collimator lens; P, prism; T, telescope lens; F, curve along which the various parts of the spectrum are in focus; B, blue or short wavelength part; R, red or long wavelength part.

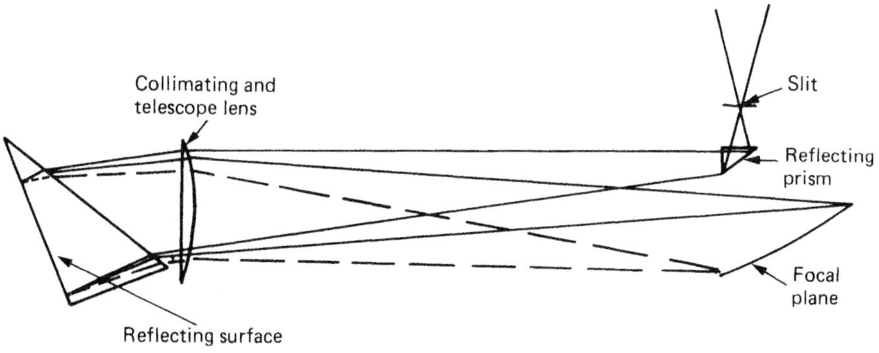

Figure 3.11 Diagram of the optical system of a Littrow spectrograph. The lens has been reversed to reduce scattered light.

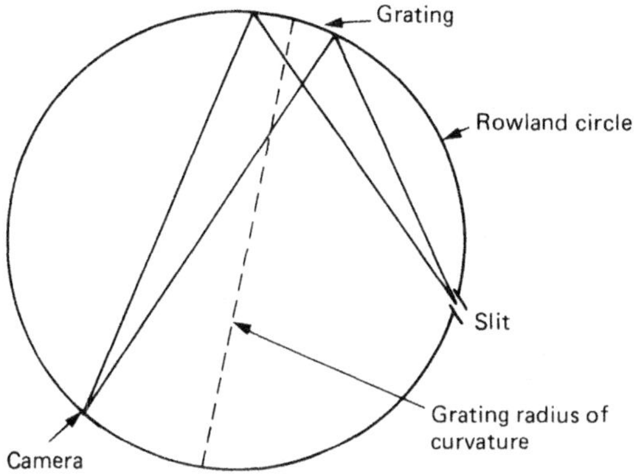

Figure 3.12 Elements of Rowland circle.

more than longer ones. The resulting dispersion is greater for the UV than for IR wavelengths.

Gratings Gratings may be considered as a large number of parallel, close, equidistant slits or diffracting lines. The equation $n\lambda = 2d \sin \theta$ shows the dependence of θ upon the wavelength of the incident light, where n is an integer, λ is the wavelength of incident light, d is the distance between the lines, and θ is the angle between the diffracted beam and the normal incident beam.

Modern gratings offer the spectroscopist uniform dispersion and coverage of a wide spectral range. Today, nearly all manufacturers have turned almost exclusively to grating instruments.

3.2.1.2 Vacuum spectrographs

Many elements, particularly the non-metallic ones, have their most persistent lines in the spectral region 150–220 nm. Light of these wavelengths is absorbed by air, and instruments are manufactured in which the optical paths are evacuated to overcome this problem.

3.2.1.3 Excitation – spectroscopic sources

Many factors are considered in the choice of a source. Sample form, necessary sensitivity and the elements which must be determined are the most critical. The main sources used are (a) a d.c. arc, (b) a high voltage condensed spark, (c) an arc triggered by a high voltage spark, (d) flames, (e) plasma jets and (f) inductively coupled plasmas. A recent form of excitation consists of evaporating a non-conducting sample by means of a laser and exciting the vapour with a high voltage spark.

3.2.1.4 Standards

In order to achieve a quantitative estimation of the impurity concentrations, some form of standard sample of known purity must be analysed under exactly the same conditions as the unknown samples and the intensity of the spectral lines compared. Thus a spectrochemical laboratory may have many thousands of standards covering the whole range of materials likely to require analysis.

3.2.1.5 Applications

There are very few on-line instruments employing atomic emission techniques, but mention should be made of a continuous sodium monitor for boiler/feed water. The water is nebulized into a flame, the sodium emission is isolated by means of a monochromator and the intensity measured by means of a photomultiplier and associated electronics. Standard solutions are automatically fed into the instrument from time to time to check the calibration.

In both the steel and non-ferrous alloy industries, large grating spectroscopes are used to control the composition of the melts before they are finally poured. A complete analysis for some 30–40 elements can be made within 2 minutes of a small sample being taken. Suitable additions are then made to the melt to satisfy the required composition specification. In these cases the output from the instrument is fed to a computer which is programmed to produce actual elemental concentrations and also the necessary amounts required to be added to known weights of melts in the furnaces for them to be of the correct composition. Analysis of water samples or samples in solution can be carried out using an inductively coupled plasma direct reading spectrometer. Some 60 elements can be determined in each sample every two minutes. The source is ionized argon pumped inductively from an r.f. generator into which the sample is nebulized. Temperatures of about 8500 °C are achieved. Many instruments of this type are now manufactured and have been of great value to the water industry and to environmental chemists generally – in particular, those instruments manufactured by A.R.L., Philips and Javell Ash. Limits of detection are of the order of 1 ppb (parts per 10^9) with an accuracy of about 10 per cent.

3.2.2 Atomic absorption spectroscopy

In emission spectroscopy, as we have already seen, the sample is excited, the emitted radiation dispersed and the intensities of the selected lines in the emission spectrum measured. If self-absorption and induced emission are neglected, then the integrated intensity of emission of a line is given by

$$\int I_v \, dv = CN_j F$$

where N_j is the number of atoms in the higher-energy level involved in the transition responsible for the line, F is the oscillation strength of the line, and C is a constant dependent upon the dispersing and detecting systems. Assuming that the atoms are in thermal equilibrium at temperature T, then the number of atoms in the excited state, of excitation energy E_j is given by

$$N_j = N_0 \frac{P_j}{P_0} \exp(-E_j/KT)$$

where N_0 is the number of atoms in the ground state, P_j and P_0 are statistical weights of the excited and ground states respectively, and K is Boltzmann's constant. For a spectral term having a total quantum number J_1, P is equal to $2J_1+1$. From the above equations, it can be seen that the emitted intensity depends on T and E_j. Examples of the variation of N_j/N_0 with temperature are given in Table 3.3.

In nearly all cases, the number of atoms in the lowest excited state is very small compared with the number of atoms in the ground state and the ratio only becomes appreciable at high temperatures. The strongest resonance lines of most elements have wavelengths less than 600 nm and as temperatures in the flames used are normally less than 3000 °K, the value of N_j will be negligible compared with N_0.

In absorption, consider a parallel beam of radiation of intensity I_0, frequency v incident on an atomic vapour of thickness l cm, then if I_v is the intensity of the transmitted radiation, and K_v is the absorption coefficient of the vapour at frequency v, then

$$I_v = I_0 \exp(-E_v l)$$

From classical dispersion theory

$$\int K_v dv = \frac{\pi e^2}{mc} N_v f$$

where m and e are the electronic mass and charge respectively, c is the velocity of light, N_v the number of atoms/cm^3 capable of absorbing radiation of frequency v, and f the oscillator strength (the average number of electrons per atom capable of being excited by the incident radiation). Thus, for a transition initiated from the ground state, where N_v is for all practical purposes equal to N_0 (the total number of atoms/cm^3), the integrated absorption is proportional to the concentration of free atoms in the absorbing medium. The theoretical sensitivity is therefore increased because all the atoms present will take part in the absorption whereas in the emission techniques only a very small number are excited and are used for detection.

In practice, the flame, into which the solution is nebulized, is treated as if it were the cell of absorbing solution in conventional spectrophotometry. The absorbance in the flame of light of a resonant wave-

Table 3.3 Values of N_j/N_0 for various resonance lines

Resonance line	*Transition*	P_j/P_0	N_j/N_0			
			T = 2000 K	*T* = 3000 K	*T* = 4000 K	*T* = 5000 K
Cs 852.1 nm	$2S_{1/2}$–$2P_{3/2}$	2	4.4×10^{-4}	7.24×10^{-3}	2.98×10^{-2}	6.82×10^{-2}
K 766.5 nm	$2S_{1/2}$–$2P_{3/2}$	2	2.57×10^{-4}	4.67×10^{-3}	1.65×10^{-2}	3.66×10^{-2}
Na 589.0 nm	$2S_{1/2}$–$2P_{3/2}$	2	9.86×10^{-6}	5.88×10^{-4}	4.44×10^{-3}	1.51×10^{-2}
Ca 422.7 nm	$1S_0$–$1P_1$	3	1.21×10^{-7}	3.69×10^{-5}	6.03×10^{-4}	3.33×10^{-3}
Zn 213.8 nm	$1S_0$–$1P_1$	3	7.29×10^{-15}	5.58×10^{-10}	1.48×10^{-7}	4.32×10^{-6}

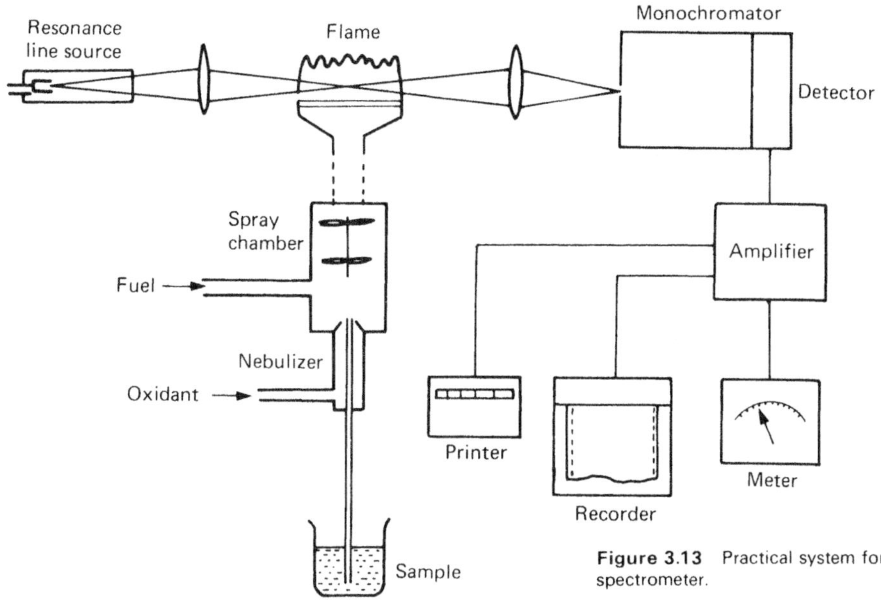

Figure 3.13 Practical system for atomic absorption spectrometer.

length of a particular element, is a direct measure of the concentration of atoms of that element in solution being nebulized into the flame. A practical system for an atomic absorption spectrometer is shown in Figure 3.13.

When only small volumes of sample are available the flame may be replaced by a graphite tube or rod furnace. Small volumes (10 μl) are placed on the graphite and the latter is heated resistively in stages to about 3000 °C and the absorption of a resonant wavelength measured as a pulse. The sensitivity of this technique is such that very low concentrations of some elements may be determined ($\sim$0.001 ppm). The limit of detection using a flame varies from element to element from less than 1 ppm up to about 50 ppm. The technique has found wide use in analysis of solutions in virtually every industry – from 'pure' water analysis to the analysis of plating solutions, from soil extracts to effluent from a steel works.

There are many manufacturers of atomic absorption spectrophotometers and the modern instruments are very highly automated. The resonant line source is usually a high intensity hollow cathode lamp and up to ten of these may be contained in a turret so that each is used in turn. The flames are usually air–propane, air–acetylene or nitrous oxide–acetylene – the hotter flames being necessary to atomize the more refractory elements. The output from the monochromator and detector is usually handled by a microprocessor, so that once the instrument has been calibrated, results are automatically printed out as concentrations. Another instrument based on atomic absorption is the mercury vapour detector. A mercury vapour lamp is the resonant source and the detector is tuned to the mercury line at 253.6 nm. Air to be sampled is passed through a tube located between source and detector and the absorption is a measure of the mercury vapour in the air. There are many instruments manufactured for this purpose and all are very sensitive with limits of detection of around 0.1 ppm by volume.

3.2.3 Atomic fluorescence spectroscopy

This is a technique closely allied to atomic absorption. To initiate atomic fluorescence, neutral atoms in a flame cell are excited as in atomic absorption, i.e. by absorption of a characteristic radiation. Fluorescence occurs when these atoms are de-activated by the emission of radiation at the same or a different wavelength. The fluorescent wavelength is characteristic of the atoms in question and its intensity is proportional to the atomic concentration. In practice, initiation is achieved with a high intensity source and the fluorescent signal emitted by the atomic vapour is examined at right angles by passing it into a radiation detection system. Very briefly the basic equation relating the intensity of a fluorescent signal to atomic concentration is

$$F = 2.303\phi I_0 e_A lcp$$

where F is the intensity of fluorescent radiation, ϕ the quantum efficiency (which factor has to be used to account for energy losses by processes other than a fluorescence), I_0 is the intensity of the excitation radiation, e_A the atomic absorptivity at the wavelength of irradiation, l the flame path length, c the concentration of the neutral atom absorbing species, and p a proportionality factor relating to the fraction of the total fluorescence observed by the detector. Thus, $F = K\phi I_0 c$ for a particular set of instrumental conditions, and c is proportional to F and F will increase if the intensity of the irradiating source is increased.

There are four types of atomic fluorescence.

Resonance fluorescence This is the most intense type of fluorescence and most widely used in practice. It occurs when the fluorescent and excitation wavelengths are the same, that is the atom is excited from the ground state to the first excited state and then emits fluorescent energy on de-activation to the ground state.

Direct line fluorescence Here, the valence electron is excited to an energy level above the first excited state. It is then de-activated to a lower energy level (not the ground state) and fluorescent energy is emitted. The wavelength of fluorescence is longer than the excitation wavelength, e.g. the initiation of thallium fluorescence at 535 nm by a thallium emission at 377.6 nm.

Stepwise fluorescence This entails excitation of the atom to a high energy level. The atom is then deactivated to the first excited state. There, it emits resonance radiation on returning to the ground state, e.g. the emission of sodium fluorescence at 589 nm, following excitation at 330.3 nm.

Sensitized fluorescence This occurs when the atom in question is excited by collision with an excited atom of another species and normal resonance fluorescence follows. Thallium will fluoresce at 377.6 nm and 535 nm following a collision of neutral thallium atoms with mercury atoms excited at 253.7 nm.

An instrument used to determine trace amounts of elements in solution by atomic fluorescence very simply consists of (a) an excitation source which can be a high intensity hollow cathode lamp, a microwave-excited electrodeless discharge tube, some spectral discharge lamps or more recently, a tunable dye laser, (b) a flame cell or a graphite rod as in atomic absorption, and (c) a detection system to measure the fluorescence at right angles to the line between source and flame. The detection system is usually a simple monochromator or narrow band filter followed by a photomultiplier tube, amplifier and recording device. Limits of detection are achieved which are much lower than those obtained by atomic absorption because it is easier to measure small signals against a zero background than to measure small differences in large signals as is done in atomic absorption. Detection limits as low as 0.0001 ppm are quoted in the literature.

3.3 X-ray spectroscopy

3.3.1 X-ray fluorescence spectroscopy

Many books have been written about this technique and only a brief outline is given here.

The technique is analogous to atomic emission spectroscopy in that characteristic x-radiation arises from energy transferences involved in the rearrangement of orbital electrons of the target element following ejection of one or more electrons in the excitation process. The electronic transitions involved are between orbits nearer to the nucleus (see Figure 3.14).

Thus if an atom is excited by an electron beam or a beam of x-rays, electronic transitions take place and characteristic x-radiation is emitted for that atom. If, after collimation, these x-rays fall on to a crystal lattice – which is a regular periodic arrangement of atoms – a diffracted beam will only result in certain directions, depending on the wavelength of the x-rays λ, the angle of incidence θ, and atomic spacing within the crystal d. Bragg's law for the diffraction of x-rays states that $n\lambda = 2d \sin\theta$. Thus the K_α, K_β, L_α, L_β, M_α etc. x-radiations will be diffracted at different angles. These fluorescent radiations are then collimated and detected by a variety of detectors. The intensity of these radiations is a measure of the concentration of that particular atom. Thus if a sample containing many elements is subjected to x-radiation, fluorescent radiation for all the elements present will be spread out into a

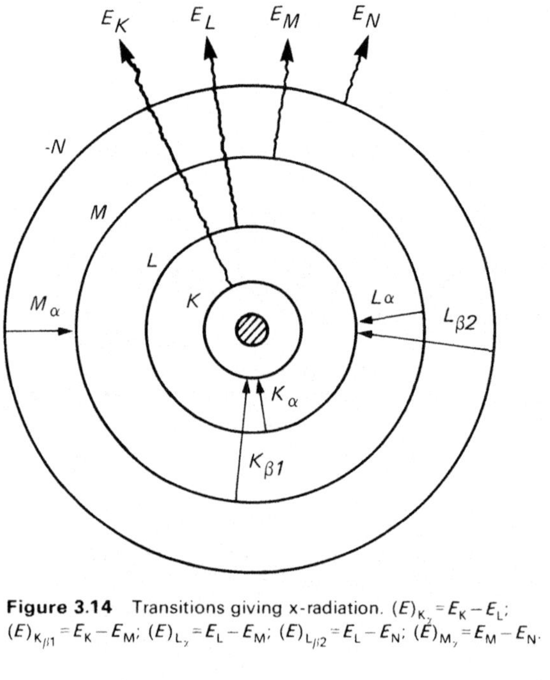

Figure 3.14 Transitions giving x-radiation. $(E)_{K_\alpha} = E_K - E_L$; $(E)_{K_{\beta 1}} = E_K - E_M$; $(E)_{L_\alpha} = E_L - E_M$; $(E)_{L_{\beta 2}} = E_L - E_N$; $(E)_{M_\alpha} = E_M - E_N$.

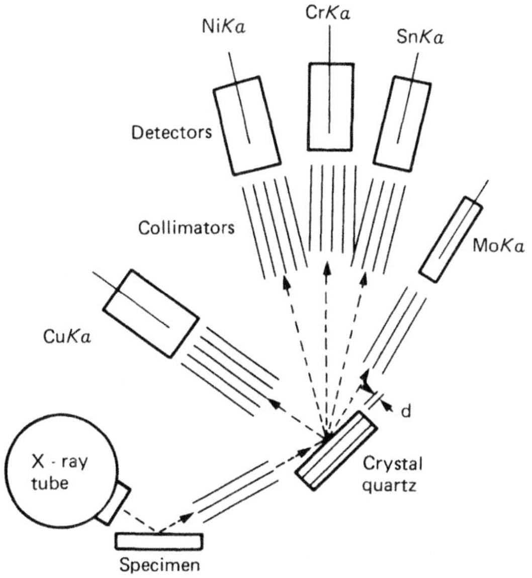

Figure 3.15 Multi-channel spectrometer having 5 collimator – detector channels arranged to receive 5 different analyte lines, each from a different crystallographic plane (*hkil*) from the same quartz crystal.

spectrum, depending on the elements present and the crystal being used (see Figure 3.15).

All modern x-ray fluorescence spectrometers use this layout. The source of x-rays is usually an x-ray tube, the anode of which is chromium, tungsten or rhodium. All types of sample can be analysed, ranging from metals, through powders to solutions. The collimator systems are based on series of parallel plates. As their purpose is to limit the divergence of the x-ray beam and provide acceptable angular resolution, the distance between the plates must be such that the divergence embraces the width of the diffraction profile of the crystal. In general, this entails a spacing between plates of 200–500 μm.

Most modern instruments can accommodate six analysing crystals, any one of which can be automatically placed in the fluorescent x-ray beam. A list of the types of crystal used is shown in Table 3.4. The detectors are either gas flow proportional counters or scintillation counters. (See Volume 3, Nuclear Instrumentation Technology.) The instruments are microprocessor-controlled and this varies the output of the x-ray source, chooses the correct crystals and controls the samples going into the instrument. A small computer analyses the output from the detectors and (having calibrated the instrument for a particular sample type) calculates the concentration of the elements being analysed – allowing for matrix and inter-element effects. Instruments of this type, made by Philips, Siemens, and ARL are widely used in the metallurgical industry as the technique – although capable of low limits of detection – is very accurate for major constituents in a sample, such as copper in brass. Analysis of atmospheric particulate pollution is carried out using x-ray fluorescence. The sample is filtered on to a paper and the paper and deposit analysed.

A portable instrument, which uses a radioactive isotope as a source is used to monitor particular elements (depending on settings) in an ore sample before processing. This instrument is now marketed by Nuclear Enterprises. See Volume 3 (Measurements employing nuclear techniques).

Electron probe microanalysis is a technique which is based on the same principle as x-ray fluorescence, electrons being the exciting source, but by using electronic lenses the electron beam can be focused onto a very small area of a sample and so analysis of areas as

Table 3.4 Analysing crystals

Crystal	*Reflection plane*	*2d spacing* (Å) (1 Å = 0.1 nm)	*Lowest atomic number detectable*	
			K series	L series
Topaz	(303)	2.712	V (23)	Ce (58)
Lithium fluoride	(220)	2.848	V (23)	Ce (58)
Lithium fluoride	(200)	4.028	K (19)	In (49)
Sodium chloride	(200)	5.639	S (16)	Ru (44)
Quartz	$(10\bar{1}1)$	6.686	P (15)	Zr (40)
Quartz	$(10\bar{1}0)$	8.50	Si (14)	Rb (37)
Penta erythritol	(002)	8.742	Al (13)	Rb (37)
Ethylenediamine tartrate	(020)	8.808	Al (13)	Br (35)
Ammonium dihydrogen phosphate	(110)	10.65	Mg (12)	As (23)
Gypsum	(020)	15.19	Na (11)	Cu (29)
Mica	(002)	19.8	F (9)	Fe (26)
Potassium hydrogen phthalate	$(10\bar{1}1)$	26.4	O (8)	V (23)
Lead stearate		100	B (5)	Ca (20)

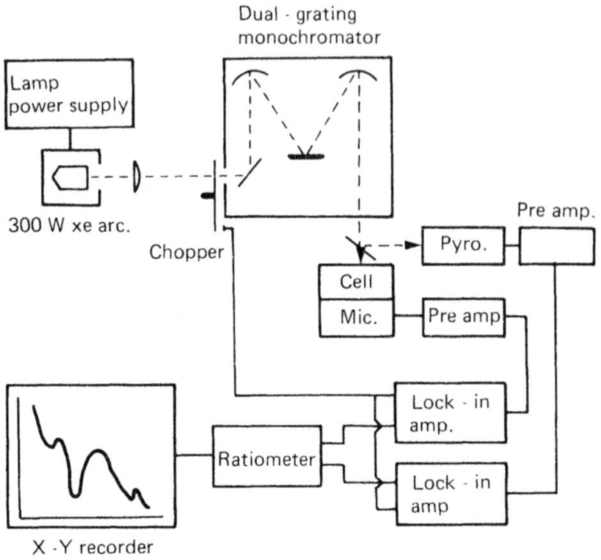

Figure 3.16 Photo-acoustic spectrometer layout.

small as 0.1 μm diameter can be carried out. The technique can be used for looking at grain boundaries in metallurgical specimens and plotting elemental maps in suspected heterogeneous alloys. Again this is a technique which is very specialized.

A further allied technique is photoelectron spectroscopy (PES) or Electron Spectroscopy for Chemical Analysis (ESCA). In Figure 3.14, showing the transitions within an atom to produce x-rays, it is seen that some electrons are ejected from the various shells in the atom. The energy of these electrons is characteristic of that atom and so by producing an energy spectrum of electrons ejected from a sample when the latter is subjected to x-ray or intense UV radiation, the presence of different elements and their concentrations can be determined. It should be pointed out that this technique is essentially a surface technique and will only analyse a few monolayers of sample. Instruments are manufactured by Vacuum Generators.

3.3.2 X-ray diffraction

This is a technique which is invaluable for the identification of crystal structure. In the section on x-ray fluorescence it was seen that crystals diffract x-rays according to Bragg's law:

$$n\lambda = 2d \sin \theta$$

Thus if a small crystal of an unidentified sample is placed in an x-ray beam, the x-rays will be diffracted equally on both sides of the sample to produce an x-ray pattern on a film placed behind the sample. The position of the lines on the film (i.e. their distance from the central beam) is a function of the crystal lattice structure and by reference to standard x-ray diffraction data, the crystals in the sample are identified. Again this is a specialized technique and beyond the scope of this book.

Manufacturers of x-ray fluorescence spectrometers also make x-ray diffraction spectrometers. Typical uses for an instrument are the identification of different types of asbestos, and corrosion deposit studies.

3.4 Photo-acoustic spectroscopy

An instrument marketed by EDT Research makes use of this technique to study both liquid and solid samples. Figures 3.16 and 3.17 give schematic diagrams of the instrument and cell. Radiation from an air-cooled high pressure xenon arc source, fitted with an integral parabolic mirror, is focused onto a variable speed rotating light chopper mounted at the entrance slit of a high radiance monochromator. The monochromator has two gratings to enable optical acoustic spectra to be obtained in the UV, visible and near-infrared. The scanning of the monochromator is completely automatic over the spectral range covered and a range of scan rates can be selected. The exit and entrance slits provide variable band passes of width 2–16 nm in the UV and 8–64 nm in the IR. A reflective beam-splitter passes a fraction of the dispersed radiation to a pyroelectric detector to provide source compensation and a reference signal. Source radiation is then focused onto the specially designed opto-acoustic cell and sample-holder assembly. The sample cell contains a sensitive microphone and pre-amplifier. Special cells are used for different applications. Absorption of the radiation by the molecular species in the sample occurs and is converted to kinetic energy. The sample temperature fluctuates and causes a variation in the pressure of the gas surrounding the sample. This pressure variation is monitored by the microphone. The amplitude of the microphone signal is recorded as a function of the wavelength of the incident radiation to give an absorption spectrum of

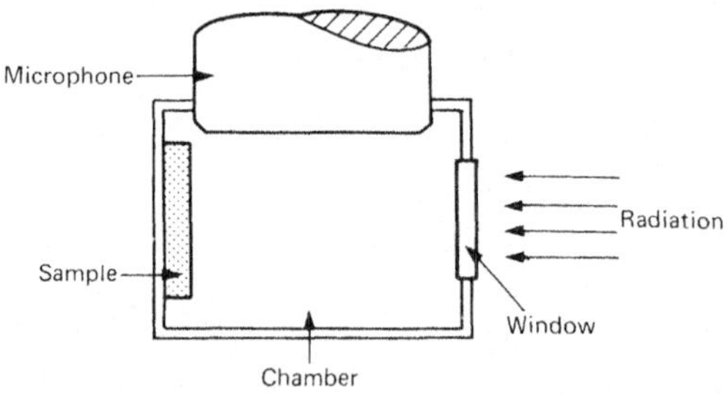

Figure 3.17 Schematic representation of a photo-acoustic cell employed for the examination of solid materials.

the sample. Typical applications include the identification of foodstuffs, blood and bloodstains, paints and inks, papers and fabrics, and pharmaceutical materials.

3.5 Microwave spectroscopy

The portion of the electromagnetic spectrum extending approximately from 1 mm (300 000 MHz) to 30 cm (1000 MHz) is called the microwave region. Spectroscopic applications of microwaves consist almost exclusively of absorption work in gaseous samples. With some exceptions, the various types of spectra are distinguished by their energy origins. As mentioned earlier, in the visible and UV regions the transitions between electronic energy states are directly measurable as characteristics of elements, and vibrational and rotational energies of molecules are observed only as perturbation effects. In the infrared region the vibrational spectra are observed directly as characteristic of functional groups with rotational energies observed as perturbation effects. In the microwave region transitions between rotational energies of molecules are observed directly as characteristic of absorbing molecules as a whole with nuclear effects as first-order perturbations. In the radio frequency (r.f.) region, the nuclear effects are directly observable. (Especially important today is the observation in the microwave region of paramagnetic resonance absorption (PMR) and also nuclear magnetic resonance. Both these techniques will be discussed briefly in a later section.) As in any other type of absorption spectroscopy, the instrument required consists of a source of radiation, a sample cell and detector. Unlike optical spectrometers, the microwave spectrometer is a completely electronic instrument requiring no dispersive components because the source is monochromatic and any frequency can be chosen and measured with very high precision. The most common type of source is the Klystron, a specially designed high-vacuum electron tube. The output is monochromatic under any given set of conditions and different types are available to cover various parts of the microwave spectrum. The sample cell is usually a waveguide and the detector could be a silicon crystal, although bolometers and other heat-type detectors are sometimes used. In addition to the three basic components a complete spectrometer includes provision for modulation of the absorption spectrum, an a.c. amplifier for the detector output, a final indicator consisting of a CRT or strip recorder, a sweep generator to vary synchronously the source frequency, a gas sample handling system and necessary power supplies.

Since the lines in a microwave spectrum are usually completely resolved, it is only necessary to compare these measured frequencies against tables of the frequencies observed for known substances in order to identify molecules. Quantitative analysis is somewhat more complex, but is based on the fact that the integrated intensity and the product of the peak height and half-width of a microwave absorption line can be directly related to the concentration of molecules per unit volume. The technique is used extensively in isotopic analysis.

3.5.1 Electron paramagnetic resonance (EPR)

This is really a special part of microwave spectroscopy because it usually involves the absorption of microwave radiation by paramagnetic substances in a magnetic field. A typical layout of a spectrometer is given in Figure 3.18. The electromagnet has a homo-

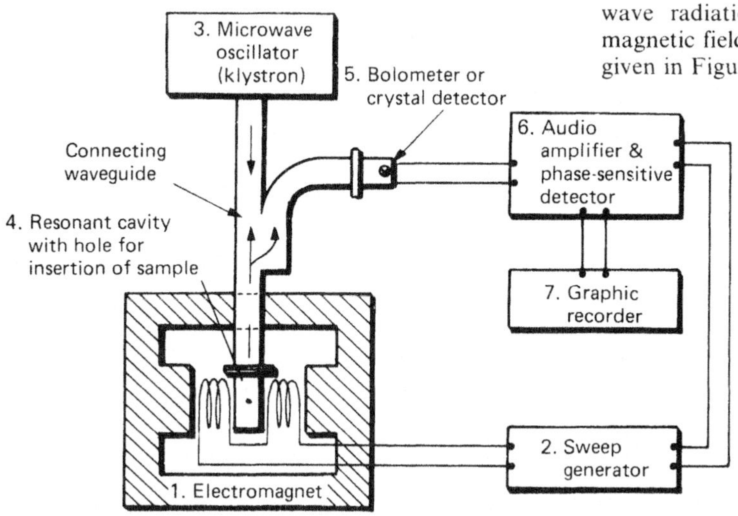

Figure 3.18 Block diagram of electron paramagnetic resonance spectrometer.

geneous gap field H which can be swept continuously from near zero to over 50 microtesla. The sweep generator produces small modulations of the main field H at the centre of the air-gap. The sample cavity resonates at the Klystron frequency.

The electron, like the proton, is a charged particle; it spins and therefore has a magnetic field. It spins much faster than a proton and so has a much stronger magnetic field. Because of this and being lighter than a proton, it precesses much more rapidly in a magnetic field. Thus when microwaves travel down a waveguide and produce a rotating magnetic field at any fixed point, it can serve to flip over electron magnets in matter, just as a rotating field in a coil flips protons. If a sample is placed on the sidewall of the waveguide and the microwave radiation, applied to the external magnetic field, causes the electrons to precess, then when the precession rate reaches a resonance value and the electrons flip, they extract energy from the microwaves and the reading on the recorder dips accordingly.

If the electron has not only a magnetic moment along its own spin axis but also one associated with its circulation in an atomic orbit, the electron will possess a total magnetic moment equal to the vector sum of the magnetic moments. The ratio of the total magnetic moment to the spin value is a constant for a given atom in a given environment and is called the gyromagnetic ratio or spectroscopic splitting factor for that particular electron. The fact that these ratios differ for various atoms and environments and that local magnetic fields depend on the structure of the matter permit spectral separation and EPR spectroscopy. Not all atoms and molecules are susceptible to this technique; in substances in which electrons are paired, magnetism is neutralized. But for unpaired electrons, electronic resonance occurs. This effect is observed in unfilled conduction bands, transition element ions, free radicals, impurities in semiconductors and, as might be expected, applications in the biological field are fruitful. The most common use is the paramagnetic oxygen analyser.

3.5.2 Nuclear magnetic resonance spectroscopy

When atomic nuclei – the hydrogen proton is the simplest – are placed in a constant magnetic field of high intensity and subjected to a radio frequency alternating field, a transfer of energy takes place between the high frequency field and the nucleus to produce a phenomenon known as 'nuclear magnetic resonance'.

If a system of nuclei in a magnetic field is exposed to radiation of frequency ν such that the energy of a quantum of radiation $h\nu$ is exactly equal to the energy difference between two adjacent nuclear energy levels, then energy transitions may occur in which the nuclei may 'flip' back and forth from one orientation to another. A quantum of energy is equally likely to tip a nucleus in either direction, so that there is a net absorption of energy from the radiation only when the number of nuclei in one energy level exceeds the number in another. Under these conditions a nuclear magnetic resonance spectrum is observed. Applications of this technique include such problems as locating hydrogen atoms in solids, measuring bond lengths, crystal imperfections, and determination of crystalline and amorphous fractions in polymers.

3.6 Neutron activation

Gamma ray spectroscopy is the technique by which the intensities of various gamma energies emanating from a radioactive source are measured. See Volume 3 (Measurements employing nuclear techniques). It can be used for qualitative identification of the components of radionuclide mixtures and for quantitative determination of their relative abundance. Such a situation arises in neutron activation analysis. This is a technique of chemical analysis for extremely minute traces down to ppb (parts per 10^9) of chemical elements in a sample. It employs a beam of neutrons for activation of isotopes which can then be identified, with counters, by the radioactive characteristics of the new nuclear species. This technique has been applied for the trace analysis of many elements in a variety of materials, from coal ash to catalysts, halides in phosphors, and trace impurities in many metals.

3.7 Mass spectrometers

The mass spectrometer is capable of carrying out quick and accurate analysis of a wide variety of solids, liquids and gases and has a wide range of application in process monitoring and laboratory research. When combined with the gas chromatograph it provides an extremely powerful tool for identifying and quantifying substances which may be present in extremely small quantities.

While the optical spectrometer resolves a beam of light into components according to their wavelengths, a mass spectrometer resolves a beam of positive ions into components according to their mass/charge ratio, or if all carry single elementary charges, according to their masses. As with the optical spectrometer the mass spectrometer may be used to identify substances and to measure the quantity present.

The original mass spectrometer was devised by F. W. Aston about 1919 to measure the mass of individual positive ions. The accuracy of the instrument enabled the different masses of what appeared to be chemically identical atoms to be

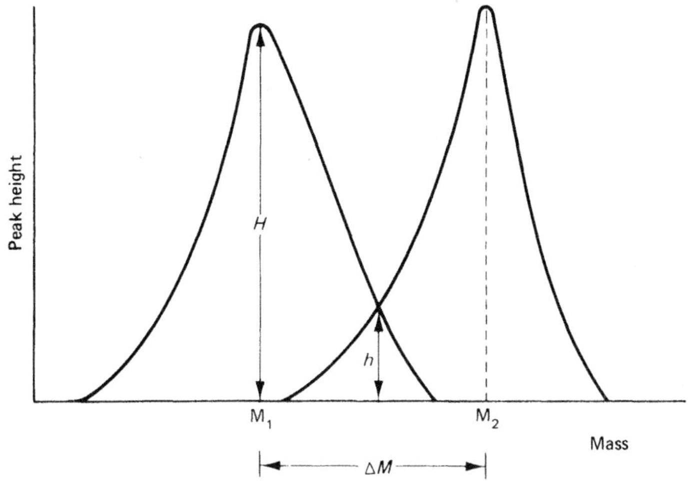

Figure 3.19 Peak separation for a mass spectrometer.

measured, resulting in the discovery of isotopes. Considerable development has taken place over the years, resulting in very versatile instruments having very high resolving power and sensitivity.

The resolving power of a mass spectrometer is a measure of its ability to separate ions having a very small difference in mass. If two ions of masses M_1 and M_2 differing in mass by ΔM give adjacent peaks in their spectrum as shown in Figure 3.19 and the height of peak is H above the baseline, then on the 10 per cent valley definition the peaks are said to be resolved if the height of the valley h is less than or equal to 10 per cent of the peak H, i.e.

$$(h/H) \leqslant 10 \text{ per cent}$$

The resolution is then $M_1/\Delta M$, e.g. if the peaks representing two masses 100.000 and 100.005 are separated by a 10 per cent valley, the resolution of the instrument is 100.000/0.005, i.e. 20 000. Instruments with a resolution of 150 000 are readily available. The sensitivity on the other hand is a measure of the smallest detectable quantity of the substance being identified. An example of the extreme sensitivity of modern instruments is that at a resolution of 1000, 3 ng/s of a compound, relative molecular mass 300, will give a spectrum with a signal-to-noise ratio of 10:1 for a peak having an intensity of 5 per cent of the base peak when a mass range of 10:1 is scanned in 3 s.

The mass spectrometer has a very wide range of use in process monitoring and laboratory research. It is used in refineries for trace element survey, analysis of lubricating oils and identifying and quantifying the substances in mixtures of organic compounds. Its use in detecting and measuring the concentration of pollutants in air, water and solids is rapidly increasing, also its use in biochemical analysis in medicine and other fields, particularly the analysis of drugs in biological extracts.

By means of a double-beam instrument an unknown sample may be compared with a standard so that the unknown components are readily identified and the concentration measured. By suitable modifications an instrument can be made to provide an energy analysis of electrons released from the surface of a sample by x-radiation, or ultraviolet light.

3.7.1 Principle of the classical instrument

There are many different types of mass spectrometers but the ones described here are the most commonly used.

In all types the pressure is reduced to about 10^{-5} N/m^2 in order to reduce collisions between particles in the system. The spectrometer consists of an inlet system by which the sample is introduced into the region in which ions of the sample are produced. The separation of ions according to their mass-to-charge ratio may be achieved by magnetic or electric fields or by a combination of both. The differences between the various types of mass spectrometer lie in the manner in which the separation is achieved. In the instrument illustrated in Figure 3.20 the ions are accelerated by an electrical potential through accelerating and defining slits into the electrostatic analyser, where ions having energies within a restricted band are brought to a focus at the monitor slit which intercepts a portion of the ion beam. They then enter the electromagnetic analyser which gives direction and mass focusing. This double focusing results in ions of all masses being focused simultaneously along a given plane. The ions can be recorded photographically on a plate over a period of time to give a very high sensitivity and reduction of the effects of ion-beam fluctuation.

Alternatively, the accelerating or deflecting field may be arranged so that ions of a given mass are focused on a detector which may consist of a plate or, if initial amplification of the charge is required, onto an electron multiplier or scintillation detector. By arranging the deflecting field to change in a pre-determined manner, the instrument may be arranged to scan a range of masses and so record the abundance of ions of each particular mass. Such a record is known as a 'mass spectrum' and mathematical analysis of this mass spectrum enables the composition of the sample to be determined. Mass spectra obtained under constant conditions of ionization depend upon the structure of the molecules from which the ions originate. Each substance has its own characteristic mass spectrum, and the mass spectrum of a mixture may therefore be analysed in terms of the spectra of the pure components, and the percentage of the different substances in the mixture calculated.

Analysis of the mass spectrum of a mixture may involve the solution of a large number of simultaneous equations, which can be accomplished using a micro-processor or a small computer.

3.7.2 Inlet systems

The mode of introduction of the sample into the ion source is dependent upon the nature of the sample and in particular its volatility.

The simplest system designed to introduce reference compounds into the ion source includes a 35 cm^3 reservoir into which the compound is injected through a septum. Flow into the ion source is through a molecular leak and a shut-off valve is provided. Facilities for pumping out the system and obtaining temperatures up to 100 °C are provided.

Relatively volatile gases and liquids may be introduced by a probe attached to a small reservoir into which the sample is injected and from which it flows to the ion source at a controlled rate. The temperature of the system may be controlled between ambient and 150 °C.

For less volatile substances an all-glass heated system may be used. Glass is used for the system so that catalytic decomposition of the sample is reduced to a minimum. The system can be operated at temperatures up to 350 °C and incorporates its own controlled heating and temperature-monitoring facilities. It includes both large and small reservoirs to enable a wide range of quantities of liquid or solid samples to be introduced.

To introduce less volatile and solid samples into the ion chamber a probe may be used. The sample is loaded onto the tip of the probe, which is inserted into the ion source through a two-stage vacuum lock.

The probe may be heated or cooled independently of the ion chamber as required from −50 to +350 °C. The temperature is measured by a platinum resistance thermometer, forming part of the temperature control system, which enables the temperature to be set from the instrument control panel.

Effluents from a gas chromatograph column usually flow at about 50 cm^3/min and consist mainly of carrier gas. In order to reduce the flow, the gas is passed through a molecular separator designed to remove as much as possible of the carrier gas but permitting the significant components to pass into the mass spectrometer.

3.7.3 Ion sources

In the system shown the ions are produced by a spark passed between electrodes formed from the sample by applying a controlled pulsed r.f. voltage. Positive ions representative of the sample are produced in the discharge and are accelerated through a simple ion gun. This beam is defined by resolving slits before it passes into the analyser section.

Other methods may be employed in order to produce ions of the sample which are impelled towards the exit slit by a small positive potential in the ion chamber. These methods involve increasing the energy of the sample by some form of radiation. Organic compounds require photons of energy up to 13 eV to produce ionization so that a high energy beam of short wavelength radiation is sufficient. Where energies greater than 11 eV are required window materials become a problem so that the photon source has to emit radiation directly into the ion source. A helium discharge at 21.21 eV provides a convenient source of photons capable of ionizing all organic compounds.

Electrons emitted by a heated filament and accelerated by about 70 eV and directed across the ion chamber may also be used to ionize many substances. While 70 eV produces the maximum ion yield, any voltage down to the ionization voltage of the compound studied may be used.

The electric field production near a sharp point or edge at a high potential will have a high potential gradient and may be used to produce ions. Ions can also be formed by the collision of an ion and a molecule. This method can produce stable but unusual ions, e.g.

$$CH_4^+ + CH_4 \rightarrow CH_5^+ + CH_3$$

and is most efficient at pressures of about 10^{-2} N/m^2.

It is most important to realize that the process of producing ions from molecules will in many cases split up the original molecule into a whole range of ions of simpler structure and the peak of maximum height in the spectrum does not necessarily represent the ion of

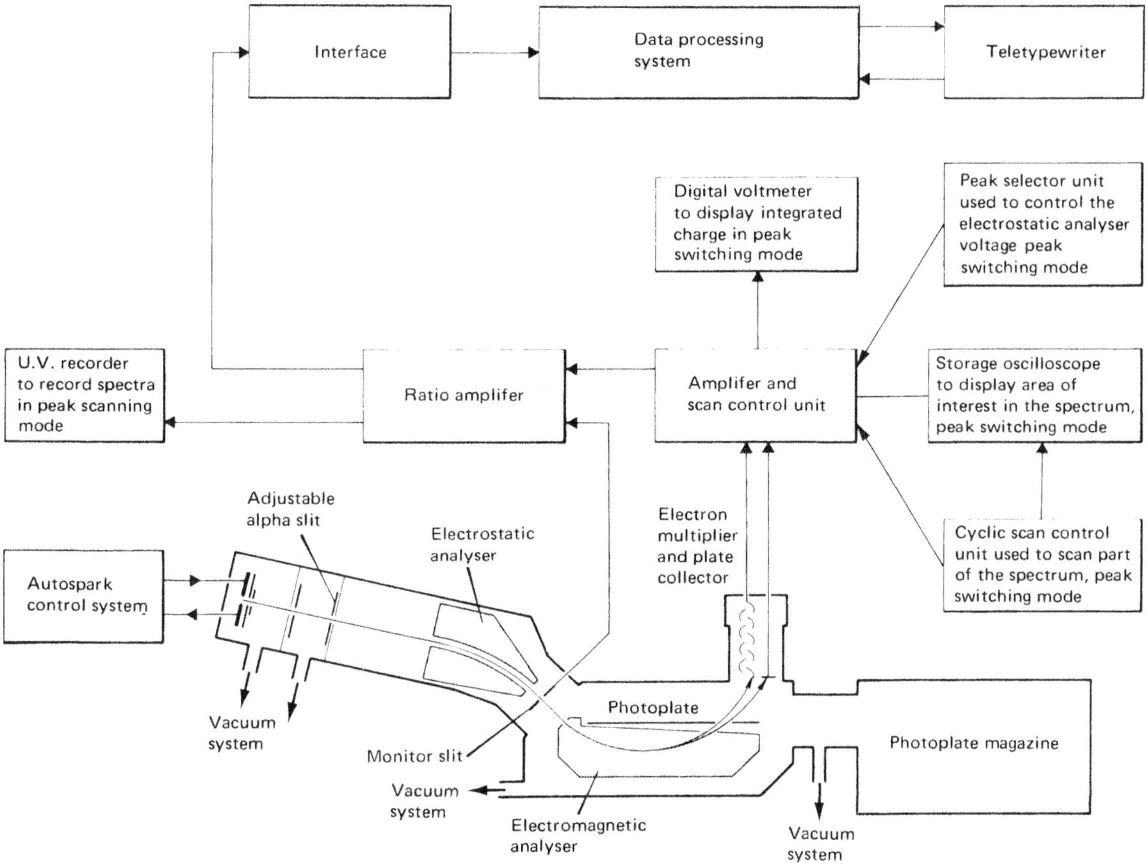

Figure 3.20 Schematic diagram of the complete system of a spark source mass spectrometer (courtesy Kratos Ltd.).

the original molecule. For example the mass spectrum of *m*-xylene $C_6H_4(CH_3)_2$ may contain 22 peaks of different m/e values, and the peak of maximum height represents a m/e ratio of 91 while the ions having the next highest peak have a m/e ratio of 106.

3.7.4 Separation of the ions

The mass spectrometer shown in Figure 3.20 employs the Mattauch-Herzog geometry but other forms of geometry achieve a similar result.

The positive ions representative of the sample produced in the ion source are accelerated by a controlled electrostatic field in a simple gun, the spread of the ions being controlled by the resolving slits. If an ion of mass m and charge e can be regarded as starting from rest, then its velocity v after falling through a potential V volts will be represented by the equation

$$\tfrac{1}{2}mv^2 = eV$$

The ion beam then passes through the electrostatic analyser where it passes between two smooth curved plates which are at different potentials, such that an electrostatic field B exists between them which is at right angles to the path of the ions. The centrifugal force on the ions will therefore be given by

$$mv^2/r = eB$$

Combining the equations we see that the radius of curvature r of the path will be given by

$$r = mv^2/eB = 2eV/eB = 2V/B$$

Thus, the curvature of the path of all ions will be dependent upon the accelerating and deflecting fields only and independent of the mass/charge ratio. Therefore, if the field B is kept constant the electrostatic analyser focuses the ions at the monitor slit in accordance with their translational energies. The monitor slit can be arranged to intercept a given portion of the beam. The energy-focused ion beam is then passed through the electromagnetic analyser where a magnetic field at right angles to the electrostatic field is applied (i.e. at right angles to the plane of the diagram). Moving electric charges constitute an electric current so that if each carries a charge e, and moves with a velocity v, at right angles to a uniform magnetic field H, each particle will be subject to a force F where $F=Hev$ in a direction given by Fleming's left-hand rule, i.e. in a direction mutually at right angles to the magnetic field and the direction of the stream. Thus the ions will move in a curved path radius r such that

$$mv^2/r=Hev$$

or

$$r=mv^2/Hev=mv/He$$

but

$$mv^2=2eV \quad \text{or} \quad v=\sqrt{(2eV/m)}$$

$$\therefore \quad r=(m/eH)\sqrt{(2eV/m)}$$

or

$$r^2=(m^2/e^2H^2)(2eV/m)$$
$$=(2V/H^2)(e/m)$$

or

$$m/e=(H^2r^2)/2V$$

At constant values of the electrostatic and electromagnetic fields all ions of the same m/e ratio will have the same radius of curvature. Thus, after separation in the electromagnetic analyser, ions having a single charge will be brought to a focus along definite lines on the photographic plate according to their mass, starting with the lowest mass on the left-hand edge of the plate and increasing to the highest mass on the right.

The ions will therefore give rise to narrow bands on the photographic plate and the density of these bands will be a measure of the number of ions falling on the band. The sensitivity range of the plate is limited, and it is necessary to make several exposures for increasing periods of time to record ions which have a large ratio of abundance. By using long exposure, ions which are present in very low abundances may be accurately measured. The intensity of the photographic lines after development of the plate may be compared with a microphotometer similar to that used with optical spectrometers.

As all ions are recorded simultaneously, ion beam fluctuations affect all lines equally and the photographic plate also integrates the ions over the whole of the exposure.

The instantaneous monitor current may be measured and used to control the sparking at the electrodes at optimum by adjusting the gap between the electrodes.

The integrated monitor current is a guide to the exposure, and the range of masses falling on the photographic plate may be controlled by adjustment of the value of the electrostatic and magnetic fields.

The plate collector and the electron multiplier detection systems enable quantitative analysis to be carried out with greater speed and precision than with the photographic plate detector. For high sensitivity the ions may be caused to fall on the first dynode of the electron multiplier and the final current further amplified, and recorded on the ultraviolet sensitive strip recorder. The logarithmic ratio of the monitor and collector signals is used in recording spectra in order to minimize the errors due to variations in the ion beam.

In the peak switching mode the operator can select the peaks of interest and display them on an oscilloscope and examine them with greater precision. Increasing the resolving power of the instrument will enable what may initially appear to be a single peak to be split up into its components representing ions differing in mass by a small amount.

Provision is made for changing the amplification in logarithmic steps so that a wide range of abundances may be measured. Where a rapid qualitative and semi-quantitative analysis is required for a very wide range of masses, consecutive masses are swept across the multiplier collector by allowing the magnet current to decay from a preset value at a preset rate while the accelerating voltage is kept constant. Values of ion current from the individual ion species received at the detector are amplified and instantaneously compared with a fraction of the total ion current at the monitor by means of two logarithmic amplifiers which feed into a summing amplifier. This gives a signal proportional to the relative ion concentrations, which can be recorded on the ultraviolet sensitive strip recorder and has the form shown in Figure 3.21.

Where large amounts of data are generated the output from the ratio detector of the electrical detection system can be fed through a suitable interface into a data acquisition and processing system. If necessary this system can be programmed to print out details of the elements present in the sample and an indication of their concentration.

3.7.5 Other methods of separation of ions

3.7.5.1 *Time-of-flight mass spectrometer*

This type of instrument is shown schematically in Figure 3.22. It has a relatively low resolution but a very fast response time.

In this instrument, the ions are accelerated through a potential V, thus acquiring a velocity v given by:

$$\tfrac{1}{2}mv^2=eV \quad \text{or} \quad v=[2V(e/m)]^{1/2}$$

If the ions then pass through a field-free (drift) region of length d, to the detector the time of transit t will be d/v. That is,

$$t=d/[2V(e/m)]^{1/2}=[(e/m)2d^2V]^{-1/2}$$

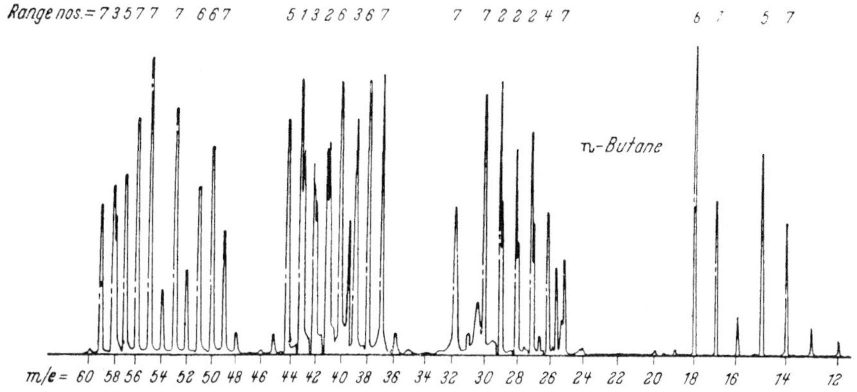

Figure 3.21 Ultraviolet-sensitive strip recording.

Thus, the ions will arrive at the detector after times proportional to $(m/e)^{1/2}$. The time intervals between the arrival of ions of different mass at the detector is usually very short and the mass spectrum is most conveniently displayed on a cathode ray tube. The time-of-flight mass spectrometer occupies a unique place in mass spectrometry as it provides a simple rapid measurement of the abundance of various isotopes or elements comprising a sample. In practice, 10 000 to 100 000 spectra can be scanned per second. With the aid of suitable electronic circuitry it is possible to monitor reaction rates and to investigate reaction profiles of only 100 μs duration. Longer length drift tubes have also contributed to improved mass resolution. It is also possible to scan from 0 to 900 atomic mass units in 1.5 seconds and also, to prevent multiplier saturation when very large ion peaks are present near smaller peaks, appropriate 'gating' peaking can be applied to the multiplier. Thus, it is possible to suppress mass 40 without interfering with the recording of mass 39 or 41. This has extended the practical range of sensitivity in identifying gas chromatograph effluent by orders of magnitude.

3.7.5.2 Quadrupole mass spectrometer

This type of instrument is particularly suited to vacuum system monitoring and to a wide range of gas analysis. Although it has a relatively modest resolving power (about 16 000 maximum) it has the advantages of compactness, robustness and relatively low cost.

Ions, produced by bombarding the sample with electrons from a filament assembly, are extracted electrostatically from the ionizer and focused by electrostatic lenses into the quadrupole mass filtering system. The latter consists of two pairs of metal rods, precisely aligned and housed in a chamber at a pressure of 2.6×10^{-4} N/m^2. One pair is connected to a source of d.c. voltage, while the other is supplied by a radio frequency voltage. Combination of the d.c. and r.f. voltages creates a hyperbolic potential distribution. The applied voltages increase uniformly from zero to a given maximum and then drop to zero again – a voltage sweep which is then repeated. Most ions entering the quadrupole field will undergo an oscillating trajectory of increasing amplitude so that they will eventually be collected on one of the electrodes. However, at any given time, ions of one

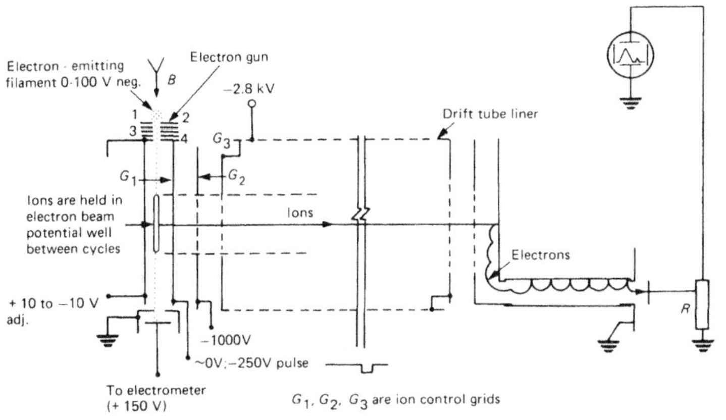

Figure 3.22 Time-of-flight spectrometer.

specific mass/charge ratio are deflected as much to one electrode as to another and are passed by the filter.

As the voltages are swept from zero to their maximum values, the entire mass range is scanned. After passing through the mass filter, the ions impinge on an electron multiplier and a signal proportional to the collected ion current can be displayed on an oscilloscope or recorder. As the voltages increase, the position of the mass peaks is linearly related to mass, making the spectrum easy to interpret. The instrument covers mass ranges up to about 400 amu. Modern instruments are able to detect partial pressures in the 10^{-13} torr range. They are equipped with variable mass scanning sweeps so that rapidly changing concentrations of gases can be monitored on a continuing basis. There are many other types of ion separators, for details on these, the reader should consult textbooks devoted to mass spectroscopy. Among these types are multiple magnet systems, the cycloidal mass spectrometer, cyclotron resonance types and r.f. mass filters.

3.8 Bibliography

Bertin, Eugene P., *Principles and Practice of X-ray Spectrographic Analysis*, Plenum Press, (1970)

Ebdon, L., *An Introduction to Atomic Absorption Spectroscopy – A Self Teaching Approach*, Heyden, (1982)

Jenkins, Ron, Gould, R. W. and Gedcke, Dale, *Quantitative X-ray Spectrometry*, Marcel Dekker, (1981)

Price, W. J., *Spectrochemical Analysis by Atomic Absorption*, Heyden, (1979)

Royal Society of Chemistry, Annual Reports on Analytical Atomic Spectroscopy

Slavin, W., *Atomic Absorption Spectroscopy* (2nd edn), Wiley, (1978)

Tertian, R. and Claisse, F., *Principles of Quantitative X-ray Fluorescence Analysis*, Heyden, (1982)

Welvy, E. L. (ed.), *Modern Fluorescence Spectroscopy*, Plenum Press, (1981)

White, Frederick A., *Mass Spectrometry in Science and Technology*, Wiley, (1968)

4 Chemical analysis – electrochemical techniques

W. G. CUMMINGS and K. TORRANCE

4.1 Acids and alkalis

In order to appreciate electrochemical techniques of chemical analysis it is necessary to have an understanding of how substances dissociate to form ions.

All acids dissociate when added to water to produce hydrogen ions in the solution, e.g. nitric acid:

$$HNO_3 \rightleftharpoons H^+ + NO_3^-$$

The extent to which dissociation takes place varies from acid to acid, and increases with increasing dilution until, in very dilute solutions, almost all the acid is dissociated.

According to the ionic theory, the characteristic properties of acids are attributed to the hydrogen ions (H^+) which they produce in solution. Strong acids (nitric, sulphuric, hydrochloric) are those that produce a large concentration of hydrogen ions when added to water. As a result the solutions are excellent conductors of electricity. Weak acids like carbonic acid (H_2CO_3) and acetic acid (CH_3COOH) when dissolved in water produce small concentrations of hydrogen ions and their solutions are poor conductors of electricity.

The strength of a weak acid is indicated by its dissociation constant K which is defined as

$$K = \frac{[A^-][H^+]}{[HA]}$$

where $[A^-]$ is the molar concentration of the acidic ions, $[H^+]$ is the concentration of hydrogen ions, and $[HA]$ is the concentration of undissociated acid.

The dissociation constant K varies with temperature but, at a given temperature, if a little more acid is added to the solution, a portion of it dissociates immediately to restore the relative amount of ions and undissociated acid to the original value.

Similarly the typical properties of alkalis in solution are attributed to hydroxyl ions (OH^-). Strong alkalis such as sodium hydroxide (NaOH) produce large concentrations of hydroxyl ions when added to water but weak alkalis such as ammonium hydroxide (NH_4OH) are only slightly ionized in water and produce much smaller concentrations of hydroxyl ions.

As with weak acids, the strength of a weak base is indicated by its dissociation constant

$$K = \frac{[B^+][OH^-]}{[BOH]}$$

where $[B^+]$ is the concentration of alkaline ions, $[OH^-]$ is the concentration of hydroxyl ions, and $[BOH]$ is the concentration of undissociated alkali.

Strong electrolytes have no dissociation constant; the expression for strong acids $[A^-][H^+]/[HA]$, and the corresponding expression for alkalis, vary considerably with change in concentration. With strong acids and alkalis the apparent degree of ionization can be taken as a measure of the strength of the acid or base.

So far it has been assumed that the effective concentrations or active masses could be expressed by the stoichiometric concentrations but, according to modern thermodynamics, this is not strictly true. For a binary electrolyte $AB \rightleftharpoons A^+ + B^-$ the correct equilibrium equation is:

$$K_a = \frac{a_{A^+} \times a_{B^-}}{a_{AB}}$$

where a_{A^+}, a_{B^-} and a_{AB} represent the activities of A^+, B^- and AB and K_a is the thermodynamic dissociation constant. The thermodynamic quantity 'activity' is related to concentration by a factor called the activity coefficient, i.e. activity = concentration × activity coefficient.

Using this concept, the thermodynamic activity coefficient is

$$K_a = \frac{[A^+][B^-]}{[AB]} \times \frac{f_{A^+} \times f_{B^-}}{f_{AB}}$$

where f refers to the activity coefficients and the square brackets to the molar concentrations.

The activity coefficients of un-ionized molecules do not differ much from unity and so for weak electrolytes in which the ionic concentration, and therefore the ionic strength is low, the error introduced by neglecting the difference between the actual values of the activity coefficients of the ions, f_{A^+} and f_{B^-}, and unity is small (less than 5 per cent). Hence for weak electrolytes, the constants obtained by using the simpler equation $K=[A^+][B^-]/[AB]$ are sufficiently precise for the purposes of calculation in quantitative analysis. Strong electrolytes are assumed to be completely dissociated and no correction for activity coefficients needs to be made for dilute solutions.

However, the concept of activity is important in potentiometric techniques of analysis (described later). The activity coefficient varies with concentration and for ions it varies with the charge and is the same for all *dilute* solutions having the same ionic strength. The activity coefficient depends upon the total ionic strength of a solution (a measure of the electrical field existing in the solution) and for ion-selective work it is often necessary to be able to calculate this. The ionic strength I is given by

$$I=0.5\sum C_i Z_i^2$$

where C_i is the ionic concentration in moles per litre of solution at Z_i is the charge of the ion concerned. Thus the ionic strength of 0.1 M nitric acid solution (HNO_3) containing 0.2 M barium nitrate [$Ba(NO_3)_2$] is given by

$$\begin{aligned}0.5[0.1\ (\text{for } H^+)&+0.1\ (\text{for } NO_3^-)\\&+0.2\times 2^2\ (\text{for } Ba^{++})\\&+0.4\times 1\ (\text{for } NO_3^-)]\\=0.5[1.4]=0.7&\end{aligned}$$

4.2 Ionization of water

As even the purest water possesses a small but definite electrical conductivity, water itself must ionize to a very slight extent into hydrogen and hydroxyl ions:

$$H_2O \rightleftharpoons H^+ + OH^-$$

This means that at any given temperature

$$\frac{a_{H^+}\times a_{OH^-}}{a_{H_2O}}=\frac{[H^+].[OH^-]}{[H_2O]}\times\frac{f_{H^+}.f_{OH^-}}{f_{H_2O}}=K$$

where a_x, [X] and f_x refer to the activity, concentration and activity coefficient of the species X, and K is a constant.

As water is only slightly ionized, the ionic concentrations are small and the activity coefficients of the ions can therefore be regarded as unity. The activity coefficient of the un-ionized molecule H_2O may also be taken as unity and the above expression therefore reduces to

$$\frac{[H^+]\times[OH^-]}{[H_2O]}=K$$

In pure water too, because there is only very slight dissociation into ions, the concentration of the undissociated water [H_2O] may also be considered constant and the equation becomes $[H^+]\times[OH^-]=K_w$. The constant K_w is known as the ionic product of water.

Strictly speaking, the assumptions that the activity coefficient of water is constant and that the activity coefficients of the ions are unity are only correct for pure water and for very dilute solutions where the ionic strength is less than 0.01. In more concentrated solutions the ionic product for water will not be constant but, as activity coefficients are generally difficult to determine, it is common usage to use K_w.

The ionic product of water, K_w, varies with temperature and is given by the equation

$$\begin{aligned}\log_{10}K_w=&14.00-0.0331(t-25)\\&+0.000\,17(t-25)^2\end{aligned}$$

where t is the temperature in °C.

Conductivity measurements show that, at 25 °C, the concentration of hydrogen ions in water is 1×10^{-7} mol litre^{-1}. The concentration of hydroxyl ions equals that of the hydrogen ions therefore $K_w=[H^+]\times[OH^-]=10^{-14}$. If the product of $[H^+]$ and $[OH^-]$ in aqueous solution momentarily exceeds this value, the excess ions will immediately recombine to form water. Similarly if the product of the two ionic concentrations is momentarily less than 10^{-14}, more water molecules will dissociate until the equilibrium value is obtained. Since the concentrations of hydrogen and hydroxyl ions are equal in pure water it is an exactly neutral solution. In aqueous solutions where the hydrogen ion concentration is greater than 10^{-7}, the solution is acid; if the hydrogen ion concentration is less than 10^{-7} the solution is alkaline.

4.3 Electrical conductivity

4.3.1 Electrical conduction in liquids

As early as 1833, Faraday realized that there are two classes of substances which conduct electricity. In the first class are the metals and alloys, and certain non-metals such as graphite, which conduct electricity without undergoing any chemical change. The flow of the current is due to the motion of electrons within the conductor, and the conduction is described as metallic, or electronic.

In the second class are salts, acids and bases which, when fused or dissolved in water, conduct electricity owing to the fact that particles, known as ions, carrying positive or negative electric charges move in opposite directions through the liquid. It is this motion of electrically charged particles which constitutes the current. Liquids which conduct electricity in this manner are known as electrolytes.

4.3.2 Conductivity of solutions

The passage of current through an electrolyte generally obeys Ohm's law and the current-carrying ability of any portion of electrolyte is termed its conductance and has the units of reciprocal resistance ($1/\Omega$), siemens (S). The specific current-carrying ability of an electrolyte is called its conductivity and consequently has the units of $S\ m^{-1}$.

The conductivity of electrolytes varies greatly with their concentration because dilution (a) increases the proportion of the dissolved electrolyte which forms ions in solution but (b) tends to reduce the number of these ions per unit of volume. In order to measure the first effect alone another term, molar conductivity, Λ, is defined,

$$\Lambda\ (S\ m^2/mol) = \kappa/c,$$

where κ is the conductivity and c is the concentration in $mol\ m^{-3}$. Although these are the basic SI units most work is reported using volume units of cm^3 since the litre is a convenient volume for laboratory use and Λ is usually in units of $S\ cm^2/mol$.

Table 4.1 Limiting ionic conductivities at 25°C

Cation	λ° *S cm²/mol*	*Anion*	λ° *S cm²/mol*
H^+	349.8	OH^-	199.1
Li^+	38.7	F^-	55.4
Na^+	50.1	Cl^-	76.4
K^+	73.5	Br^-	78.1
NH_4^+	73.6	I^-	76.8
$(CH_3)_2NH_2^+$	51.9	NO_3^-	71.5
$\frac{1}{2}Mg^{2+}$	53.1	ClO_4^-	64.6
$\frac{1}{2}Ca^{2+}$	59.5	Acetate	40.9
$\frac{1}{2}Cu^{2+}$	53.6	$\frac{1}{2}SO_4^{2-}$	80.0
$\frac{1}{2}Zn^{2+}$	52.8	$\frac{1}{2}CO_3^{2-}$	69.3

At infinite dilution the ions of an electrolyte are so widely separated by solvent molecules that they are completely independent and the molar conductivity is equal to the sum of the ionic conductivities, λ°, of the cation and anion, i.e.

$$\Lambda_\infty = \lambda^\circ_- + \lambda^\circ_+$$

The values of λ° are the values for unit charge, referred to as equivalent ionic conductivities at infinite dilution. The general case is

$$\Lambda_\infty = z_+ n_+ \lambda^\circ_+ + z_- n_- \lambda^\circ$$

where z is the charge on the ion and n the number of these ions produced by dissociation of one molecule of the salt, e.g.

$$\lambda_\infty(LaCl_3) = 3 \times 1 \times \lambda^\circ_{La} + 1 \times 3 \times \lambda^\circ_{Cl}$$

Since, for example, the ionic conductivity of the chloride ion is the same in all chloride salts then the molar conductivity at infinite dilution of any chloride salt can be calculated if the corresponding value for the cation is known. Values of ionic conductivities at infinite dilution at 25 °C are given in Table 4.1.

Providing the concentration of a fully dissociated salt is less than about 10^{-4} mol/l then the conductivity κ at 25 °C can be calculated from

$$\kappa\ (S\ cm^{-1}) = zn(\lambda^\circ_+ + \lambda^\circ_-)c\ 10^{-3}$$

or

$$\kappa\ (\mu S\ cm^{-1}) = zn(\lambda^\circ_+ + \lambda^\circ_-)c\ 10^3$$

where c is the concentration in mol/l.

Values of limiting ionic conductivities in aqueous solution are highly temperature-dependent and in some cases the value increases five- or sixfold over the temperature range 0–100 °C (see Table 4.2). These changes are considered to be due mainly to changes in the viscosity of water and the effect this has on the mobility and hydration of the ions.

Table 4.2 Ionic conductivities between 0 and 100°C (S cm²/mol)

Ion	*0°*	*5°*	*15°*	*18°*	*25°*	*35°*	*45°*	*55°*	*100°*
H^+	225	250.1	300.6	315	349.8	397.0	441.4	483.1	630
OH^-	105	—	165.9	175.8	199.1	233.0	267.2	301.4	450
Li^+	19.4	22.7	30.2	32.8	38.7	48.0	58.0	68.7	115
Na^+	26.5	30.3	39.7	42.8	50.1	61.5	73.7	86.8	145
K^+	40.7	46.7	59.6	63.9	73.5	88.2	103.4	119.2	195
Cl^-	41.0	47.5	61.4	66.0	76.4	92.2	108.9	126.4	212
Br^-	42.6	49.2	63.1	68.0	78.1	94.0	110.6	127.8	—
I^-	41.4	48.5	62.1	66.5	76.8	92.3	108.6	125.4	—
NO_3^-	40.0	—	—	62.3	71.5	85.4	—	—	195
ClO_4^-	36.9	—	—	58.8	67.3	—	—	—	185
Acetate	20.1	—	—	35	40.9	—	—	—	—
$\frac{1}{2}Mg^{2+}$	28.9	—	—	44.9	53.0	—	—	—	165
$\frac{1}{2}Ca^{2+}$	31.2	—	46.9	50.7	59.5	73.2	88.2	—	180
$\frac{1}{2}SO_4^-$	41	—	—	68.4	80.0	—	—	—	260

4.3.3 Practical measurement of electrical conductivity

From the foregoing, it can be seen that measurement of electrical conductivity enables concentration to be determined.

4.3.3.1 Alternating current cells with contact electrodes

Conductivity cells provide the means of conducting a small, usually alternating, current through a precise volume of liquid whose conductivity we wish to know. At its simplest, this process involves the measurement of the resistance between two electrodes of fixed shape and constant distance apart. The relationship between the specific conductivity κ of the solution and the resistance R across the electrodes includes a cell constant 'a' such that

$$\kappa = a/R$$

If we express the conductivity in units of S cm^{-1} then the cell constant has the dimension of cm^{-1}. In order to simplify the electrical circuits of the measuring instruments it is customary to maintain the resistance of conductivity cells between the limits of 10 and 100 000 Ω. The conductivity of aqueous solutions varies from pure water with a conductivity of about 5 μS/m to those of concentrated electrolytes with conductivities as high as 1000 S/m. In order to keep within these resistance limits it is necessary, therefore, to have cells with a range of cell constants from 0.01 to 100 cm^{-1}. A working guide to the most appropriate value of cell constant for any given range of conductivity is shown in Table 4.3.

Table 4.3 Guide to cell constant for known conductivity range

Conductivity range, μS cm^{-1}	*Cell constant cm^{-1}*
0.05 to 20	0.01
1 to 200	0.1
10 to 2000	1
100 to 20 000	10
100 to 200 000	50

In order to measure the conductivity accurately it is necessary to know the cell constant accurately. It is usual to determine the cell constant by preferably (a) measuring the conductance when the cell is filled with a solution whose conductivity is accurately known or, failing that, (b) comparing the measured conductance with that obtained from a cell of known cell constant when both cells contain the same solution at the same temperature.

The only solutions whose conductivities are known with sufficient accuracy to be used for reference purposes are aqueous solutions of potassium chloride. This salt should be of the highest purity, at least analytical reagent grade and dried thoroughly in an oven at 120 °C before preparing solutions by dissolving in de-ionized water whose conductivity is less than 2 μS/cm at room temperature. The most accurate reference solutions are prepared by weight and the two most useful solutions are given in Table 4.4.

Table 4.4 Standard solutions for cell calibration

*Solution g KCl/1000 g solution**	*κ at 18 °C S m^{-1}*	*κ at 25 °C S m^{-1}*
(A) 7.4191	1.1163	1.2852
(B) 0.7453	0.12201	0.14083

* All values are 'mass in vacuo'.

For many purposes a simpler procedure can be followed. This involves weighing only the potassium chloride and preparing solutions by volume at 20 °C, these details are given in Table 4.5.

Table 4.5 Standard solutions (volumetric) for cell calibration

Solution	*κ at 18 °C*	*κ at 25 °C*
(A′) 7.4365 g KCl/l at 20 °C	1.1167 S m^{-1}	1.2856 S m^{-1}
(B′) 0.7440 g KCl/l at 20 °C	0.1221 S m^{-1}	0.1409 S m^{-1}
(C′) 100 ml of solution B′ made up to 1 litre at 20 °C	—	146.93 μS cm^{-1}*

* For the highest accuracy the conductivity of the dilution water should be added to this value.

Calibration of conductivity cells by these solutions requires considerable care if accurate values of cell constants are to be determined. The importance of temperature control cannot be over-emphasized since the conductivity of the potassium chloride solution will change by over 2 per cent per kelvin. Alternatively the cell constant can be determined by the comparison technique with identical, rather than standard conditions in both the 'known' and 'unknown' cell. Equally important as the effect of temperature, is that of polarization in these cells where the electrodes contact the solution and conduct a significant current.

The extent of polarization depends on a number of factors, the most important of which are the nature of the electrode surface and the frequency of the a.c. signal applied to the cell. The restrictions that polarization errors, arising from electrode material, impose on the choice of cell means that cells with

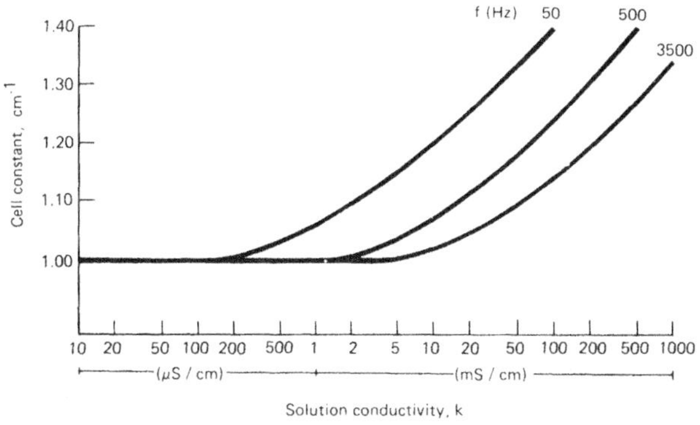

Figure 4.1 Effect of frequency on the useful range of a cell with titanium carbide coated stainless steel electrodes (courtesy F. Oehme, Polymetron).

bright metal electrodes are best suited for measurements of low conductivities where the proportion of the total resistance due to polarization is very small. Treated or coated electrodes are suitable for low ($\sim 0.05\ \mu S\ cm^{-1}$) to intermediate ($\sim 0.1\ S\ m^{-1}$) conductivities provided that the frequency of the a.c. voltage is in the range normally found in commercial instruments (50–1000 Hz).

Polarization in all the cells we have been discussing can be reduced by increasing the frequency of the applied voltage. This can best be appreciated by considering Figure 4.1 in which the apparent cell constant over a range of conductivities is plotted against three values of a.c. frequency. The true value of the cell constant was $1\ cm^{-1}$ and it can be seen that the highest frequency, 3.5 kHz, gave the true value for the cell constant over the widest concentration range. Unfortunately increase of frequency can introduce capacitative errors into the measurement, particularly from the signal cable and in many applications the choice of operating frequency is a compromise. Although variable frequency conductivity meters are available as laboratory instruments (e.g. Philips Model PW 9509, High Performance Conductivity Meter) such a facility is not usually found on industrial instruments. In this case it is necessary to consider the range of conductivities to be measured, together with the chemical and physical nature of the solutions to be measured before specifying the operating frequency. All determinations of cell constant should be carried out at this frequency.

Cell construction The materials used in cell construction must be unaffected by the electrolyte and the insulation between the electrodes must be of a high quality and not absorb anything from the process liquid.

A wide range of materials are at present available covering a wide range of pressures, temperatures and process fluids. The body may be made of glass, epoxy resins, plastics such as PTFE, pure or reinforced, PVC, Perspex, or any other material suitable for the application, but it must not be deformed in use by temperature or pressure, otherwise the cell constant will change.

The electrodes may be parallel flat plates or rings of metal or graphite cast in the tube forming the body, or in the form of a central rod with a concentric tubular body.

One common form of rod-and-tube conductivity cell consists of a satinized stainless steel rod-electrode surrounded by a cylindrical stainless steel electrode, having holes to permit the sample to flow freely through the cell. This is surrounded by an intermediate cylinder also provided with holes, and two O-rings which together with the tapered inner end form a pressure-tight seal onto the outer body when the inner cell is withdrawn for cleaning, so that the measured solution can continue to flow and the cell be replaced without interruption of the process. The outer body is screwed into the line through which the measured solution flows. Figure 4.2(a) shows the inserted cell as it is when in use, and (b) the withdrawn measuring element with the intermediate sleeve forming a seal on the outer body. The cell may be used at 110 °C up to 7 bar pressure.

Many manufacturers offer a type of flow-through conductivity cell with annular graphite electrodes, one

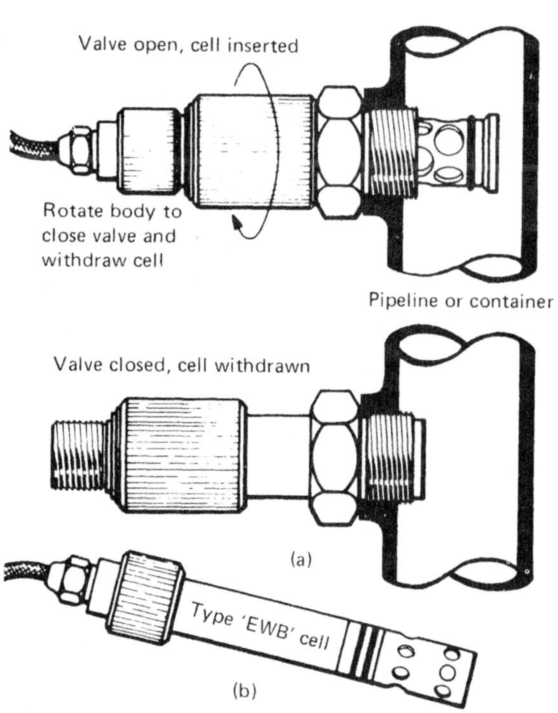

Figure 4.2 Retractable conductivity cell (courtesy Kent Industrial Measurements Ltd., Analytical Instruments).

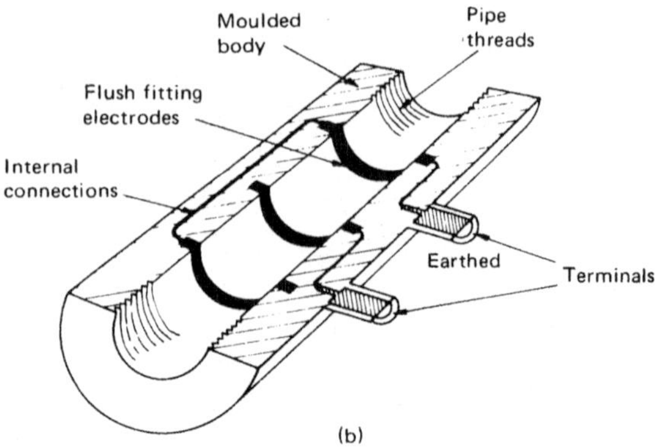

(b)

Figure 4.3 Flow-through cell (courtesy Kent Industrial Measurements Ltd., Analytical Instruments).

form of which is shown in Figure 4.3. It consists of three annular rings of impervious carbon composition material equally spaced within the bore of an epoxy resin moulded body. Conduction through the solution within the cell takes place between the central electrode and the two outer rings, which are connected to the earthed terminal of the measuring instrument; thus electrical conduction is confined entirely within the cell, where it is uninfluenced by the presence of adjoining metal parts in the pipe system. This pattern of cell, having a simple flow path, is ideally suited to the exacting requirements of dialysate concentration monitoring in the artificial kidney machine. Screw-in patterns of this cell are also generally available.

The use of an impervious carbon composition material for the electrodes substantially eliminates polarization error and provides conducting surfaces that do not require replatinization or special maintenance, other than periodic, but simple and infrequent cleaning by means of a bottle brush. Typical operating temperature and pressure limits for this type of cell are 100 °C and 7 bar.

Measuring cells should be installed in positions where they are adequately protected from mechanical shock by passing traffic, dampness and extremes of temperature. Where a flow-line cell is connected directly in the electrolyte pipe, suitable support should be given to the pipes to ensure that the cell is under no mechanical strain, and that the pipe threads in a rigid system are straight and true. Dip pattern cells should be installed so that moving parts in a tank, e.g. agitators, are well clear of the cells.

Where measuring cells are installed in pipework, it is essential that they are positioned in a rising section of the system to ensure that each cell is always full of electrolyte, and that pockets of air are not trapped. Alternatively, they may be installed in the bottom member of a U formed in horizontal pipework. In this case, screw-in cells should be in the top or side of the pipe so that sediment cannot settle in them.

Cleaning and maintenance of cells Periodic inspection and cleaning of conductivity cells is essential to ensure that the electrode surfaces are free from contamination which would otherwise alter the electrode area and effective cell constant. The frequency of such procedures is mainly dependent on the nature of the samples but the design of the cells and the accuracy required for the measurement will also have to be taken into consideration. All new cells should be thoroughly cleaned before installation and these cleaning procedures depend on the design of the cell and the electrode material.

Platinized electrodes Cleaning of these electrodes constitutes a major drawback in their application because no form of mechanical cleaning should be attempted. A suitable cleaning solution consists of a stirred mixture of 1 part by volume isopropyl alcohol, 1 part of ethyl ether and 1 part hydrochloric acid (50%). Alternatively, the sensitivity of the electrodes can frequently be restored by immersion in a 10–15% solution of hydrochloric or nitric acid for about two minutes. The electrodes should be thoroughly rinsed with water before being returned to service.

Annular graphitic electrodes Cleaning should be carried out with a 50% solution of water/detergent using a bottle brush. After thorough brushing with this solution, the cell bore should be rinsed several times in distilled water and then viewed. Looking through the bore towards a source of illumination, the surface

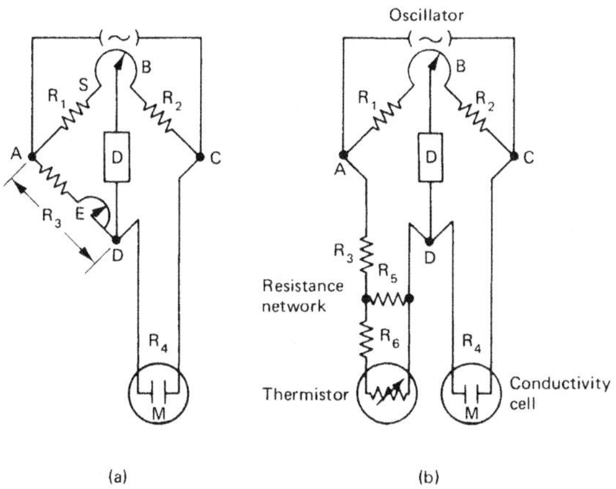

Figure 4.4 Measurement of conductance using Wheatstone bridge: (a) simple circuit, (b) thermistor temperature-corrected circuit.

should be evenly wetted with no dry patches where the water has peeled away. If dry patches appear rapidly, indicating that a thin film of grease is present, the surface is not clean.

Stainless steel and Monel A feature of many stainless steel cells is the frosted appearance of the electrodes which is essential to reduce polarization. It is most important that this frosting is not polished away by the regular use of abrasive cleaners. This type of cell may be cleaned with a 50% water detergent solution and a bottle brush.

In the case of screw-in cells the outer electrode may be removed to facilitate cleaning, but on no account should the central electrode be disturbed, as this will impair the accuracy of the electrical constant of the cell. In cases where metal cells have become contaminated with adherent particulate matter, such as traces of magnetite or other metal oxides, ultrasonic cleaning in the detergent solution has been shown to be effective.

In all cleaning processes care should be taken to keep the external electrical contact, cable entries and plugs dry.

Instruments for conventional a.c. measurement The conductance of a cell may be measured (a) by Wheatstone bridge methods or (b) by direct measurement of the current through the cell when a fixed voltage is applied.

Wheatstone bridge methods The actual conductance of the cell is usually measured by means of a self-balancing Wheatstone bridge of the form shown in Figure 4.4 and described in detail in Volume 3.

Direct measurement of cell conductance The conductance of a cell may be measured directly by the method indicated in Figure 4.5. The current is directly proportional to the conductance so the output from the current amplifier is applied to the indicator and recorder. Temperature compensation is achieved by connecting a manual temperature compensator in the amplifier circuit, or a resistance bulb may be used to achieve automatic compensation.

Multiple-electrode cells From the foregoing discussion on errors introduced by polarization together with the importance of constancy of electrode area, it can be appreciated that two-electrode conductivity cells have their limitations. In circumstances where accurate measurements of conductivity are

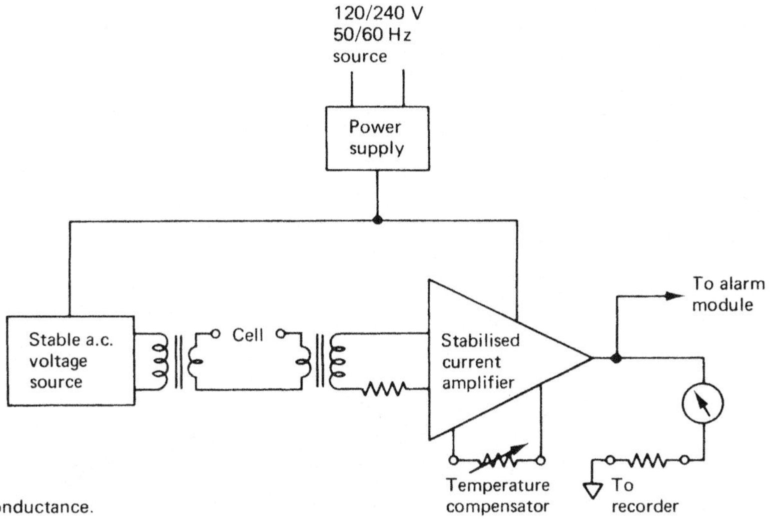

Figure 4.5 Direct measurement of cell conductance.

required in solutions of moderate or high conductivity or in solutions which can readily contaminate the electrode surfaces, multiple-electrode cells should be considered.

In its simplest form, a multiple-electrode cell has four electrodes in contact with the solution. An outer pair operate similarly to those in a conventional two-electrode cell and an a.c. current is passed through the solution via these electrodes. The voltage drop across a segment of the solution is measured potentiometrically at a second or inner pair of the electrodes, and this drop will be proportional to the resistivity or inversely proportional to the conductivity of the solution. Four-electrode cells can be operated in either the constant-current or constant-voltage mode but the latter is the more popular and will be described further. In this form of measurement the voltage at the inner electrode pair is maintained at a constant value by varying the current passed through the solution via the outer electrodes. The current flowing in the cell will be directly proportional to the conductivity, and can be measured as indicated in Figure 4.6.

The circuit shown in the figure is considerably simplified and there are multiple-electrode cells available from a number of manufacturers which contain additional electrodes whose function is to minimize stray current losses in the cell, particularly for solutions flowing through earthed metal pipework.

Since there is imperceptible current flowing through the voltage sensing electrodes, cells of this type are free from the restrictions imposed by polarization. Therefore multiple-electrode cells can be standardized with any of the potassium chloride solutions given in Tables 4.4 and 4.5. The precaution previously stated about constancy of temperature during any determination of cell constant must still be observed.

Multiple-electrode cells are available with cell constants from 0.1 to 10 cm^{-1} and can therefore be used over a wide range of solution conductivities. However, their most valuable applications are when contamination or polarization is a problem.

Temperature compensation The conductivity of a solution is affected considerably by change of temperature, and each solution has its own characteristic conductivity–temperature curve. Figure 4.7 shows how different these characteristics can be. When it is required to measure composition rather than absolute conductivity it is therefore essential to use a temperature compensator to match the solution.

Manual compensators consist of a variable and a fixed resistor in series. The temperature scale showing the position of the contact on the variable resistance is calibrated so that the resistance of the combined resistors changes by the same percentage of the value of conductivity of the solution at 25 °C as does the solution. The scale becomes crowded at the upper end, thus limiting the span of the compensator to about 70 °C.

Aqueous solutions containing very low ($\mu g\,l^{-1}$) concentrations of electrolytes must have more elaborate compensation to allow for the non-linear conductivity–temperature characteristic of pure water. This type of compensation system is applied in all conductivity transmitters (either with two-electrode or multiple-electrode cells) designed for accurate operation in the range up to 0.5 $\mu S\,cm^{-1}$.

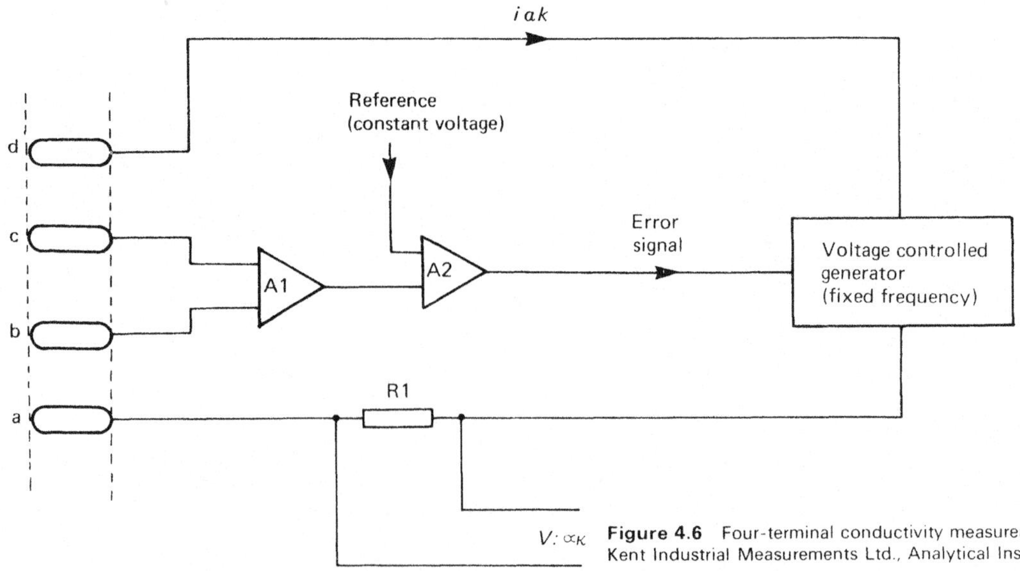

Figure 4.6 Four-terminal conductivity measurement (courtesy Kent Industrial Measurements Ltd., Analytical Instruments).

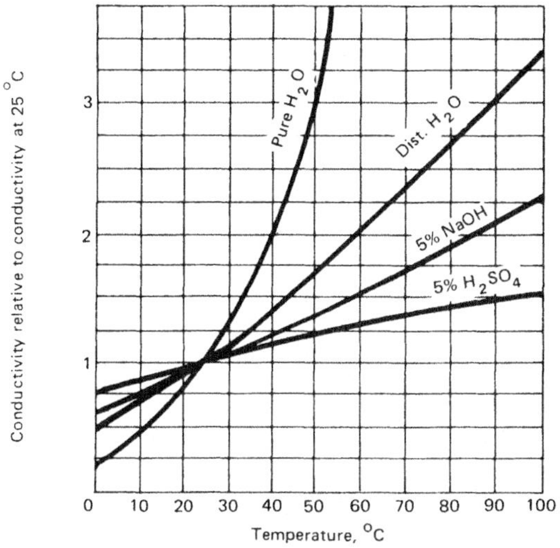

Figure 4.7 Variation of solution conductivity with temperature.

4.3.3.2 Electrodeless method of measuring conductivity

The principle of the method is to measure the resistance of a closed loop of solution by the extent to which the loop couples two transformer coils. The liquid to be measured is enclosed in a non-conducting pipe, or a pipe lined with a non-conducting material. Three forms of measuring units are available, as shown in Figure 4.8. As the method is most successful with full scale resistances of 10–1000 Ω, relatively large bore pipe may be used, reducing the possible errors due to solid deposition or film formation.

Figure 4.8(a) shows the form used for immersion in a large volume of solution. For measurements on a solution flowing through a pipe the arrangement shown in Figure 4.8(b) is used. If the liquid contains suspended solids or fibres, wide bore non-conducting pipe fitted with metallic end-pieces connected together with a length of wire to complete the circuit may sometimes be used (Figure 4.8(c)).

The principle of the measuring system is shown in Figure 4.9. Figure 4.9(a) shows the simple circuit which consists of two transformers. The first has its primary winding, the input toroid, connected to an oscillator operating at 3 or 18 kHz and as its secondary the closed loop of solution. The closed loop of solution forms the primary of the second transformer and its secondary is the output toroid. With constant input voltage the output of the system is proportional to the conductivity of the solution. The receiver is a high impedance voltage measuring circuit which amplifies and rectifies the output and displays it on a large indicator.

In order to eliminate effects of source voltage and changes in the amplifier characteristics a null balance system may be provided as shown in Figure 4.9(b). An additional winding is provided on each toroid and the position of the contact is adjusted on the main slide-wire to restore the system to the original balanced state by means of the balancing motor operated by the amplified out-of-balance signal in the usual way.

The electrodeless measurement of conductivity has obvious advantages in applications where the solution is particularly corrosive or has a tendency to foul or mechanically abrade the electrodes. Typical of these applications are measurements in oleum, hot concentrated sodium hydroxide and slurries. In addition, this technique is ideal for applications in concentrated electrolytes (not necessarily aggressive) such as estuarine or sea waters where polarization errors

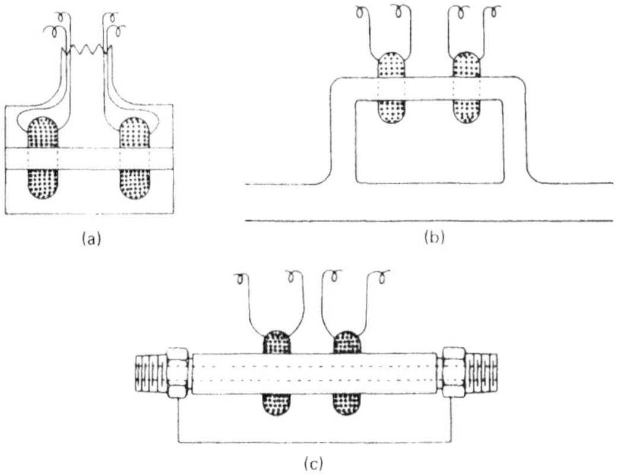

Figure 4.8 Electrodeless conductivity cells.

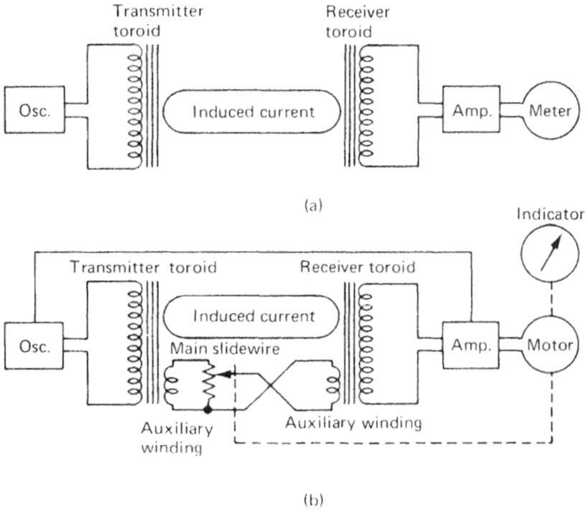

Figure 4.9 Measuring circuits for use with electrodeless cells (courtesy Beckman Instruments Inc.): (a) direct reading, (b) balanced bridge.

would be considerable in a conventional cell. Temperature compensation is normally incorporated.

4.3.4 Applications of conductivity measurement

The measurement of electrical conductivity is the simplest and probably the most sensitive method of providing a non-specific indication of the dissolved solids, or more correctly the ionic, content of a solution. If the number of ionic species in solution are few then it may be possible to use conductivity as a measure of the concentration of a particular component. Undoubtedly the robust nature of conductivity measurements has led to its use in circumstances where its non-specific response gives rise to errors in interpretation of concentration. Consequently, any successful instrumental application of conductivity as a concentration sensor has to ensure that the species of interest is the dominating ion or the only ion (together with its counter-ion of opposite charge) whose concentration is changing. With these restrictions it can be appreciated that determinations of concentrations by conductivity measurements are often supported by additional analyses or preceded by a physical or chemical separation of the species of interest.

4.3.4.1 Conductivity and water purity

Water of the highest purity is increasingly being used for industrial purposes, for example, the manufacture of electronic components and the preparation of drugs. Other examples of large-scale uses include process steam and feedwater for high pressure boilers. In all these cases conductivity provides the most reliable measurement of water purity in circumstances where contamination from non-conducting impurities is considered to be absent. The conductivity of pure water is highly temperature-dependent due to the increase in the dissociation of water molecules into hydrogen and hydroxyl ions of water, K_w, with temperature. The extent of this can be seen in Table 4.6.

The conductivity of pure water can be calculated at any temperature provided values of λ°_{OH}, λ°_H, K_w, the dissociation constant of water, and the density of water d are known at the appropriate temperature.

$$\kappa(\mu S\ cm^{-1}) = (\lambda^\circ_H + \lambda^\circ_{OH})d \cdot \sqrt{K_w} \cdot 10^3$$

In the application under consideration here (i.e. the use of pure water) the exact nature of the ionic species giving rise to a conductivity greater than that of pure water are of no interest but it is useful to note how little impurity is required to raise the conductivity. For example, at 25 °C only about 10 $\mu g\ l^{-1}$ of sodium (as sodium chloride) are required to increase the conductivity to twice that of pure water.

Table 4.6 Pure water, conductivity from 0 to 100 °C

Temperature (°C)	*Conductivity* ($\mu S\ cm^{-1}$)	*Resistivity* (°C)
0	0.0116	86.0
5	0.0167	60.0
10	0.0231	43.3
15	0.0314	31.9
20	0.0418	23.9
25	0.0548	18.2
30	0.0714	14.0
35	0.0903	11.1
40	0.1133	8.82
45	0.1407	7.11
50	0.1733	5.77
60	0.252	3.97
70	0.346	2.89
80	0.467	2.14
90	0.603	1.66
100	0.788	1.27

4.3.4.2 Condensate analyser

The purity of the water used in the steam–water circuit of power stations is particularly important for the prevention of corrosion. An essential component of such a circuit is the condenser wherein the steam from the turbines is condensed before returning to the boiler. On one side of the condenser tubes is the highly pure steam and water from the turbines and on the other is cooling water chosen for its abundance (e.g. river water or estuarine water) rather than its chemical purity. Any leakage of this cooling water through the condenser tubes leads to the ingress of unwanted impurities into the boiler and therefore must be immediately detected. Direct measurement of conductivity would detect significant ingress of say sodium chloride from estuarine water but it would not be capable of detecting small leakages since the conductivity of the condensate would be dominated by the alkaline additives carried over in the steam from the boiler. A better method of detection of leakage is to pass the condensate through a cation exchange column in the H^+-form, then measuring the conductivity. Using this procedure, all cations in the condensate are exchanged for hydrogen ions and the solution leaving the column will be weakly acidic if any salts have entered through the condenser. Otherwise, the effluent from the column will ideally be pure water since the cations of the alkaline boiler water additives (NH_4OH, NaOH) will be exchanged and recombine as,

$$H^+ + OH^- \rightleftharpoons H_2O$$

A secondary advantage of such a system is the enhancement of the conductivity due to replacement of cations by hydrogen ions which gives about a fivefold enhancement in ionic conductance. This is particularly important with very low leak rates.

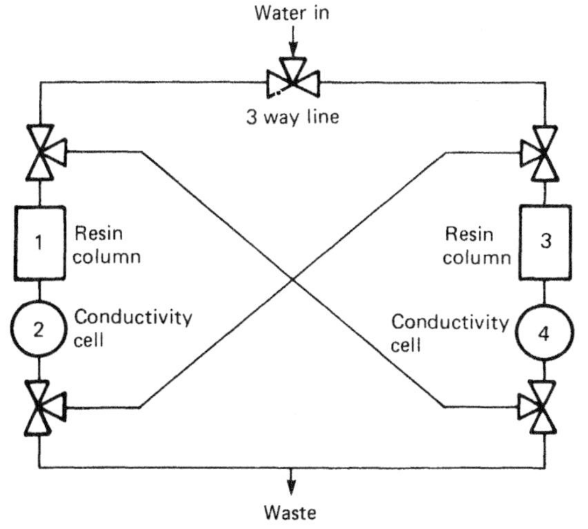

Figure 4.10 Condensate analyser.

A schematic diagram of an instrument based on the above principles is given in Figure 4.10. The incoming sample flows at about 400 ml min^{-1} through a H^+-form cation exchange column (1), 500 mm deep and 50 mm in diameter, and then to a flow-through conductivity cell (2). The effluent from the cell flows to waste via an identical column/cell system (3 and 4) which is held in reserve. Since there will be no exchange on this second column it will not be depleted and the constant flow of water or weak acid keeps it in constant readiness for instant replacement of column (1) when the latter becomes exhausted. The measured conductivity can be recorded and displayed and, where necessary, alarms set for notification of specific salt ingress levels. In the case of power stations using estuarine water for cooling the condensers the condensate analyser can be used to give a working guide to the salt going forward to the boiler (see Table 4.7).

Table 4.7 Relationship between conductivity and salt fed to the boiler

Conductivity at 25 °C (μS cm^{-1})	*Chloride in condensate* (ppm)	*Salt going forward to boiler* (g NaCl/Tonne)
0.137	0.01	0.0165
0.604	0.05	0.0824
1.200	0.10	0.1649
1.802	0.15	0.2473
2.396	0.20	0.3298
6.003	0.50	0.8265

4.3.4.3 Conductivity ratio monitors

These instruments measure the conductivities at two points in a process system continuously and compare the ratio of the measurements with a preset ratio. When the measured ratio reaches the preset value, a signal from the monitor can either operate an alarm or initiate an action sequence or both.

One application of this type of dual conductivity measurement is to control the regeneration frequency of cation exchange units (usually in the H^+-form) in water treatment plants. The conductivity at the outlet of such a unit will be higher than at the inlet since cations entering the ion exchange bed will be replaced by the much more conductive hydrogen ion ($\lambda^\circ_H = 350$, $\lambda^\circ_{Na} = 50$). For example, an inlet stream containing 10^{-4} mol l^{-1} of sodium chloride will have ratios of 3.5, 3.3 and 2.3 for 100, 90 and 50 per cent exchange respectively. A value corresponding to the acceptable extent of exchange can then be set on the instrument. Reverse osmosis plants use ratio monitors to measure the efficiency of their operation and these are usually calibrated in percentage rejection or passage.

This type of operational control is most effective when the chemical constituents of the inlet stream do not vary greatly, otherwise the ratio will be subject to errors from unconsidered ionic conductivities.

4.3.4.4 Ion chromatography

Although conductivity measurements have a non-specific response they can, when combined with a separation technique, provide extremely sensitive and versatile detectors of chemical concentration. The best example of this is in ion chromatography which in recent years has been shown to be an invaluable instrumental technique for the identification and measurement of the concentration of ions, particularly at low levels, in aqueous solution.

The general principles of chromatography are outlined in Chapter 2. In an ion chromatograph, a small volume of sample is injected into a carrier or eluent electrolyte stream. The eluent together with the sample is carried forward under high pressure (5–50 bar) to an ion exchange column where chromatographic separation of either the cations (+ve) or anions (−ve), depending on the nature of the exchanger, takes place. The ion exchange material in these chromatographic separator columns is fundamentally the same as conventional ion exchange resins but the exchange sites are limited to the surface of very fine resin beads. This form of exchanger has been shown to have the characteristics required for rapid separation and elution of the ionic components in the order expected from the general rules of ion exchange (e.g. Cl^- before Br^- before SO_4^{2-}). At this stage the conductivity can be monitored and the elution peaks corresponding to the separated ionic components measured as increases superimposed on the relatively high background conductivity of the eluent. This is the procedure used in the ion chromatograph manu-

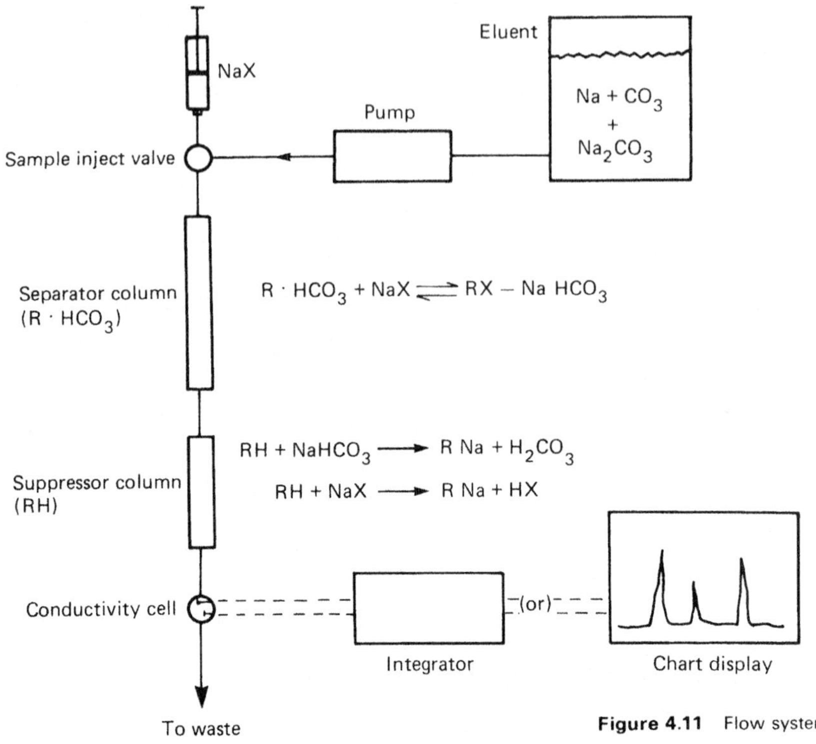

Figure 4.11 Flow system for anion chromatography.

factured by Wescan Instruments Inc. In another instrument manufactured by The Dionex Corporation, the eluent from the separator column passes through a second ion exchange column where the ions of opposite charge to those which have been separated chromatographically are all converted to a common form. This second column, termed a 'suppressor column', reduces the background conductivity of the eluent and thus ensures that conductivity changes due to the sample constitute a significant portion of the total measured conductivity. With a system such as this, the retention time identifies the elution peak and the area under the peak is a measure of the concentration of the ionic species giving rise to it. In many cases peak heights rather than areas can be used as the indicators of concentration, thus simplifying the measurement since an integrator is not required. For most purposes this is adequate since sharp elution peaks are obtained by keeping mixing to a minimum by use of very narrow bore transmission tubing combined with a conductivity cell whose volume is of the order of 6 μl. In cells of this size polarization resistance can be considerable due to the proximity of the electrodes.

A schematic outline of the main features of a typical system for the determination of anions is given in Figure 4.11.

In this particular example the eluent consisting of a mixture of 2.4×10^{-3} mol l^{-1} sodium carbonate and 3×10^{-3} mol l^{-1} sodium bicarbonate has a conductivity of about 700 μS cm^{-1}. The separator column consists of a strong base anion exchanger (R . HCO_3) mainly in the bicarbonate form and the suppressor column is a strong acid cation exchanger in the H^+-form (R . H). After the eluent has passed through the cation exchange it will be weakly acid carbonic acid (H_2CO_3) having a conductivity level of about 25 μS cm^{-1} and with this much reduced base conductivity level it is possible to detect quantitatively the small changes due to the acids (H . X) from the sample anions.

4.3.4.5 Sulphur dioxide monitor

A technique used to measure the concentration of sulphur dioxide in air in the parts per hundred million (pphm) range is based on the measurement of the change in the conductivity of a reagent before and after it has absorbed sulphur dioxide. The principle of the measurement is to absorb the sulphur dioxide in hydrogen peroxide solution, thus forming sulphuric acid which increases the electric conductivity of the absorbing reagent.

Continuous measurements can be made by passing air upwards through an absorption column down which the hydrogen peroxide absorbing solution is

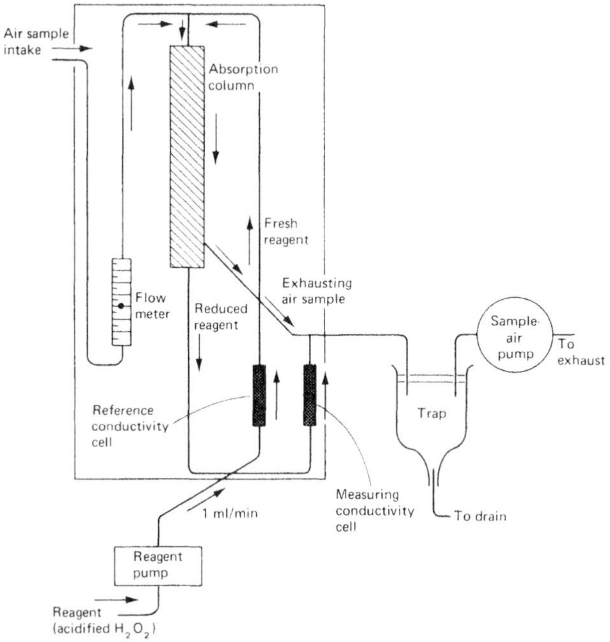

Figure 4.12 Continuous sulphur dioxide monitor.

flowing. Provided flow rates of air and hydrogen peroxide reagent are maintained constant, the sulphur dioxide concentration is proportional to the measured conductivity of the hydrogen peroxide reagent. Figure 4.12 is a diagram of suitable apparatus.

4.3.4.6 Salt-in-crude-oil monitor

A rapid continuous measurement of the salt in crude oil before and after desalting is based on the measurement of the conductivity of a solution to which a known quantity of crude oil has been added. The sample of crude oil is continuously circulated through a loop in the measurement section of the 'salt-in-crude monitor'. When the test cycle is initiated, solvent (xylene) is introduced from a metering cylinder into the analyser cell. A sample is then automatically diverted from the sample circulating loop into a metering cylinder calibrated to deliver a fixed quantity of crude oil into the analysis cell. A solution containing 63% n-butanol, 37% methanol, 0.25% water is then metered into the analysis cell from another calibrated cylinder.

The cell contents are thoroughly mixed by a magnetic stirrer, then the measuring circuit is energized and an a.c. potential is applied between two electrodes immersed in the liquid. The resulting a.c. current is displayed on a milliammeter in the electrical control assembly, and a proportional d.c. millivolt signal is transmitted from the meter to a suitable recorder.

At the end of the measuring period, a solenoid valve is opened automatically to drain the contents of the measuring cell to waste. The minimum cycle time is about 10 minutes.

Provision is made to introduce a standard sample at will to check the calibration of the instrument. Salt concentrations between 1 and 200 kg salt per 1000 m^3 crude oil can be measured with an accuracy of $\pm 5\%$ and a repeatability of 3% of the quantity being measured.

4.4 The concept of pH

4.4.1 General theory

Ionic concentrations were discussed in Section 4.2. The range of hydrogen ion concentrations met in practice is very wide; also when dealing with small concentrations it is inconvenient to specify hydrogen or hydroxyl concentrations. A method proposed by S. P. L. Sörenson in 1909 is now used universally – this is the concept of a hydrogen ion exponent or pH defined as:

$$\mathrm{pH} = -\log_{10}[\mathrm{H}^+] = \log_{10}\frac{1}{[\mathrm{H}^+]}$$

Thus pH is the logarithm to base 10 of the reciprocal of the hydrogen ion concentration. The advantage of this nomenclature is that all values of acidity and alkalinity between those of solutions molar with respect to hydrogen and hydroxyl ions can be expressed by a series of positive numbers between 0 and 14. Thus a neutral solution with $[\mathrm{H}^+] = 10^{-7}$ has a pH of 7. If the pH is less than 7 the solution is acid, if greater than 7, the solution is alkaline.

It must be realized that pH measuring devices measure the effective concentration, or activity, of the hydrogen ions and not the actual concentration. In very dilute solutions of electrolyte the activity and concentration are identical. As the concentration of electrolyte in solution increases above 0.1 mol/litre however, the measured value of pH becomes a less reliable measure of the concentration of hydrogen ions. In addition, as the concentration of a solution increases the degree of dissociation of the electrolyte decreases.

A dilute solution of sulphuric acid is completely dissociated and the assumption that pH = $-\log 2(H_2SO_4)$ is justified. (The 2 occurs because each molecule of acid provides two hydrogen ions.) Anhydrous sulphuric acid is only slightly dissociated, the degree of dissociation rising as the pure acid is diluted.

A maximum hydrogen ion concentration occurs in the neighbourhood of 92% H_2SO_4, but, at this concentration, the difference between actual hydrogen ion concentration and the activity of the hydrogen ions is large, and the measured pH minimum of about -1.4

occurs at a much lower sulphuric acid content.

A more reliable indication of the ionic behaviour of a solution will be obtained if we define pH in terms of the hydrogen ion activity aH^+ so that

$$pH = \log_{10}(1/aH^+) = -\log_{10} aH^+$$

where aH is related to the hydrogen ion concentration cH^+ by the equation

$$aH^+ = fH^+cH^+$$

where fH^+ is the activity coefficient, see Section 4.1. The pH values of common acids, bases and salts are given in Table 4.8.

Table 4.8 pH values of common acids, bases and salts

Compound	*Molarity*	*pH*
Acid benzoic	(Saturated)	2.8
Acid boric	0.1	5.3
Acid citric	0.1	2.1
Acid citric	0.01	2.6
Acid hydrochloric	0.1	1.1
Acid oxalic	0.1	1.3
Acid salicylic	(Saturated)	2.4
Acid succinic	0.1	2.7
Acid tartaric	0.1	2.0
Ammonia, aqueous	0.1	11.3
Ammonium alum	0.05	4.6
Ammonium chloride	0.1	4.6
Ammonium oxalate	0.1	6.4
Ammonium phosphate, primary	0.1	4.0
Ammonium phosphate, secondary	0.1	7.9
Ammonium sulphate	0.1	5.5
Borax	0.1	9.2
Calcium hydroxide	(Saturated)	12.4
Potassium acetate	0.1	9.7
Potassium alum	0.1	4.2
Potassium bicarbonate	0.1	8.2
Potassium carbonate	0.1	11.5
Potassium dihydrogen citrate	0.1	3.7
Potassium dihydrogen citrate	0.02	3.8
Potassium hydrogen oxalate	0.1	2.7
Potassium phosphate, primary	0.1	4.5
Sodium acetate	0.1	8.9
Sodium benzoate	0.1	8.0
Sodium bicarbonate	0.1	8.3
Sodium bisulphate	0.1	1.4
Sodium carbonate	0.1	11.5
Sodium carbonate	0.01	11.0
Sodium hydroxide	0.1	12.9
Sodium phosphate, primary	0.1	4.5
Sodium phosphate, secondary	0.1	9.2
Sodium phosphate, tertiary	0.01	11.7
Sulphamic acid	0.01	2.1

4.4.2 Practical specification of a pH scale

As the value of pH defined as $-\log_{10}$ (hydrogen ion activity) is extremely difficult to measure it is necessary to ensure that, when different workers state a pH value they mean the same thing. An operational definition of pH has been adopted in British Standard 1647:1961. The e.m.f. E_X of the cell

Pt H_2/soln. X/conc. KCl soln./ref. electrode

is measured and likewise the e.m.f. E_S of the cell

Pt H_2/soln. S/conc. KCl soln./ref. electrode

both cells being at the same temperature throughout and the reference electrodes and bridge solutions being identical in the two cells.

The pH of the solution X denoted by pH(X) is then related to the pH of the solution S denoted by pH(S) by the definition:

$$pH(X) - pH(S) = (E_X - E_S)/(2.3026\,RT/F)$$

where R is the gas constant, T is temperature in kelvins, and F is the Faraday constant. Thus defined, pH is a pure number.

To a good approximation, the hydrogen electrodes in both cells may be replaced by other hydrogen-responsive electrodes, e.g. glass or quinhydrone. The two bridge solutions may be of any molarity not less than 3.5 mol/kg provided they are the same.

4.4.3 pH standards

The difference between the pH of two solutions having been defined as above, the definition of pH can be completed by assigning at each temperature a value of pH to one or more chosen solutions designated as standards. In BS 1647 the chosen primary standard is a solution of pure potassium hydrogen phthalate having a concentration of 0.05 mol/litre.

This solution is defined as having a pH value of 4.000 at 15 °C and the following values at other temperatures between 0 and 95 °C:

Between 0 and 55 °C

$$pH = 4.000 + 1/2[(t-15)^2/100]$$

Between 55 and 95 °C

$$pH = 4.000 + 1/2[(t-15)^2/100] - (t-55)/500$$

Other standard buffer solutions are given on p. 120.

The e.m.f. E_X is measured and likewise the e.m.f. E_1 and E_2 of similar cells with solution X replaced by standard solutions S_1 and S_2, so that E_1 and E_2 are on either side of and as near as possible to E_X. The pH of the solution X is then obtained by assuming linearity between pH and E, i.e.

$$(pHX - pH\,S_1)/(pH\,S_2 - pH\,S_1) = (E_X - E_1)/(E_2 - E_1)$$

4.4.4 Neutralization

When acid and base solutions are mixed, they combine to form a salt and water, e.g.

hydrochloric acid		sodium hydroxide		sodium chloride		water HOH
H^+Cl^-	+	Na^+OH^-	=	Na^+Cl^-	+	(largely
(dissociated)		(dissociated)		(dissociated)		undissociated)

Thus, if equal volume of equally dilute solutions of strong acid and strong alkali are mixed, they yield neither an excess of H^+ ions nor of OH^- ions and the resultant solution is said to be neutral. The pH value of such a solution will be 7.

4.4.5 Hydrolysis

Equivalent amounts of acid and base when mixed will produce a neutral solution only when the acids and bases used are strong electrolytes. When a weak acid or base is used, hydrolysis occurs. When a salt such as sodium acetate, formed by a weak acid and a strong base, is present in water, the solution is slightly alkaline because some of the H^+ ions from the water are combined with acetic radicals in the relatively undissociated acetic acid, leaving an excess of OH^- ions, thus:

odium acetate + water→acetic acid + sodium hydroxide

Na^+Ac^- + HOH→ HAc + Na^+OH^-

(dissociated) (largely undissociated) (dissociated)

The pH value of the solution will therefore be greater than 7. Experiment shows it to be 8.87 in 0.1 mol/litre solution at room temperature.

Similarly, ammonium chloride (NH_4Cl), the salt of a weak base and a strong acid, hydrolyses to form the relatively undissociated ammonium hydroxide (NH_4OH), leaving an excess of H^+ ions. The pH value of the solution will therefore be less than 7. Experiment shows it to be 5.13 at ordinary temperatures in a solution having a concentration of 0.1 mol/litre.

A neutralization process therefore does not always produce an exactly neutral solution when one mole of acid reacts with one mole of base.

4.4.6 Common ion effect

All organic acids and the majority of inorganic acids are weak electrolytes and are only partially dissociated when dissolved in water. Acetic acid, for example, ionizes only slightly in solution, a process represented by the equation

$$HAc \rightleftharpoons H^+ + Ac^-$$

Its dissociation constant at 25 °C is only 1.8×10^{-5}, i.e.

$$([H^+][Ac^-])/[HAc] = 1.8 \times 10^{-5} \text{ mol/litre}$$

or

$$[H^+][Ac^-] = 1.8 \times 10^{-5}[HAc]$$

Therefore in a solution of acetic acid of moderate concentration, the bulk of the acid molecules will be undissociated, and the proportion present as acetic ions and hydrogen ions is small. If one of the salts of acetic acid, such as sodium (NaAc) is added to the acetic acid solution, the ionization of the acetic acid will be diminished. Salts are, with very few exceptions, largely ionized in solution, and consequently when sodium acetate is added to the solution of acetic acid the concentration of acetic ions is increased. If the above equation is to continue to hold, the reaction $H^+ + Ac^- \rightarrow HAc$ must take place, and the concentration of hydrogen ions is reduced and will become extremely small.

Most of the acetic ions from the acid will have recombined; consequently the concentration of un-ionized acid will be practically equal to the total concentration of the acid. In addition, the concentration of acetic ions in the equilibrium mixture due to the acid will be negligibly small, and the concentration of acetic ions will, therefore, be practically equal to that from the salt. The pH value of the solution may, therefore, be regulated by the strength of the acid and the ratio [salt]/[acid] over a wide range of values.

Just as the ionization of a weak acid is diminished by the addition of a salt of the acid, so the ionization of a weak base will be diminished by the addition of a salt of the base, e.g. addition of ammonium chloride to a solution of ammonium hydroxide. The concentration of hydroxyl ions in the mixture will be given by a similar relationship to that obtained for hydrogen ions in the mixture of acid and salt, i.e.

$$[OH^-] = K[\text{alkali}]/[\text{salt}]$$

4.4.7 Buffer solutions

Solutions of a weak acid and a salt of the acid such as acetic acid mixed with sodium acetate and solutions of a weak base and one of its salts, such as ammonium hydroxide mixed with ammonium chloride (as explained above in Section 4.4.6) undergo relatively little change of pH on the further addition of acid or alkali and the pH is almost unaltered on dilution. Such solutions are called buffer solutions; they find many applications in quantitative chemical analysis. For example, many precipitations are made in certain ranges of pH values and buffer solutions of different values are used for standardizing pH measuring equipment.

Buffer solutions with known pH values over a wide range can be prepared by varying the proportions of the constituents in a buffer solution; the value of the pH is given by

$$pH = \log_{10}\left(\frac{1}{K}\right) + \log_{10}\frac{[\text{salt}]}{[\text{acid}]}$$

The weak acids commonly used in buffer solutions include phosphoric, boric, acetic, phthalic, succinic and citric acids with the acid partially neutralized by alkali or the salt of the acid used directly. Their preparation requires the use of pure reagents and careful measurement and weighing but it is more important to achieve correct proportions of acid to salt than correct concentration. An error of 10 per cent in the volume of water present may be ignored in work correct to 0.02 pH units.

National Bureau of Standards (USA) standard buffer solutions have good characteristics and for pH 4, pH 7 and pH 9.2 are available commercially, as pre-weighed tablets, sachets of powder or in solution form. Those unobtainable commercially are simple to prepare provided analytical grade reagents are used dissolved in water with a specific conductance not exceeding 2 μS/cm.

4.5 Electrode potentials

4.5.1 General theory

When a metallic electrode is placed in a solution, a redistribution of electrical charges tends to take place. Positive ions of the metal enter the solution leaving the electrode negatively charged, and the solution will acquire a positive charge. If the solution already contains ions of the metal, there is a tendency for ions to be deposited on the electrode, giving it a positive charge. The electrode eventually reaches an equilibrium potential with respect to the solution, the magnitude and sign of the potential depending upon the concentration of metallic ions in the solution and the nature of the metal. Zinc has such a strong tendency to form ions that the metal forms ions in all solutions of its salts, so that it is always negatively charged relative to the solution. On the other hand, with copper, the ions have such a tendency to give up their charge that the metal becomes positively charged even when placed in the most dilute solution of copper salt.

This difference between the properties of zinc and copper is largely responsible for the e.m.f. of a Daniell cell (Figure 4.13). When the poles are connected by a wire, sudden differences of potential are possible (a) at the junction of the wires with the poles, (b) at the junction of the zinc with the zinc sulphate, (c) at the junction of the zinc sulphate with the copper sulphate, (d) at the junction of the copper with the copper sulphate. The e.m.f. of the cell will be the algebraic sum of these potential differences.

In the measurement of the electrode potential of a metal, a voltaic cell similar in principle to the Daniell cell is used. It can be represented by the scheme

Metal 1	Solution containing ions of metal 1	Solution containing ions of metal 2	Metal 2

Under ordinary conditions, when all the cell is at the same temperature, the thermoelectric e.m.f. at the junctions of wires and electrodes will vanish.

The potential difference which arises at the junction of the solutions, known as the 'liquid junction potential', or 'diffusion potential', is due to the difference in rate of diffusion across the junction of the liquids of the cations and anions. If the cations have a greater rate of diffusion than the anions then the solution into which the cations are diffusing will acquire a positive charge, and the solution which the cations are leaving will acquire a negative charge. Therefore there is a potential gradient across the boundary. If the anions have the greater velocity, the direction of the potential gradient will be reversed. The potential difference at the junction of the two liquids may be reduced to a negligible value either by having present in the two solutions relatively large and equal concentrations of an electrolyte, such as potassium nitrate, which produces ions which diffuse with approximately equal velocities or by inserting between

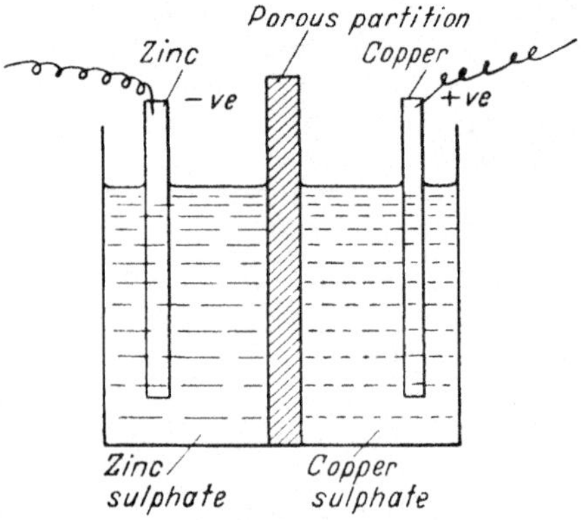

Figure 4.13 Daniell cell.

the two solutions a 'salt bridge' consisting of a saturated solution of potassium chloride or of ammonium or potassium nitrate. These salts produce ions whose diffusion rates are approximately equal.

When salt bridges are used in pH work, the liquid junction potentials are reduced to less than 1 mV unless strong acids or alkalis are involved. If an excess of neutral salt is added to the acid or alkali, the liquid junction potential will be reduced. Thus the error involved is rarely measurable on industrial instruments.

All measurements of the e.m.f. of cells give the potential of one electrode with respect to another. In the Daniell cell, all that can be said is that the copper electrode is 1 volt positive with respect to the zinc electrode. It is not possible to measure the potential of a single electrode as it is impossible to make a second contact with the solution without introducing a second metal–solution interface. Practical measurement always yields a difference between two individual electrode potentials.

In order to assign particular values to the various electrode potentials an arbitrary zero is adopted; all electrode potentials are measured relative to that of a standard hydrogen electrode (potential taken as zero at all temperatures). By convention, the half cell reaction is written as a reduction and the potential designated positive if the reduction proceeds spontaneously with respect to the standard hydrogen electrode, otherwise the potential is negative.

The standard hydrogen electrode consists of a platinum electrode coated with platinum black, half immersed in a solution of hydrogen ions at unit activity (1.228 M HCl at 20 °C) and half in pure hydrogen gas at one atmosphere pressure. In practice, however, it is neither easy nor convenient to set up a hydrogen electrode, so subsidiary reference electrodes are used, the potential of which relative to the standard hydrogen electrode has previously been accurately determined. Practical considerations limit the choice to electrodes consisting of a metal in contact with a solution which is saturated with a sparingly soluble salt of the metal and which also contains an additional salt with a common anion. Examples of these are the silver/silver chloride electrode ($Ag/AgCl_{(s)}KCl$) and the mercury/mercurous chloride electrode ($Hg/Hg_2Cl_{2s}KCl$) known as the calomel electrode. In each case the potential of the reference electrode is governed by the activity of the anion in the solution, which can be shown to be constant at a given temperature.

4.5.2 Variation of electrode potential with ion activity (the Nernst equation)

The most common measurement of electrode potential is in the measurement of pH, i.e. hydrogen ion activity, and selective ion activity, p(ion). The circuit involved is as shown in Figure 4.14.

The measured potential is the algebraic sum of the potentials developed within the system, i.e.

$$E = E_{\text{Int.ref.}} + E_s + E_j - E_{\text{Ext.ref.}}$$

where $E_{\text{Int.ref.}}$ is the e.m.f. generated at the internal reference inside the measuring electrode, E_s is the e.m.f. generated at the selective membrane, E_j is the e.m.f. generated at the liquid junction, and $E_{\text{Ext.ref.}}$ is the e.m.f. generated at the external reference electrode.

At a fixed temperature, with the reference electrode potentials constant and the liquid junction potentials zero the equation reduces to

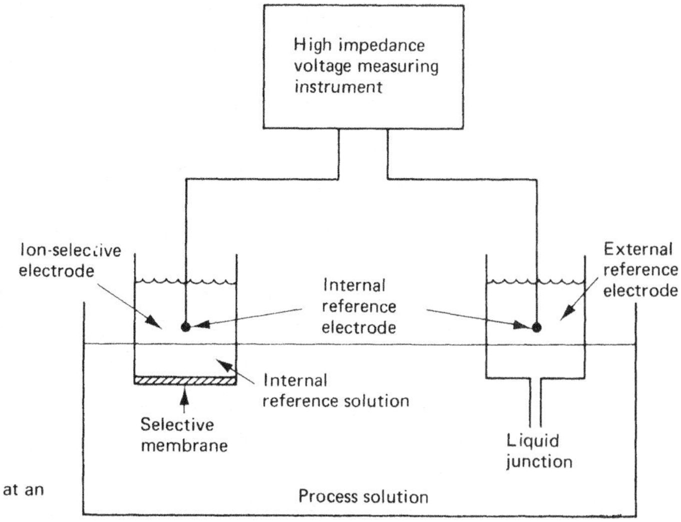

Figure 4.14 Method of measuring potential developed at an ion-selective membrane.

$$E = E' + E_s$$

where E' is a constant.

The electrode potential generated is related to the activities of the reactants and products that are involved in the electrode reactions.

For a general half cell reaction

oxidized form + n electrons → reduced form

or

$$aA + bB + \ldots + ne^- \rightarrow xX + yY + \ldots$$

the electrode potential generated can be expressed by the Nernst equation

$$E = E_0 + \frac{RT}{nF} \ln \frac{\text{OXID}}{\text{RED}} \text{ volts}$$

or

$$E = E_0 + 2.303 \frac{RT}{nF} \log_{10} \frac{[A]^a . [B]^b}{[X]^x . [Y]^y} \text{ volts}$$

where R is the molar gas constant (8.314 joule . mol^{-1} K^{-1}), T is absolute temperature in kelvins, F is the Faraday constant (96 487 coulomb . mol^{-1}), and n is the number of electrons participating in the reaction according to the equation defining the half cell reaction. The value of the term $2.303RT/nF$ is dependent upon the variables n and T and reduces to $0.059/n$ volts at 25 °C and $0.058/n$ volts at 20 °C.

An ion-selective electrode (say selective to sodium ions) is usually constructed so that the ion activity of the internal reference solution inside the electrode is constant and the Nernst equation reduces at constant temperature to

$$E = E_0 + \frac{RT}{nF} \ln a$$

where E_0 includes all the constants and a is the activity of the sodium ion. As sodium is a positive ion with one charge

$$E = E_0 + 59.16 \log_{10}(a) \text{ mV at } 25\,°\text{C}$$

This equation shows that a tenfold increase in ion activity will increase the electrode potential by 59.16 mV.

If the ion being measured is doubly charged the equation becomes

$$E = E_0 + \frac{59.16}{2} \log_{10}(a) \text{ mV at } 25\,°\text{C}$$

The applicability of these equations assume that the ion-selective electrode is sensitive uniquely to one ion. In most cases in practice, the electrode will respond to other ions as well but at a lower sensitivity. The equation for electrode potential thus becomes:

$$E = E_0 + 59.16 \log_{10}(a_1 + K_2 a_2 + \ldots) \text{ mV}$$

where $K_2 a_2$ etc. represents the ratio of the sensitivity of the electrode of the ion 2 to that of ion 1. The literature on ion-selective electrodes provided by manufacturers usually gives a list of interfering ions and their sensitivity ratios.

4.6 Ion-selective electrodes

Whereas, formerly, ion-selective electrodes were used almost exclusively for measuring hydrogen ion activity (pH), many electrodes have now been developed to respond to a wide range of selected ions. These electrodes are classified into five groups according to the type of membrane used.

4.6.1 Glass electrodes

The glass electrode (Figure 4.15(a)) used for pH measurement is designed to be selective to hydrogen ions but, by choosing the composition of the glass membrane, glass electrodes selective to sodium, potassium, ammonium, silver and other univalent cations can be made.

4.6.2 Solid state electrodes

In these electrodes the membrane consists of a single crystal or a compacted disc of the active material. In Figure 4.15(b) the membrane isolates the reference solution from the solution being measured. In Figure 4.15(c) the membrane is sealed with a metal backing with a solid metal connection. A solid state electrode selective to fluoride ions employs a membrane of lanthanum fluoride (LaF_3). One which is selective to sulphide ions has a membrane of silver sulphide. There are also electrodes available for measurement of Cl^-, Br^-, I^-, Ag^+, Cu^{2+}, Pb^{2+}, Cd^{2+}, and CN^- ions.

4.6.3 Heterogeneous membrane electrodes

These are similar to the solid state electrodes but differ in having the active material dispersed in an inert matrix. Electrodes in this class are available for Cl^-, Br^-, I^-, S^{2-}, and Ag^+ ions.

4.6.4 Liquid ion exchange electrodes

In this type of electrode (Figure 4.15(d)) the internal reference solution and the measured solution are separated by a porous layer containing an organic liquid of low water solubility. Dissolved in the organic phase are large molecules in which the ions of interest are incorporated. The most important of these electrodes is the calcium electrode, but other electrodes in this class are available for the determination of Cl^-, ClO_4^-, NO_3^-, Cu^{2+}, Pb^{2+}, and BF_4^- ions. The liquid

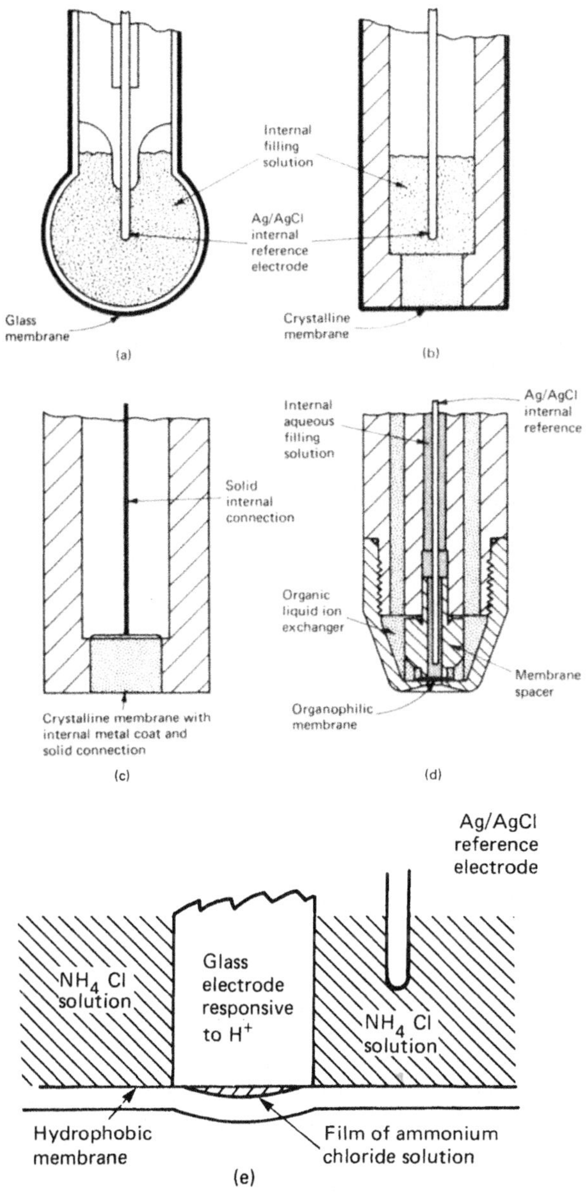

Figure 4.15 Ion-selective electrodes: (a) glass, (b) crystalline membrane with internal reference electrode, (c) crystalline membrane with solid connection, (d) liquid ion exchange, (e) gas sensing membrane (courtesy Orion Research Inc.).

ion exchange electrodes have more restricting chemical and physical limitations than the glass or solid state electrodes but they may be used to measure ions which cannot yet be measured with a solid state electrode.

4.6.5 Gas-sensing membrane electrodes

These electrodes are not true membrane electrodes as no current passes across the membrane. They are complete electrochemical cells, monitored by an ion-selective electrode as the internal chemistry is changed by the ion being determined passing from the sample solution across the membrane to the inside of the cell.

An example is an ammonia electrode (Figure 4.15(e)). The sensing surface of a flat-ended glass pH electrode is pressed tightly against a hydrophobic polymer membrane which is acting as a seal for the end of a tube containing ammonium chloride solution. A silver/silver chloride electrode is immersed in the bulk solution. The membrane permits the diffusion of free ammonia (NH_3), but not ions, between the sample solution and the film of ammonium chloride solution. The introduction of free ammonia changes the pH of the internal ammonium chloride solution which is sensed by the internal glass pH electrode.

4.6.6 Redox electrodes

In elementary chemistry a substance is said to be oxidized when oxygen is combined with it and said to be reduced when oxygen is removed from it. The definition of oxidation and reduction may, however, be extended. Certain elements, e.g. iron and tin, can exist as salts in more than one form. Iron, for example can be combined with sulphuric acid in the form of ferrous iron, valency 2, or ferric iron, valency 3.

Consider the reaction:

ferrous				ferric		ferric
sulphate	+	chlorine	=	chloride	+	sulphate
$6FeSO_4$	+	$3Cl_2$	=	$2FeCl_3$	+	$2Fe_2(SO_4)_3$

The ferrous sulphate is oxidized to ferric sulphate; chlorine is the oxidizing agent. In terms of the ionic theory, the equation may be written

$$6Fe^{2+} + 3Cl_2 = 6Fe^{3+} + 6Cl^-$$

i.e. each ferrous ion loses an electron and so gains one positive charge. When a ferrous salt is oxidized to a ferric salt each mole of ferrous ions gains one mole (1 faraday) of positive charges or loses one mole of negative charges, the negative charge so lost being taken up by the oxidizing agent (chlorine). Oxidation, therefore, involves the loss of electrons; reduction, the gain of electrons. Thus the oxidation of a ferrous ion to ferric ion can be represented by the equation

$$Fe^{2+} - e = Fe^{3+}$$

When a suitable electrode, such as an inert metal which is not attacked by the solution and which will not catalyse side reactions, is immersed in a solution containing both ferrous and ferric ions, or some other substance in the reduced and oxidized state, the electrode acquires a potential which will depend upon the tendency of the ions in the solution to pass from a higher or lower state of oxidation. If the ions in solution tend to become oxidized (i.e. the solution has reducing properties) the ions tend to give up electrons to the electrode which will become negatively charged relative to the solution. If, on the other hand, the ions in solution tend to become reduced (i.e. the solution has oxidizing properties), then the ions will tend to take up electrons from the electrode and the electrode will become positively charged relative to the solution. The sign and magnitude of the electrode potential, therefore, gives a measure of the oxidizing or reducing power of the solution, and the potential is called the oxidation–reduction or redox potential of the solution, E_h. The potential E_h may be expressed mathematically by the relationship

$$E_h = E_0 + (RT/nF)\log_{10}(a_o/a_r)$$

where a_o is the activity of the oxidized ion and a_r is the activity of the reduced ion.

To measure the oxidation potential it is necessary to use a reference electrode to complete the electrical circuit. A calomel electrode is often used for this (see Section 4.7 below).

The measuring electrode is usually either platinum or gold, but other types are used for special measurements: as, for example, the hydrogen electrode for use as a primary standard and the quinhydrone electrode for determining the pH of hydrofluoric acid solutions. However, the latter two electrodes do not find much application in industrial analytical chemistry.

4.7 Potentiometry and specific ion measurement

4.7.1 Reference electrodes

All electrode potential measurements are made relative to a reference electrode and the e.m.f. generated at this second contact with the solution being tested must be constant. It should also be independent of temperature changes (or vary in a known manner), be independent of the pH of the solution and remain stable over long periods.

Standard hydrogen electrodes are inconvenient (p. 130) and in practice three types of reference are commonly used.

Silver/silver chloride electrode This consists of a silver wire or plate, coated with silver chloride, in contact with a salt bridge of potassium chloride saturated with silver chloride. The concentration of the potassium chloride may vary from one type of electrode to another but concentrations of 1.00 or 4.00 mol per litre or a saturated solution are quite common. This saturated type of electrode has a potential of −0.199 V relative to a hydrogen electrode. It has a variety of physical forms which are discussed below.

Mercury/mercurous chloride or calomel electrode The metal used is mercury which has a high resistance to corrosion and being fluid at ambient temperature cannot be subject to strain. The mercury is in contact with either mercurous chloride or in some electrodes with mercurous chloride and potassium chloride paste. Contact with the measured solution is through a salt bridge of potassium chloride whose concentration may be 3.8 mol per litre or some other concentration appropriate to the application. Contact with the mercury is usually made by means of a platinum wire which may be amalgamated. The calomel, saturated potassium chloride, electrode has a potential relative to the hydrogen electrode of −0.244 V.

Where the use of potassium salt is precluded by the condition of use, it may be replaced by sodium sulphate, the bridge solution having a concentration of 1 mol per litre.

Whatever the type of the reference electrode, contact must be made between the salt bridge and the measured solution. Two common methods are through a ceramic plug whose shape and porosity govern the rate at which the salt bridge solution diffuses out and the process solution diffuses into and contaminates the bridge solution. If the plug is arranged to have a small cross-sectional area relative to its length, the rate of diffusion is very small (say less than 0.02 cm^3/day) and the electrode can be considered to be sealed and is used until it becomes unserviceable. It is then replaced by a similar electrode.

Where the application warrants it a high rate of diffusion from the electrode has to be tolerated (say 1 or 2 cm^3/day), so the relative dimensions and porosity of the plug are changed, or it is replaced by a glass sleeve which permits relatively fast flow of salt bridge solution, thus reducing the rate and degree of fouling of the junction. In these circumstances, the electrode is refilled on a routine basis, or a continuous supply of bridge solution is arranged into the electrode at the appropriate pressure for the application.

A wide range of electrodes are illustrated in Figures 4.16–4.19. The choice of the appropriate reference electrode for the application is vital, and consideration must be given to the pressure, temperature and nature of the process stream. The accuracy of the measurement and the frequency of maintenance depends upon the correct choice of electrode. The e.m.f. of the reference electrode will only remain constant provided satisfactory contact is made by the salt bridge, so the junction must not become plugged by suspended solids, viscous liquids, or reaction products of the

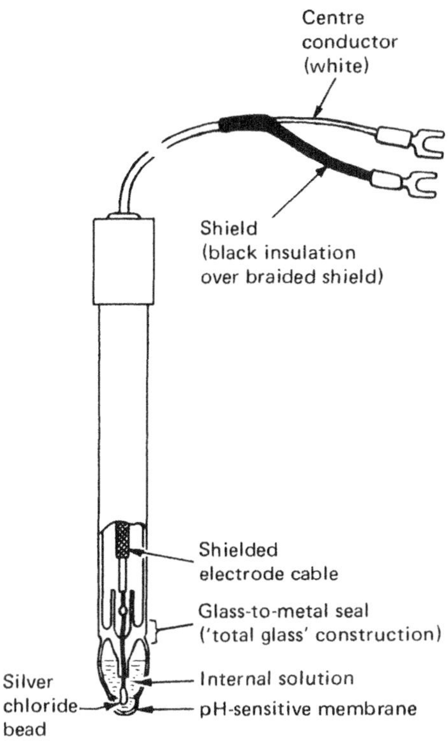

Figure 4.16 pH measuring electrode (courtesy The Foxboro Company).

process stream. Where this is a danger, the faster flow type of plug must be used. Many routine measurements can, however, be made with the non-flowing electrode, thus avoiding the necessity of refilling, or arranging a pressurized continuous supply. Flowing types of junctions are usually required where an accuracy of ± 0.02 pH units (± 1 or 2 mV) is required, where frequent or large temperature or composition changes occur, or where the process fluid is such that it is prone to foul the junction.

The temperature of operation will influence the choice of concentration of the filling solutions. Potassium chloride solution having a concentration of 4 mol per litre saturates and starts to precipitate solids at about 19 °C, and will freeze at -4 °C, while if the concentration is reduced to 1 mol per litre the solution will freeze at -2 °C without becoming saturated. Thus, no precipitation will take place in the solution of lower concentration. Although not damaging, precipitated potassium chloride and associated silver chloride will tend to clog reference junctions and tubes, decreasing electrolyte flow rate, and increasing the risk of spurious potentials. For these reasons, flowing reference electrodes are not recommended for low temperature applications unless provision is made to prevent freezing or precipitation in the electrode and any associated hardware.

When materials such as sulphides, alkali phosphates or carbonates, which will react with silver, are present in the process stream, either non-flowing electrodes, or electrodes containing potassium chloride at 1 mol per litre should be used. The diffusion rate of silver can be neglected in the non-flowing type, and the solubility of silver chloride in potassium chloride at a concentration of 1 mol per litre is only 1 or 2 per cent of that in a solution at 4 mol per litre.

High temperatures with wide fluctuations are best handled by potassium chloride solution at 1 mol per litre.

4.7.2 Measurement of pH

Glass electrode Almost all pH measurements are best made with a glass electrode (the earliest of the ion-selective electrodes), the e.m.f. being measured relative to a reference electrode. The glass electrode can be made to cover practically the whole of the pH scale and is unaffected by most chemicals except hydrofluoric acid. It can also be used in the presence of oxidizing or reducing agents without loss of measuring accuracy.

The electrode consists of a thin membrane of

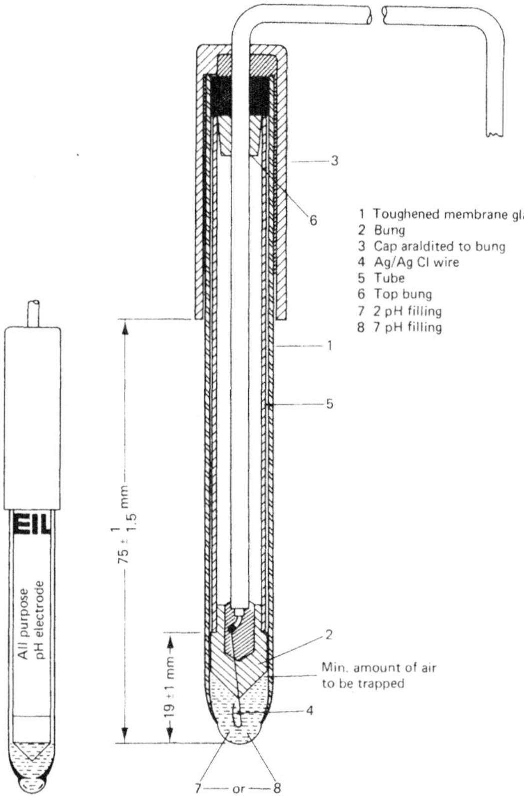

Figure 4.17 Industrial electrode (courtesy Kent Industrial Measurements Ltd., Analytical Instruments).

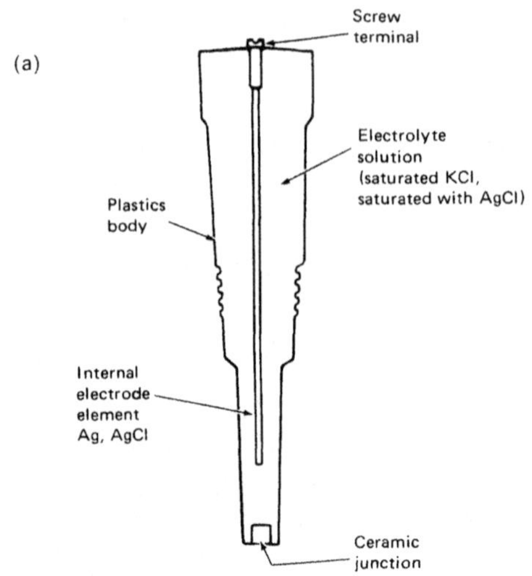

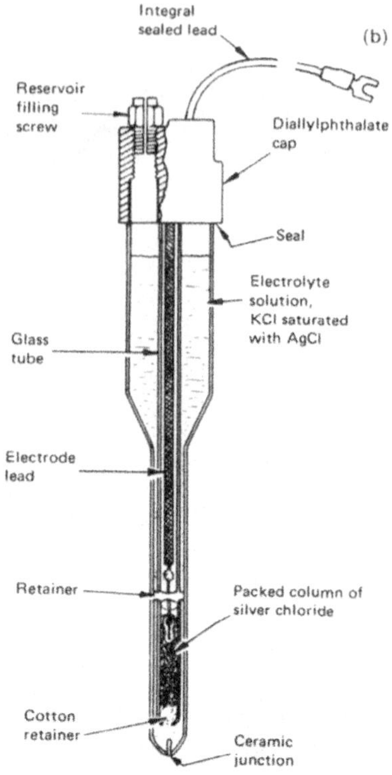

Figure 4.18 Reference electrodes (courtesy The Foxboro Company): (a) sealed electrode, (b) flowing type.

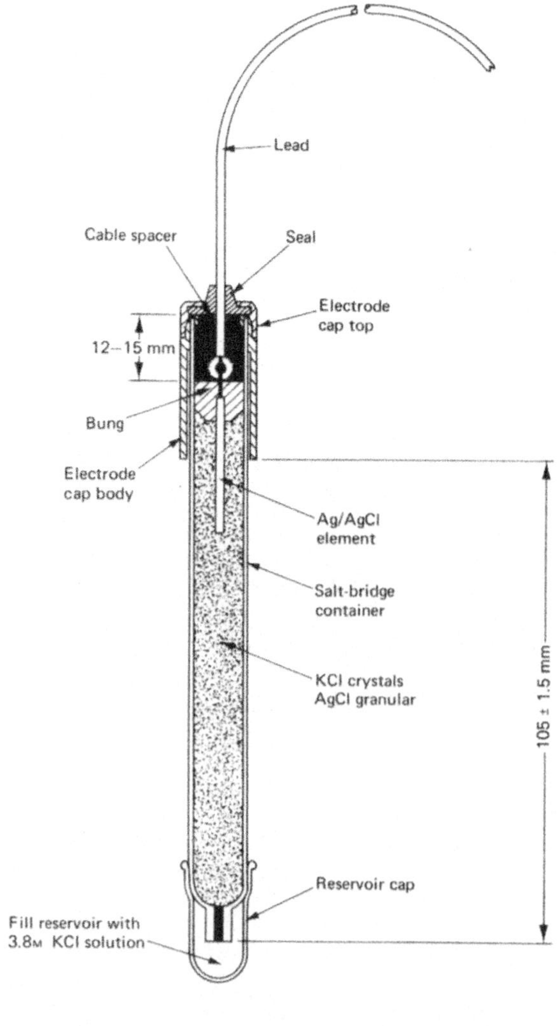

sodium-ion-selective glass sealed onto the end of a glass tube that has no ion-selective properties. The tube contains an internal reference solution in which is immersed the internal reference electrode and this is connected by a screened lead to the pH meter. The internal reference electrode is almost always a silver/silver chloride electrode although, recently, Thalamid electrodes* have sometimes been used. The internal reference solution contains chloride ions to which the internal silver/silver chloride reference electrode responds and hydrogen ions to which the electrode as a whole responds. The ion to which the glass electrode responds, hydrogen in the case of pH electrodes, is determined by the composition of the glass membrane.

* The Thalamid electrode is a metal in contact with a saturated solution of the metallic chloride. Thallium is present as a 40 per cent amalgam and the surface is covered with solid thallous chloride. The electrode is immersed in saturated potassium chloride solution. Oxygen access is restricted to prevent the amalgam being attacked. The advantage of the Thalamid electrode is that there is scarcely any time lag in resuming its electrode potential after a temperature change.

Figure 4.19 Sealed silver/silver chloride electrode (courtesy Kent Industrial Measurements Ltd., Analytical Instruments).

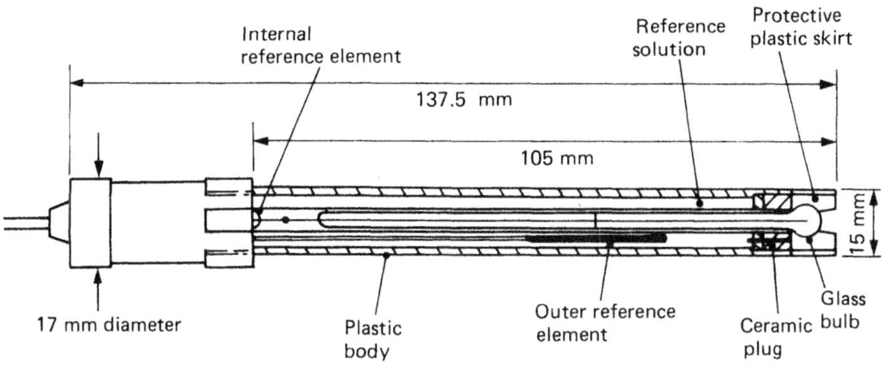

Figure 4.20 Combined reference electrode and glass electrode for pH measurement (courtesy Kent Industrial Measurements Ltd., Analytical Instruments).

A glass pH electrode can be represented as

reference electrode	test solution $^{a}H^{+}$	glass membrane	internal reference solution $^{a'}H^{+a'}Cl$	AgCl	Ag

Glass electrodes for pH measurement are of three main types (a) general-purpose for wide ranges of pH over wide ranges of temperature, (b) low temperature electrodes (less than 10 °C) which are low resistance electrodes and are generally unsuitable for use above pH 9 to 10, (c) high pH and/or high temperature electrodes (greater than 12 pH units). Glass electrodes are manufactured in many forms some of which are shown in Figures 4.16 and 4.20. Spherical membranes are common but hemispherical or conical membranes are available to give increased robustness where extensive handling is likely to occur. Electrodes with flat membranes can be made for special purposes such as measurement of the pH of skin or leather and micro-electrodes are available but at great expense. Combination glass and reference electrodes (Figure 4.20) can be obtained and some electrodes can be steam-sterilized.

New electrodes supplied dry should be conditioned before use as the manufacturer recommends or by leaving them overnight in 0.1 mol/litre^{-1} hydrochloric acid. Electrodes are best not allowed to dry out and they should be stored in distilled or demineralized water at temperatures close to those at which they are to be used. The best treatment for pH electrodes for high pH ranges is probably to condition them and store them in borax buffer solution.

Electrical circuits for use with glass electrodes For measurement of pH the e.m.f. in millivolts generated by the glass electrode compared with that of the reference electrode has to be converted to a pH scale, that is, one showing an increase of one unit for a decrease in e.m.f. of approximately 60 mV. The pH scale requires the use of two controls – the calibration control and the slope control. The latter may not always be identified as such on the pH meter as it often acts in the same way as the temperature compensation control. The slope and temperature compensation controls adjust the number of millivolts equivalent to one pH unit. The calibration control relates the measured e.m.f. to a fixed point on the pH scale.

A typical pH measuring system (glass electrode and reference electrode immersed in a solution) may have a resistance of several hundred megohms. To obtain an accurate measurement of the e.m.f. developed at the measuring electrode, the electrical measuring circuit must have a high input impedance and the insulation resistance of the electrical leads from the electrodes to the measuring circuit must be extremely high ($\sim 10^5$ MΩ – a 'Megger' test is useless). The latter is best achieved by keeping the electrode leads as short as possible and using the best moisture-resistant insulating materials available (e.g. polythene or silicone rubber).

The usual method of measurement is to convert the developed e.m.f. into a proportional current by means of a suitable amplifying system. The essential requirements of such a system have been met completely by modern electronic circuits and one system uses an amplifier with a very high negative feedback ratio. This means that the greater part of the input potential is balanced by a potential produced by passing the meter current through an accurately known resistor, as shown in Figure 4.21. If the p.d. V_0, developed across the feedback resistance is a very large fraction of the measured potential V_1, then the input voltage V is a very small fraction of V_1, and

$$I_0 = (V_1 - V)/R, \text{ approaches } V_1/R$$

With modern integrated circuit techniques it is possible to obtain an amplifier with a very high input impedance and very high gain, so that little or no current is drawn from the electrodes.

Such a system is employed in the pH-to-current converter shown in Figure 4.22 which employs zener

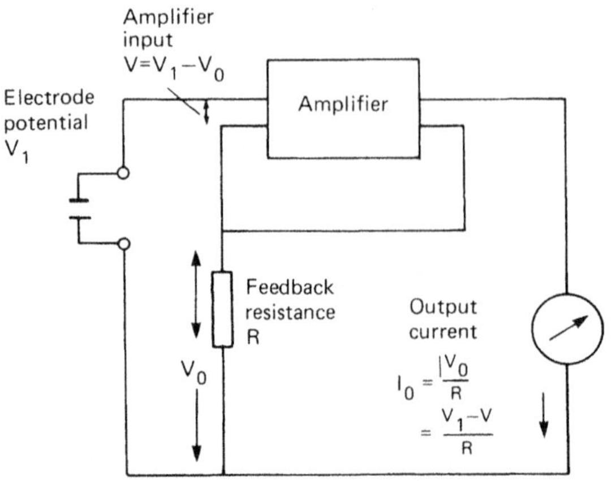

Figure 4.21 Principle of d.c. amplifier with negative feedback (courtesy Kent Industrial Measurements Ltd.).

diode stabilized supplies and feedback networks designed to give a high gain, high input impedance diode bridge amplifier.

The d.c. imbalance signal resulting from the pH signal, asymmetry correcting potential and the feedback voltage, changes the output of a capacity balance diode bridge. This output feeds a transistor amplifier which supplies feedback and output proportional to the bridge error signal. Zener diode stabilized and potentiometer circuits are used to provide continuous adjustment of span, elevation, and asymmetry potential over the entire operating range of the instrument.

The input impedance of the instrument is about $1 \times 10^{12}\,\Omega$ and the current taken from the electrodes less than 0.5×10^{-12} A.

The principle of another system which achieves a similar result is shown in Figure 4.23. It uses a matched

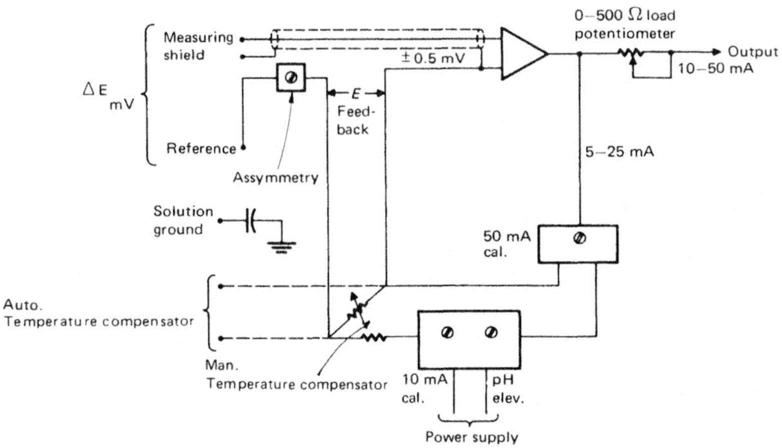

Figure 4.22 High gain, high impedance pH-to-current converter (courtesy The Foxboro Company).

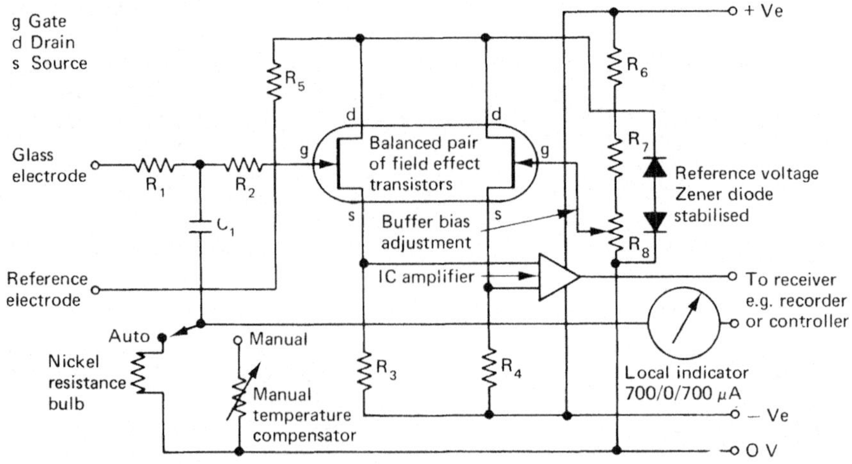

Figure 4.23 pH measuring circuit using field effect transistors.

pair of field effect transitors (FETs) housed in a single can. Here the e.m.f. produced by the measuring electrode is fed to the gate of one of the pair. The potential which is applied to one side of the high gain operational amplifier will be governed by the current which flows through the transistor and its corresponding resistance R_3. The potential applied to the gate of the second FET is set by the buffer bias adjustment which is fed from a zener stabilized potential supply. The potential developed across the second resistance R_4 which is equal in resistance to R_3 will be controlled by the current through the second of the pair of matched FETs. Thus the output of the operational amplifier will be controlled by the difference in the potentials applied to the gates of the FETs, that is, to the difference between the potential developed on the measuring electrode and the highly stable potential set up in the instrument. Thus, the current flowing through the local and remote indicators will be a measure of the change of potential of the measuring electrode.

If the e.m.f. given by the glass electrode is plotted against pH for different temperatures it will be seen that there is a particular value of the pH at which the e.m.f. is independent of temperature. This point is known as the 'iso-potential point'.

If the iso-potential point is arranged to be the locus of the slope of the measuring instrument, the pH measuring circuit can be modified to include a temperature sensor arranged to change the negative feed-back so that the circuit compensates for the change in slope of the e.m.f./pH relationship. It is important to realize that the temperature compensation only corrects for the change in the electrode response due to temperature change and the iso-potential control setting therefore enables pH electrodes calibrated at one temperature to be used at another. The iso-potential control does *not* compensate for the actual change in pH of a solution with temperature. Thus if pH is being measured to establish the composition of a solution one must carry out the measurements at constant temperature.

A few commercial pH meters have a variable iso-potential control so that they can be used with several different combinations of electrodes but it is more generally the case that pH meters have fixed iso-potential control settings and can only be used with certain combinations of pH and reference electrodes. It is strongly recommended that, with fixed iso-potential control settings, both the glass and reference electrodes be obtained from the manufacturer of the pH meter. Temperature compensation circuits generally work only on the pH and direct activity ranges of a pH meter and not on the millivolt, expanded millivolt and relative millivolt ranges.

Modern pH meters with analogue displays are scaled 0 to 14 pH units with the smallest division on the scale equivalent to 0.1 unit giving the possibility of estimating 0.02 pH units by interpolation. The millivolt scale is generally 0 to 1400 mV with a polarity switch, or -700 to $+700$ mV without one. The smallest division is 10 mV, allowing estimation to 2 mV. Many analogue meters have a facility of expanding the scale so that the precision of the reading can be increased up to 10 times. Digital outputs are also available with the most sensitive ones reading to 0.001 pH unit (unlikely to be meaningful in practice) or 0.1 mV. Instruments incorporating microprocessors are also now available – these can calculate the concentration of substances from pH measurements and give readout in concentration units. Blank and volume corrections can be applied automatically.

Precision and accuracy Measurements reproducible to 0.05 pH units are possible in well buffered solutions in the pH range 3 to 10. For routine measurements it is rarely possible to obtain a reproducibility of better than ± 0.01 pH units.

In poorly buffered solutions reproducibility may be no better than ± 0.1 pH unit and accuracy may be lost by the absorption of carbon dioxide or by the presence of suspensions, sols and gels. However, measured pH values can often be used as control parameters even when their absolute accuracies are in doubt.

Sodium ion error Glass electrodes for pH measurement are selective for hydrogen ions, not uniquely responsive to them, and so will also respond to sodium and other ions especially at alkaline pH values (more than about 11). This effect causes the pH value to be underestimated. Sodium ions produce the greatest error, lithium ions about a half, potassium ions about a fifth and other ions less than a tenth of the error due to sodium ions. One can either standardize the electrode in an alkaline buffer solution containing a suitable concentration of the appropriate salt or, better, use the special lithium and caesium glass electrodes developed for use in solutions of high alkalinity. These are less prone to interference. For a given glass electrode at a stated measuring temperature the magnitude of the error can be found from tables provided by electrode manufacturers. An example is shown in Figure 4.24.

Temperature errors The calibration slope and standard potential of ion-selective electrodes (including glass pH electrodes) are affected by temperature. If the pH is read directly off the pH scale, some form of temperature correction will be available, but often only for the calibration slope and not for the standard potential. If measurements are made at a temperature different from that at which the electrode was calibrated there will be an error. This will be small if the meter has an iso-potential setting. For the most accurate work the sample and buffer solutions should

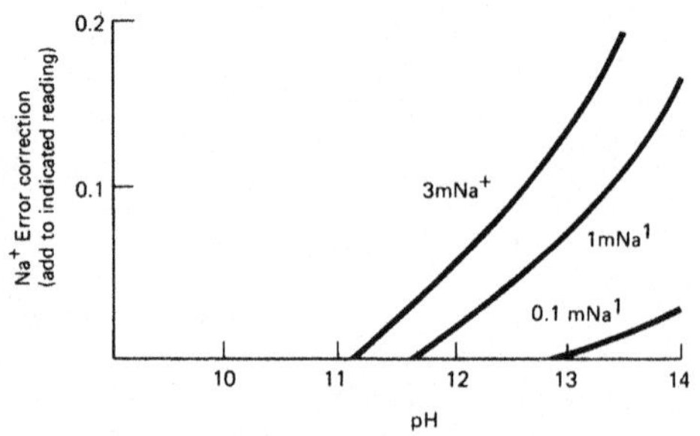

Figure 4.24 Relationship of pH and Na ion error (courtesy Kent Industrial Measurements Ltd., Analytical Instruments).

be at the same temperature, even if iso-potential correction is possible.

Stirring factor In well buffered solutions it may not be necessary to stir when making pH measurements. However, it is essential in poorly buffered solutions.

The hydrogen electrode The hydrogen electrode, consisting in practice of a platinum plate or wire coated with platinum block (a finely divided form of the metal) can measure hydrogen ion activity when hydrogen is passed over the electrode. However this electrode is neither easy nor convenient to use in practice and is now never used in industrial laboratories nor on plant.

The antimony electrode The antimony electrode is simply a piece of pure antimony rod (~12 mm diameter, 140 mm long), housed in a protective plastic body resistant to acid attack, see Figure 4.25. The protruding antimony rod when immersed in a solution containing dissolved oxygen becomes coated with antimony trioxide Sb_2O_3 and the equilibria governing the electrode potential are:

$$Sb \rightarrow Sb^{3+} + 3e^-$$

$$Sb_2O_3 + 6H^+ \rightarrow 2Sb^{3+} + 3H_2O, \quad K = \frac{[Sb^{3+}]}{[H^+]^3}$$

However, there are many possible side reactions depending on the pH and the oxidizing conditions; salt effects are large. There is therefore difficulty in calibrating with buffer solutions; stirring temperature and the amount of oxygen present all have rather large effects. A reproducibility of about 0.1 pH unit is the best that is normally attained, the response is close to Nernstian over the pH range 2 to 7 and the response time can be as short as 3 minutes but is often about 30 minutes.

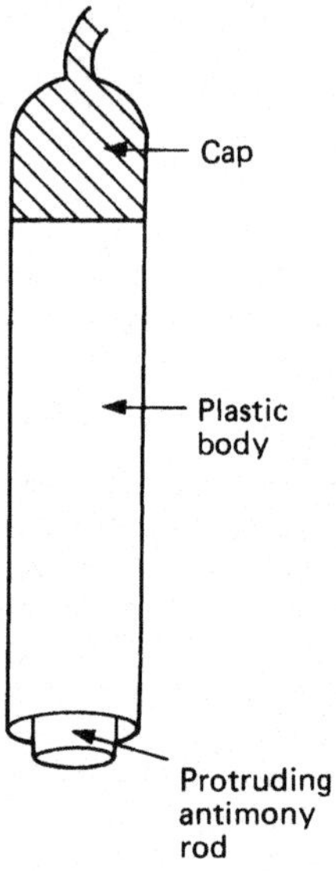

Figure 4.25 Antimony electrode.

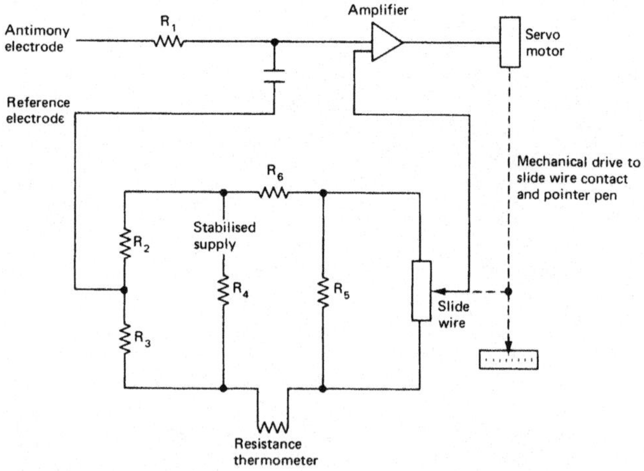

Figure 4.26 Low impedance measuring circuit for use with antimony electrodes.

The outstanding advantage of the antimony electrode is its ruggedness and for this reason it has been used for determining the pH of soils. Also, of course, it is indispensable for solutions containing hydrofluoric acid which attack glass. If the electrode becomes coated during use its performance can be restored by grinding and polishing the active surface and then reforming the oxide film by immersion in oxygenated water before using in deoxygenated solutions.

However, there is much more uncertainty to every aspect of behaviour of the antimony electrode than with the glass electrode and even the fragile glass electrodes of years ago with their limited alkaline range displaced the antimony electrode when accurate pH measurements were required. Modern glass electrodes are excellent in respect of robustness and range and antimony electrodes are not much used apart from the specialized applications already mentioned. In these, the resistance of the measuring system is low, so a simple low impedance electrical circuit can be used with them, for example a voltmeter or a potentiometric type of system as described in Volume **0**. Figure 4.26 shows the principle of such a system. Any difference between the electrode e.m.f. and that produced across the potentiometer will be amplified and applied to the servo-motor which moves the slide-wire contact to restore balance.

Industrial pH systems with glass electrodes Two types of electrode systems are in common use: the continuous-flow type of assembly, and the immersion, or dip-type of assembly.

Continuous-flow type of assembly The physical form of the assembly may vary a little from one manufacturer to another but Figure 4.27 illustrates a typical assembly designed with reliability and easy maintenance in mind. Constructed in rigid PVC throughout, it operates at pressure up to 2 bar and temperatures up to 60 °C. For higher temperatures and pressures the assembly may be made from EN 58J stainless steel, flanged and designed for straight-through flow when pressures up to 3 bar at temperatures up to 100 °C can be tolerated. It accommodates the standard measuring electrode, usually of toughened glass.

A reservoir for potassium chloride (or other electrolyte) forms a permanent part of the electrode holder. A replaceable reference element fits into the top of the reservoir, and is held in place by an easily detachable clamp nut. A microceramic plug at the lower end of the reservoir ensures slow electrolyte leakage (up to six months continuous operation without attention is usually obtained). The ceramic junction is housed in a screw-fitting plug, and is easily replaceable.

The close grouping of electrodes makes possible a small flow cell, and hence a fast pH response at low flow rates. An oil-filled reservoir built into the electrode holder houses a replaceable nickel wire resistance element, which serves as a temperature compensator. (This is an optional fitment.)

The flow through the cell creates some degree of turbulence and thus minimizes electrode coating and sedimentation.

The integral junction box is completely weatherproof and easily detachable. Electrode cables and the output cable are taken via individual watertight compression fittings into the base of the junction box. A desiccator is included to absorb moisture which may be trapped when the cover is removed and replaced.

Two turns of the lower clamp nut allow the entire electrode unit to be detached from the flow cell and hence from the process fluid. The electrodes can be immersed easily in buffer solution.

Immersion type Basically this assembly is similar to the flow type except that the flow cell is replaced by a protecting guard which protects the electrode but allows a free flow of solution to the electrodes. Also the upper cap is replaced by a similarly moulded tube which supports the electrode assembly, but brings the terminal box well above the electrode assembly so that the terminals are clear of the liquid surface when the assembly is in the measured solution. Immersion depths up to 3 m are available.

Electrode assemblies should be designed so that the electrodes can be kept wet when not in use. It is often possible to arrange for the easy removal of the assembly from the process vessel so that it can be immersed in a bucket filled with process liquid, water or buffer solution during shut-down.

The design of the assembly is often modified, to suit the use. For example, in measuring the pH of pulp in a paper beater the electrodes and resistance bulb are mounted side by side in a straight line and then inclined downstream at about 45° from the vertical so that they present no pockets to collect pulp and are self-cleaning.

When the assembly is immersed in a tank, care must be taken in the siting to ensure the instrument is measuring the properties of a representative sample; adequate mixing of the process material is essential. Sometimes it is more convenient to circulate the contents of a tank through a flow type of assembly and then return the liquid to the tank.

The main cause of trouble in electrode assemblies is the fouling of the electrodes. In order to reduce this, two forms of self-cleaning are available and the choice of method is dependent on the application. Where the main cause of trouble is deposits on the glass electrode and mechanical cleaning is required, this may be achieved by the cleaning attachment shown on a dip

(a)

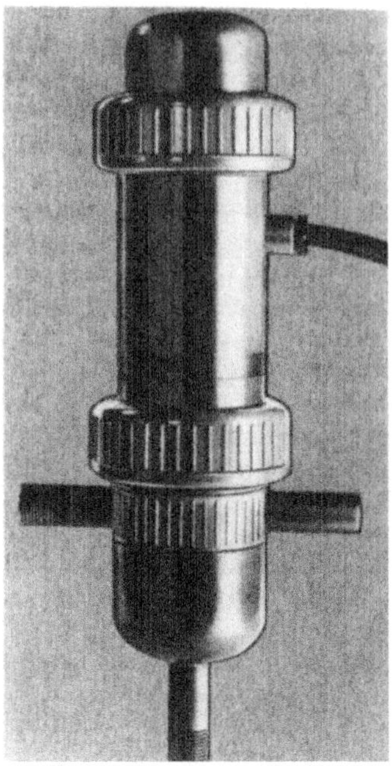

(b)

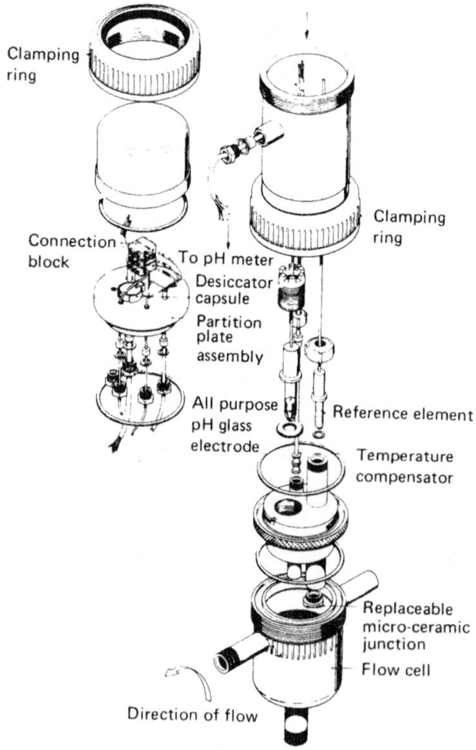

(c)

Figure 4.27 Flow-type of electrode system (courtesy Kent Industrial Measurements Ltd., Analytical Instruments): (a) external view, (b) upper section detaches for easy buffering, (c) exploded view showing the components.

system in Figure 4.28. The pneumatically driven rubber membrane wipes the electrode, providing a simple reliable cleaning action. It is driven by compressed air at preset intervals from a controller which incorporates a programmed timer mechanism that governs the frequency of the wiping action. The cleaning attachment is constructed entirely of polypropylene and 316 stainless steel, except for the rubber wiper which may be replaced by a polypropylene brush type should this be more suitable.

Alternatively an ultrasonic generator operating at 25 kHz, can be fitted to the electrode assembly, this greatly increasing the periods between necessary electrode cleaning.

4.7.3 Measurement of redox potential

When both the oxidized and reduced forms of a substance are soluble in water the old-fashioned metal redox electrode is useful – an equilibrium being set up between the two forms of the substance and the electrons in the metal electrode immersed in the solution. Again a reference electrode, generally calomel, has to be used and determinations can be made either by using the redox electrode as an indicator during titrations or by direct potentiometric determination. Arrangements are similar to those for a pH electrode. Redox electrodes, too, can be immersed directly in a liquid product stream when monitoring on plant. The high impedance e.m.f. measuring circuits as used for pH electrode systems are completely satisfactory but, as metal redox electrodes are low resistance systems, low impedance e.m.f. measuring

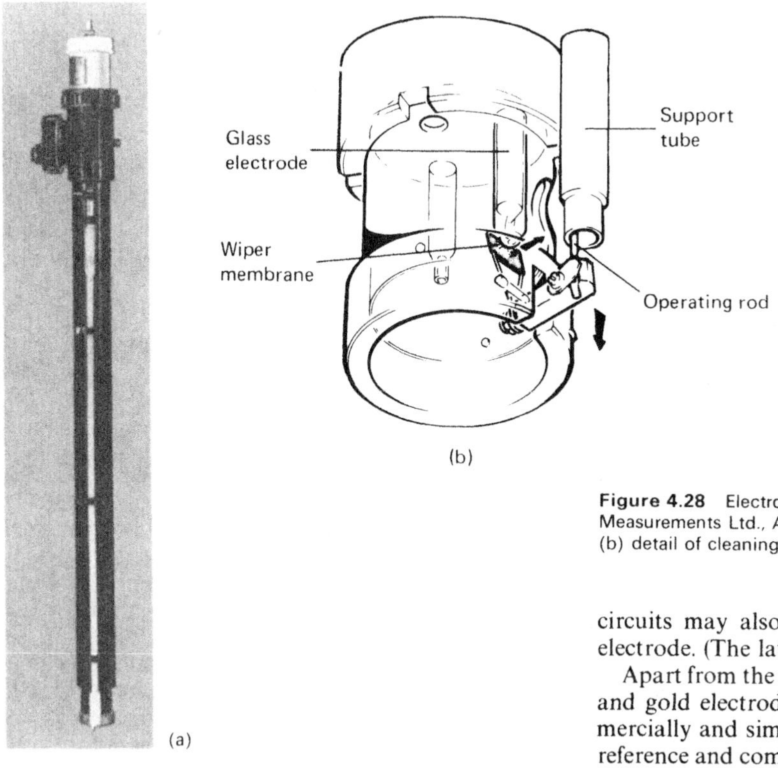

Figure 4.28 Electrode cleaning (courtesy Kent Industrial Measurements Ltd., Analytical Instruments): (a) assembly, (b) detail of cleaning attachment.

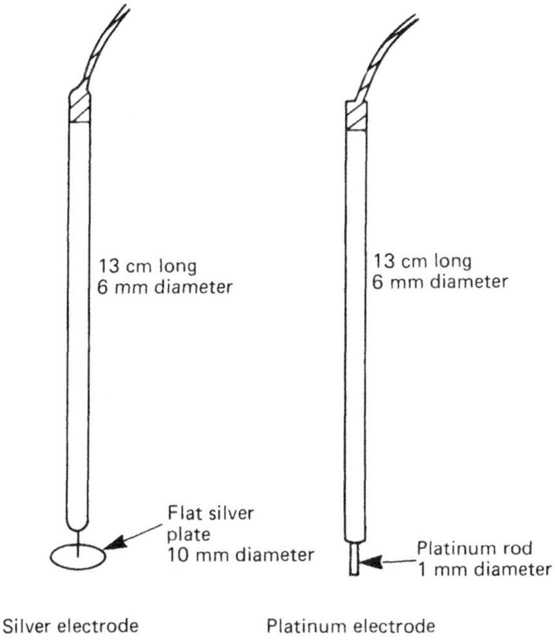

Figure 4.29 Examples of metal redox electrodes.

circuits may also be used as for the antimony pH electrode. (The latter is also a metal redox electrode.)

Apart from the antimony electrode, platinum, silver and gold electrodes (Figure 4.29) are available commercially and simple electrodes for use with separate reference and combination electrodes can be obtained both for laboratory and industrial use.

Analytical chemistry applications of redox electrodes include determination of arsenic, cyanides, hydrogen peroxide, hypochlorite or chlorine, ferrous iron, halides, stannous tin, and zinc. The silver electrode is widely used for halide determination. Platinum electrodes are suitable for most other determinations with the exception of when cyanide is being oxidized with hypochlorite (for example, in neutralizing the toxic cyanide effluent from metal plating baths). In this case a gold electrode is preferable.

4.7.4 Determination of ions by ion-selective electrodes

General considerations The measurement of the concentration or the activity of an ion in solution by means of an ion-selective electrode is as simple and rapid as making a pH measurement (the earliest of ion-selective electrodes). In principle it is necessary only to immerse the ion-selective and reference electrodes in the sample, read off the generated e.m.f. by means of a suitable measuring circuit and obtain the result from a calibration curve relating e.m.f. and concentration of the substance being determined. The difference from pH determinations is that most ion-selective electrode applications require the addition of a reagent to buffer or adjust the ionic strength of the sample before measurement of the potential. Thus, unlike measurement of pH and redox potentials, ion-selective electrodes cannot be immersed directly in a plant stream of liquid product and a sampling arrangement has to be used. However this can usually be done quite simply.

pH and pIon meters High impedance e.m.f. measuring circuits must be used with most ion-selective electrodes and are basically the same as used for measuring pH with a glass electrode. The pH meters measure e.m.f. in millivolts and are also scaled in pH units. Provided the calibration control on the pH meter (which relates the measured e.m.f. to a fixed point on the pH scale) has a wide enough range of adjustment, the pH scale can be used for any univalent positive ion, for example, measurements with a sodium-selective electrode can be read on the meter as a pNa scale (or $-\log C_{Na}$). Measurements with electrodes responding to divalent or negative ions cannot be related directly to the pH scale. However, manufacturers generally make some modifications to pH meters to simplify measurements with ion-selective electrodes and the modified meters are called 'pIon meters'. Scales are provided, analogous to the pH scale, for ions of various valencies and/or a scale that can be calibrated to read directly in terms of concentration or valency. Meters manufactured as pIon meters generally also have pH and millivolt scales. To date pIon scales only cover ions with charges of ± 1 and ± 2 because no ion-selective electrodes for determining ions of high charge are yet available commercially. Direct activity scales read in relative units only and so must be calibrated before use in the preferred measurement units.

As with pH meters, pIon meters can be obtained with analogue and digital displays, with integral microprocessors, with recorder and printer outputs and with automatic standardization. Temperature compensation can be incorporated but although ion-selective and reference electrode combinations have iso-potential points, the facility of being able to set the iso-potential control has so far been restricted to pH measurement. On dual pH/pIon meters the iso-potential control (if it exists) should be switched out on the pIon and activity scales if one wishes to make a slope correction when working with an ion-selective electrode at constant temperature.

For the best accuracy and precision pIon meters should be chosen that can discriminate 0.1 mV for direct potentiometry; 1 mV discrimination is sufficient when using ion-selective electrodes as indicators for titrimetric methods.

Practical arrangements For accurate potentiometry the temperature of the solution being analysed and the electrode assembly should be controlled and ideally all analyses should be carried out at the same temperature, e.g. by using a thermostatically controlled water bath. Solutions must also be stirred, otherwise the e.m.f. developed by the electrode may not be representative of the bulk of the solution. A wide range of stirring speeds is possible but too slow a speed may give long response times and too high a speed may generate heat in the solution. Precautions must also be taken to minimize contamination.

Taking all these items into account, best results in the laboratory can be obtained by mounting the electrodes in a flow cell through which the test solution is being pumped, see Figure 4.30. This is a mandatory arrangement for on-stream instruments and in the laboratory in cases where the ion concentration being determined is close to the limit of detection of the electrode.

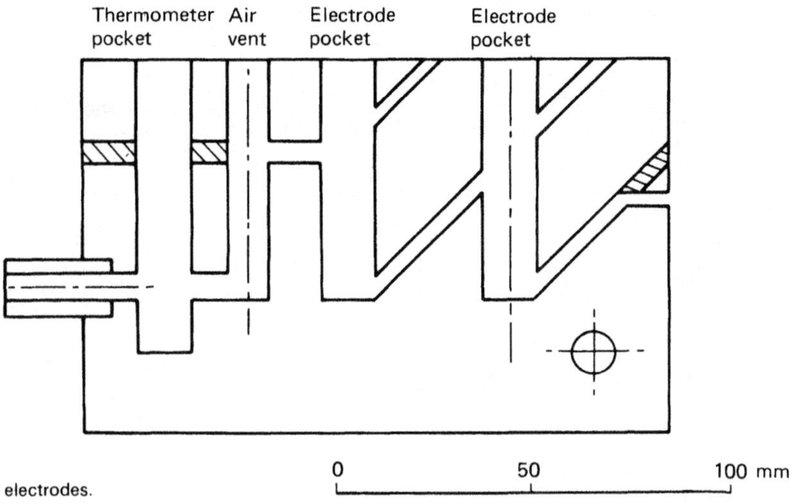

Figure 4.30 Flow cell for ion-selective electrodes.

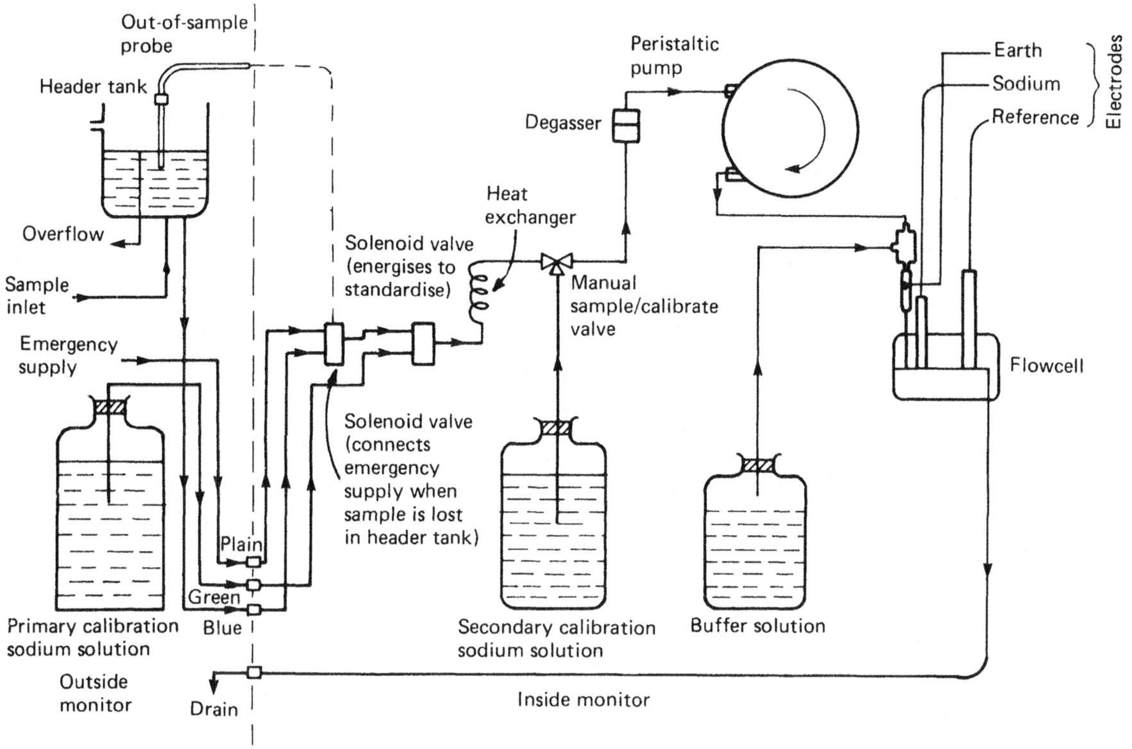

Figure 4.31 Schematic diagram for ion-selective monitor (courtesy Kent Industrial Measurements Ltd., Analytical Instruments).

Flow cells should be constructed of a material that will not contaminate a sample with the ion being determined; the flow rates of the solution must be high enough to provide 'stirring' but low enough that sample volumes are kept low. There should be good displacement of a previous sample by an incoming one, solution from the reference electrode should not reach the measuring electrode and, when liquid flow through a flow cell stops, the cell must retain liquid around the electrodes to prevent them drying out. Finally a flow cell should be water-jacketed so that its temperature can be controlled. Suitable flow cells can be machined out of Perspex and are available commercially.

Pumps used must be capable of pumping at least two channels simultaneously at different rates, the larger volume for the sample and the lesser for the reagent solution. Peristaltic pumps are the most frequently used. It follows that all interconnecting tubing and other components in contact with the sample must be inert with respect to the ion being determined.

As direct potentiometric determination of ions by ion-selective electrodes requires more frequent calibration than the more stable pH systems, industrially developed ion-selective electrode systems often incorporate automatic recalibration. This makes them more expensive than pH measuring systems. A typical scheme for an ion-selective monitor (in this case for sodium) is shown in Figures 4.31 and 4.32.

Sample water flows to the constant head unit and is then pumped anaerobically at a constant rate into the flow cell where it is equilibrated with ammonia gas obtained by pumping a stream of air through ammonia solution. (Instead of ammonia gas a liquid amine could be used and this would then be the buffer liquid delivered by the second channel of the pump.) The sample then flows through the flow cell to contact the ion-selective and reference electrodes and then to a drain.

Automatic chemical standardization takes place at preset intervals (in this case once every 24 hours) with provision for manual initiation of the sequence at any time. The standardization sequence commences by activating a valve to stop the sample flow and to allow a sodium ion solution of known strength (the standard sodium solution) to be pumped into the flow cell. When the electrodes have stabilized in the new solution, the amplifier output is compared with a

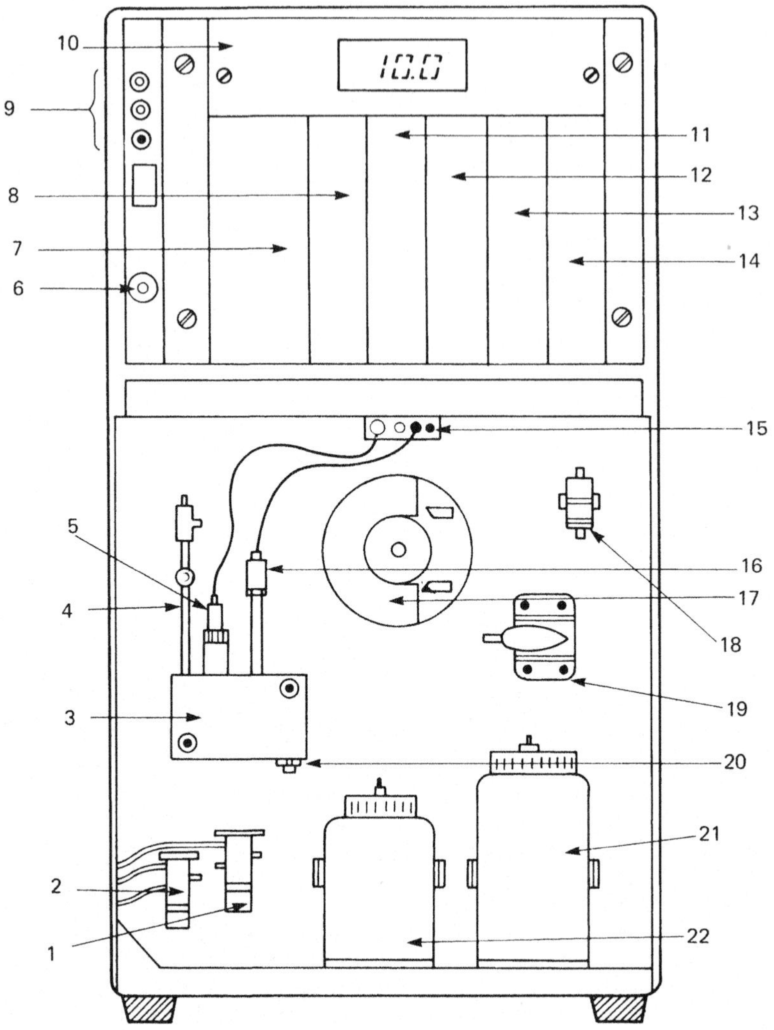

Figure 4.32 Diagrammatic arrangement of components for an ion-selective monitor (courtesy Kent Industrial Measurements Ltd., Analytical Instruments). 1, solenoid valve (energizes during standardization to connect primary standard solution); 2, solenoid valve (energizes to admit emergency sample supply when sample is lost in the header tank); 3, flow cell; 4, earthing tube; 5, sodium electrode; 6, SUPPLY ON lamp (illuminates when power is connected to the monitor); 7, 8020 100 amplifier; 8, 8033 200 current output module; 9, SERVICE lamp (red) and ON-LINE lamp (green) with push-button (optional feature); 10, digital display module (linear motor readout optional); 11, 8060 300 compensation module; 12, 8021 400 alarm and temperature control module; 13, 8020 500 power supply; 14, 8020 600 function module; 15, electrodes connection point (junction box); 16, refillable calomel reference electrode; 17, peristaltic pump; 18, gas debubbler; 19, manual SAMPLE/CALIBRATE valve; 20, flow cell drain; 21, secondary standard solution container (1 litre) (heat exchanger located behind the panel at this point); 22, buffer solution container (500 ml)

preset standard value in the auto-compensation unit and any error causes a servo-potentiometer to be driven so as to adjust the output signal to the required value. The monitor is then returned to measurement of the sample. The standardization period lasts 30 minutes, a warning lamp shows that standardization is taking place and any alarm and control contacts are disabled. It is also possible to check the stability of the amplifier and, by a manual introduction of a second sodium standard, to check and adjust the scale length.

Conditioning and storage of electrodes The manufacturer's instructions regarding storage and pretreatment of electrodes should be followed closely. The general rules are that (a) glass electrodes should not be allowed to dry out because reconditioning may not be successful, (b) solid state electrodes can be stored in deionized water, for long periods, dry-covered with protective caps and generally ready for use after rinsing with water, (c) gas-sensing membranes and liquid ion-exchange electrodes must never be allowed to dry out, (d) reference electrodes are as important as the measuring electrodes and must be treated exactly as specified by the manufacturer. The element must not be allowed to dry out, as would happen if there were insufficient solution in the reservoir.

Ion-selective electrodes available and application areas There is a very wide range of electrodes available. Not only are there many specific ion monitors but several manufacturers now market standardized modular assemblies which only need different electrodes, different buffer solutions and minor electrical adjustments for the monitors to cope with many ion determinations.

Table 4.9 Available ion-selective electrodes

Solid-state membrane electrodes	*Glass membrane electrodes*	*Liquid ion exchange membrane electrodes*	*Gas-sensing electrodes*
Fluoride	pH	Calcium	Ammonia
Chloride	Sodium	Calcium + magnesium	Carbon dioxide
Bromide	Potassium	(i.e. water hardness)	Sulphur dioxide
Iodide			Nitrous oxide
Thiocyanate			Hydrogen sulphide
Sulphide			Hydrogen fluoride
Silver		Barium	
Copper		Nitrate	
Lead		Potassium	
Cadmium			
Cyanide			
Redox			
pH (antimony)			

Table 4.9 shows the ion-selective electrodes available for the more common direct potentiometric determination of ions.

Ion-selective electrodes, as their name implies, are selective rather than specific for a particular ion. A potassium electrode responds to some sodium ion activity as well as to potassium, and this can be expressed as:

$$E_{\text{measured}} = \text{constant} \pm S \log (a_{\text{potassium}^+} + K\, a_{\text{Na}^+})$$

where K is the selectivity coefficient of this electrode to sodium and $0 < K < 1$.

Thus the fraction K of the total sodium activity will behave as though it were potassium. The *smaller* the value of K, the more selective that electrode is to potassium, i.e. the *better* it is. To identify a particular selectivity coefficient the data are best written in the form:

$$K_{\text{potassium}^+/\text{sodium}^+} = 2.6 \times 10^{-3}$$

This shows that the selectivity of potassium over sodium for the potassium electrode is about 385:1, i.e. $1/(2.6 \times 10^{-3})$. It is important to note that selectivity coefficients are not constant, but vary with the concentration of both primary and interferent ions and the coefficients are, therefore, often quoted for a particular ion concentration. They should be regarded as a guide to the effectiveness of an electrode in a particular measurement and not for use in precise calculations, particularly as quoted selectivity coefficients vary by a factor of 10 or more. For accurate work the analyst should determine the coefficient for himself for his own type of solution.

Direct potentiometric determination of ions by means of ion-selective electrodes has many applications. Examples are determination of pH, sodium and chloride in feedwater, condensate and boiler water in power stations; cyanide, fluoride, sulphide and chloride in effluents, rivers and lakes; fluoride, calcium and chloride in drinking water and sea water; bromide, calcium, chloride, fluoride, iodide, potassium and sodium in biological samples; calcium, chloride, fluoride and nitrate in soils; sulphur dioxide in wines and beer; chloride and calcium in milk; sulphide and sulphur dioxide in the paper-making industry; fluoride, calcium, chloride, nitrate and sulphur dioxide in foodstuffs, pH in water and effluents, papers, textiles, leather and foodstuffs, and calcium, chloride, fluoride and potassium in pharmaceuticals.

4.8 Common electrochemical analysers

4.8.1 Residual chlorine analyser

When two dissimilar metal electrodes are immersed in an electrolyte, and connected together, current will flow due to the build-up of electrons on the more electropositive electrode. The current will soon stop, however, owing to the fact that the cell will become polarized.

If, however, a suitable depolarizing agent is added, a current will continue to flow, the magnitude of which will depend upon the concentration and nature of the ions producing the depolarization. Thus, by choice of suitable materials for the electrodes and arranging for the addition of the depolarizing agent which is in fact the substance whose concentration is to be measured, amperometric analysers may be made to measure the concentration of a variety of chemicals. In some instruments a potential difference may be applied to the electrodes, when the current is again a linear function of the concentration of the depolarizing agent.

The sensitivity of the analyser is sometimes increased by using buffered water as the electrolyte so

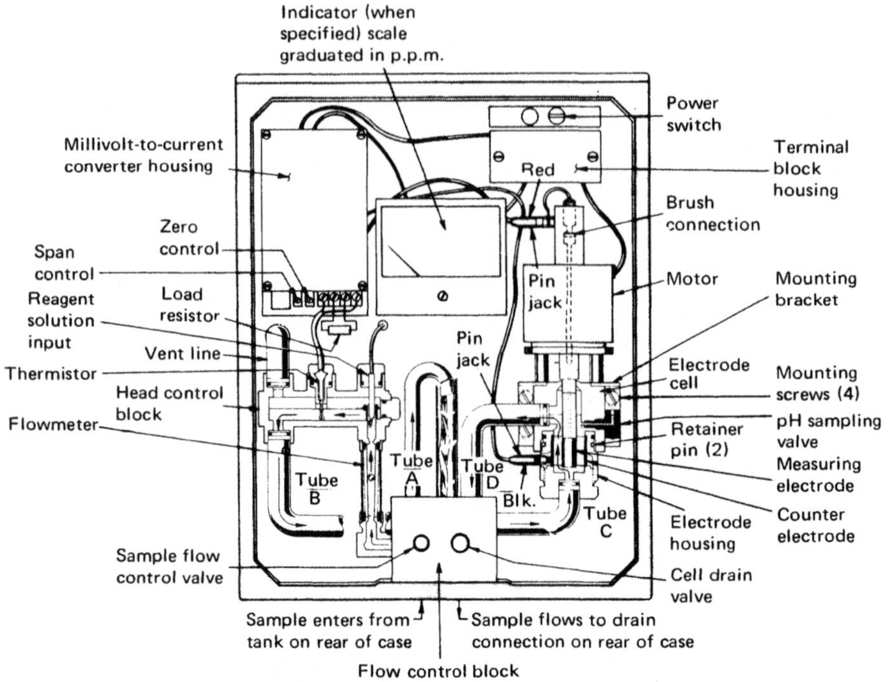

Figure 4.33 Residual chlorine analyser (courtesy Fischer & Porter).

that the cell operates at a definite pH. Amperometric instruments are inherently linear in response, but special steps have to be taken in order to make them specific to the substance whose concentration is to be measured, because other substances may act as depolarizing agents and so interfere with the measurement. When the interfering substances are known steps may be taken to remove them.

Where the instrument is intended to measure pollutants in air or gas, the gas to be tested is either bubbled through a suitable cell or arranged to impinge upon the surface of the liquid in the cell. In these cases interfering gases can be removed by chemical or molecular filters in the sampling system.

This form of instrument may be used to detect halogens, such as chlorine, in air and instruments with ranges from 0–0.5 to 0–20 ppm are available measuring with an accuracy of $\pm 2\%$ and a sensitivity of 0.01 ppm. By altering the electrolyte the instrument may be changed to measure the corresponding acid vapours, i.e. HCl, HBr and HF. One type of instrument for measuring chlorine in water is shown in Figure 4.33.

The sample stream is filtered in the tank on the back of the housing, and then enters the analyser unit through the sample flow control valve and up the metering tube into the head control block where reagent (buffer solution to maintain constant pH) is added by means of a positive displacement feed pump.

Buffered sample flows down tube B, through the flow control block and up tube C to the bottom of the electrode cell assembly. Sample flow rate is adjusted to approximately 150 millilitres per minute. Flow rate is not critical since the relative velocity between the measuring electrode and the sample is established by rotating the electrode at high speed.

In the electrode cell assembly, the sample passes up through the annular space between the concentrically mounted outer (copper) reference electrode and the inner (gold) measuring electrode and out through tube D to the drain. The space between the electrodes contains plastic pellets which are continuously agitated by the swirling of the water in the cell. The pellets keep the electrode surfaces clear of any material which might tend to adhere. The measuring electrode is coupled to a motor which operates at 1550 rev/min. The electrical signal from the measuring electrode is picked up by a spring-loaded brush on top of the motor and the circuit is completed through a thermistor for temperature compensation, precision resistors and the instationary copper electrode.

The composition of the electrodes is such that the polarization of the measuring electrode prevents current flow in the absence of a strong oxidizing agent. The presence of the smallest trace of strong oxidizer, such as chlorine (hypochlorous acid), will permit a current to flow by oxidizing the polarizing layer. The amplitude of the self-generated depolarization current is proportional to the concentration of the strong oxidizing agent. The generated current is passed through a precision resistor and the millivoltage across the resistor is then measured by the indicating or recording potentiometer. This instrument is calibrated to read in terms of the type (free or total) of residual chlorine measured. When measuring total residual chlorine, potassium iodide is added to the buffer. This reacts with the free and combined chlorine to liberate iodine in an amount equal to the total chlorine. The iodine depolarizes the cell in the same manner as hypochlorous acid, and a current directly proportional to the total residual chlorine is generated.

4.8.2 Polarographic process oxygen analyser

An instrument using the amperometric (polarographic) method of measurement is an oxygen analyser used for continuous process measurement of oxygen in flue gas, inert gas monitoring and other applications.

The key to the instrument is the rugged sensor shown in Figure 4.34. The sensor contains a silver anode and a gold cathode that are protected from the sample by a thin membrane of PTFE. An aqueous KCl solution is retained in the sensor by the membrane and forms the electrolyte in the cell (Figure 4.35).

Oxygen diffuses through the PTFE membrane and reacts with the cathode according to the equation:

$$4e^- + O_2 + 2H_2O \rightarrow 4OH^-$$

The corresponding anodic reaction is

$$Ag + Cl^- \rightarrow AgCl + e^-$$

For the reaction to continue, however, an external potential (0.7 volts) must be applied between cathode and anode. Oxygen will then continue to be reduced at the cathode, causing the flow of a current, the

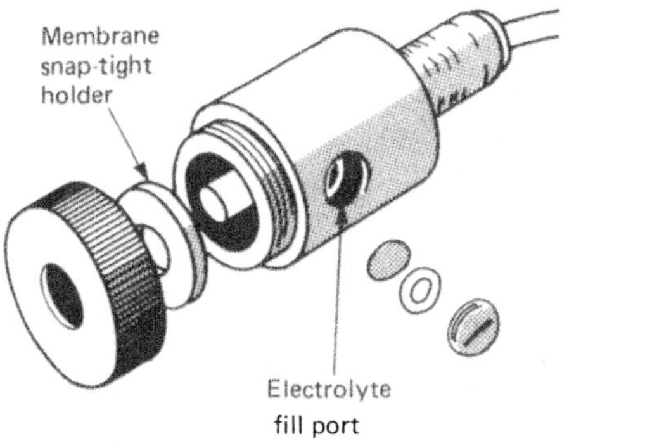

Figure 4.34 Process oxygen analyser (courtesy Beckman Instruments Inc.).

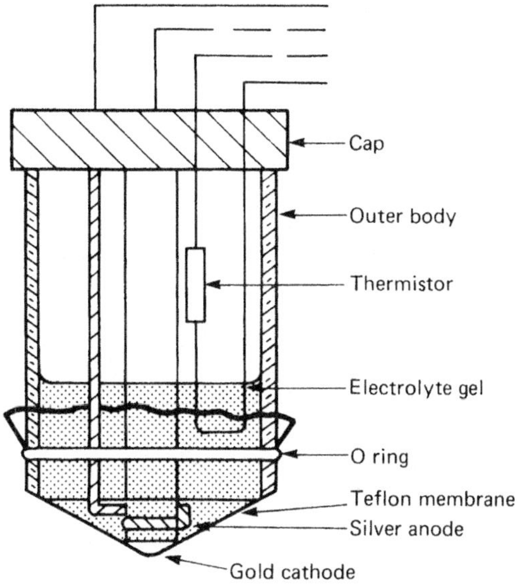

Figure 4.35 Diagram of polarographic oxygen sensor (courtesy Institute of Measurement and Control).

magnitude of which is proportional to the partial pressure of oxygen in the sample gas.

The only materials in contact with the process are PVC and PTFE and the membrane is recessed so that it does not suffer mechanical damage. The cell needs to be recharged with a new supply of electrolyte at 3- or 6-monthly intervals depending on the operating conditions and the membrane can be replaced easily should it be damaged.

The cell current is amplified by a solid state amplifier which gives a voltage output which can be displayed on an indicator or recorded. The instrument has a range selection switch giving ranges of 0–1, 0–5, 0–10 or 0–25 per cent oxygen and a calibration adjustment. The calibration is checked by using a reference gas, or air when the instrument should read 20.9 per cent oxygen on the 0–25 per cent scale. The instrument has an accuracy of ± 1 per cent of scale range at the calibration temperature but an error of ± 3 per cent of the readily will occur for a 16 °C departure in operating temperature.

When in use the sensor may be housed in an in-line type housing or in a dip-type of assembly, usually made of PVC suitable for pressures up to 3.5 bar.

4.8.3 High temperature ceramic sensor oxygen probes

Just as an electrical potential can be developed at a glass membrane which is a function of the ratio of the hydrogen concentrations on either side, a pure zirconia tube maintained at high temperature will develop a potential between its surfaces which is a function of the partial pressure of oxygen which is in

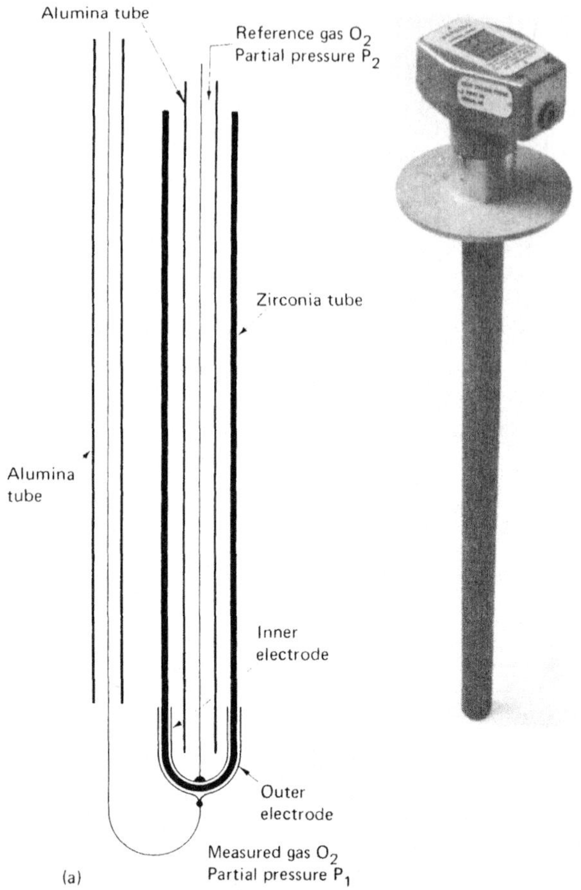

Figure 4.36 Oxygen probe (courtesy Kent Instruments).

contact with its surfaces. This is the principle involved in the oxygen meter shown in Figure 4.36.

The potential developed is given by the Nernst equation:

$$E_s = (RT/4F)\{\ln[\text{internal partial pressure of } O_2^{4-} \text{ ions}]/[\text{external partial pressure of } O_2^{4-} \text{ ions}]\}$$

Thus, if the potential difference between the surfaces is measured by platinum electrodes in contact with the two surfaces a measure may be made of the ratio of the partial pressure of the oxygen inside and outside the probe. If dry instrument air (20.9 per cent oxygen) is fed into the inside of the probe, the partial pressure of oxygen inside the tube may be regarded as constant, so that the electrical potential measured in a similar manner to that adopted in pH measurement will be a measure of the concentration of the oxygen in the atmosphere around the measuring probe. Thus by positioning the probe in a stack or flue where the temperature is above 600 °C a direct measurement of the oxygen present may be made. (In another manufacturer's instrument the probe is maintained at a temperature of 850 °C by a temperature-controlled heating element.) The instrument illustrated can operate from 600 to 1200 °C, the reading being corrected for temperature, which is measured by a thermocouple. The probe is protected by a silicon carbide sheath. The zirconia used is stabilized with calcium.

Standard instruments have ranges of oxygen con centration of 20.9–0.1 per cent, 1000–1 ppm, 10^{-5}–10^{-25} partial pressure and can measure oxygen with an accuracy of better than ±10 per cent of the reading.

As temperatures in excess of 600 °C must be used some of the oxygen in the sample will react with any combustible gas present, e.g. carbon monoxide and hydrocarbons. Thus the measurement will be lower than the correct value but will still afford a rapid means of following changes in the oxygen content of a flue gas caused by changes in combustion conditions.

4.8.4 Fuel cell oxygen-measuring instruments

Galvanic or fuel cells differ from polarographic cells and the high temperature ceramic sensors in that they are power devices in their own right, that is, they require no external source of power to drive them. One manufacturer's version is shown in Figure 4.37.

A lead anode is made in that geometric form that maximizes the amount of metal available for reaction with a convex disc as the cathode. Perforations in the cathode facilitate continued wetting of the upper surface with electrolyte and ensure minimum internal resistance during the oxygen sensing reaction. The surfaces of the cathode are plated with gold and then covered with a PTFE membrane. Both electrodes are immersed in aqueous potassium hydroxide electrolyte. Diffusion of oxygen through the membrane enables

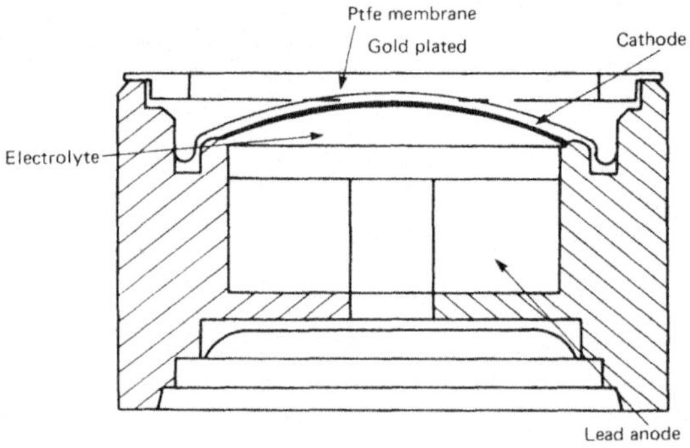

Figure 4.37 Diagrammatic micro-fuel cell oxygen sensor (courtesy Analysis Automation).

the following reactions to take place:

Cathode $4e^- + O_2 + 2H_2O \rightarrow 4OH^-$

Anode $Pb + 2OH^- \rightarrow PbO + H_2O + 2e$

Overall cell reaction $2Pb + O_2 \rightarrow PbO$

The electrical output of the cell can be related to the partial pressure of oxygen on the gas side of the membrane in a manner analogous to that described for membrane-covered polarographic cells. In this instance, however, because there is no applied potential and no resultant hydrolysis of the electrolyte, absence of oxygen in the sample corresponds to zero electrical output from the cell. There is a linear response to partial pressure of oxygen and a single point calibration, e.g. on air, is sufficient for most purposes.

The main limitation of this type of oxygen sensor is the rate of diffusion of oxygen across the membrane; this determines the speed of response and, at low oxygen partial pressure, this may become unacceptably slow. However, to overcome this, one type of fuel cell oxygen sensor has a completely exposed cathode, i.e. not covered with a PTFE membrane.

In common with all membrane cells, the response of the micro-fuel cell is independent of sample flow rate but the cell has a positive temperature-dependence. This is accommodated by incorporating negative temperature coefficient thermistors in the measuring circuit. These fuel cells have sufficient electrical output to drive readout meters without amplification. However where dual or multi-range facilities are required some amplification may be necessary.

4.8.5 Hersch cell for oxygen measurement

This galvanic cell differs from fuel cells in that a third electrode is added to the cell and a potential applied to provide anodic protection to the anode. In one manufacturer's cell (Figure 4.38) the cathode is silver and the anode cadmium. The third electrode is platinum. The anodic protection limits the cadmium current to a few microamperes and extends the life of the cadmium. However, this arrangement gives an electrical output from the cell which is non-linear with oxygen partial pressure and it is necessary for the signal to be passed through a 'shaping' circuit for the readout to be given in concentration units. Calibration is carried out by generating a predetermined concentration of oxygen in a sample by electrolysis, and

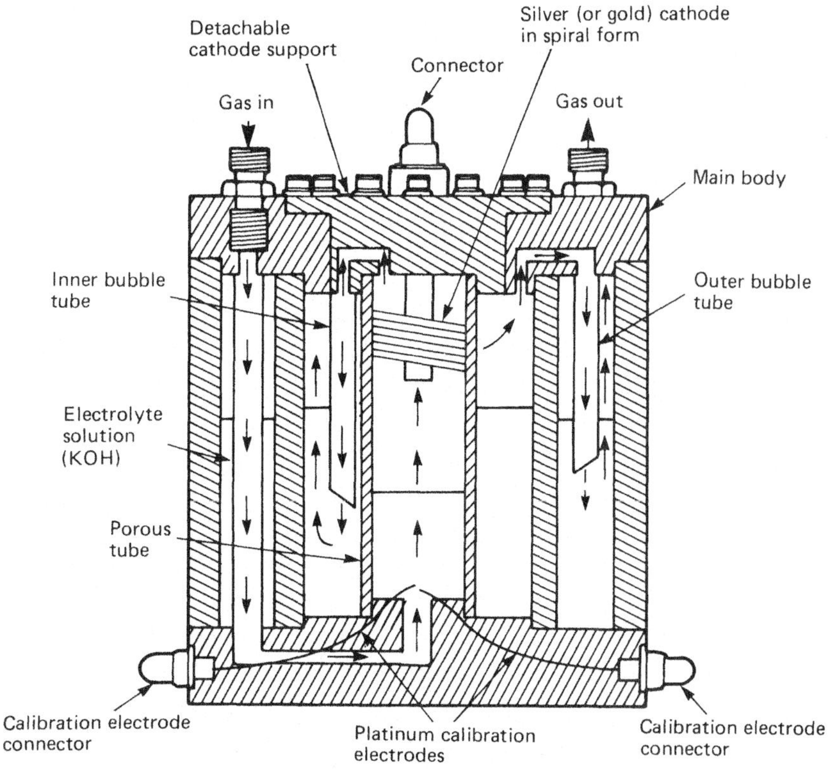

Figure 4.38 Cross-section of Hersch cell (courtesy Anacon (Instruments) Ltd.).

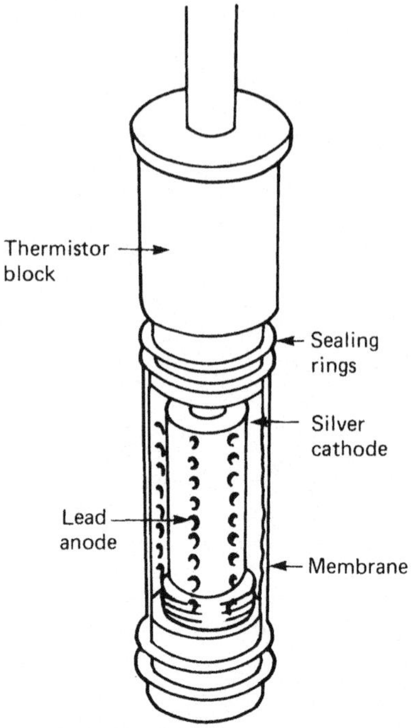

Figure 4.39 Diagram of Mackereth oxygen sensor assemblies (courtesy Kent Industrial Measurements Ltd., Analytical Instruments).

electrodes for this are incorporated in the cell. When dry gas samples are being used they must be humidified to prevent the water-based electrolyte in the cell from drying out.

4.8.6 Sensor for oxygen dissolved in water

Electrochemical sensors with membranes for oxygen determination can be applied to measuring oxygen dissolved in water; both polarographic and galvanic sensors can be used.

A most popular type of sensor is the galvanic Mackereth electrode. The cathode is a perforated silver cylinder surrounding a lead anode with an aqueous electrolyte of potassium bicarbonate (Figure 4.39). The electrolyte is confined by a silicone rubber membrane which is permeable to oxygen but not to water and interfering ions.

The oxygen which diffuses through the membrane is reduced at the cathode to give a current proportional to the oxygen partial pressure. Equations for the reactions have already been given (p. 139).

Accurate temperature control is essential (6 per cent error per degree) and thermistor- or resistance-thermometer-controlled compensation circuits are generally used. Working ranges can be from a few μg O_2/litre of water up to 200 per cent oxygen saturation. The lead anode is sacrificial and electrodes therefore

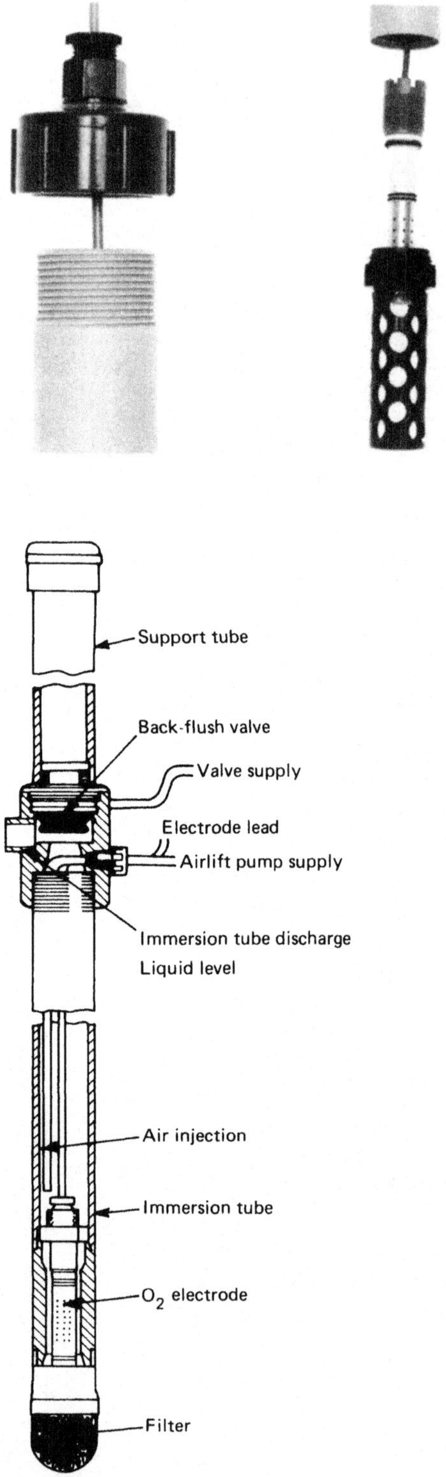

Figure 4.40 Varieties of Mackereth oxygen sensor assemblies (courtesy Kent Industrial Measurements Ltd., Analytical Instruments).

have to be refurbished according to the actual design and the total amount of oxygen that has diffused into the cell. Cells are calibrated using water containing known amounts of oxygen. Indicating meters or recorders can be connected and manufacturers offer both portable instruments and equipment for permanent installation with timing devices, water pumps, etc. There are also several variations on the basic design of electrodes to cope with oxygen determination in water plant, rivers, lakes, sewage tanks, etc. (see Figure 4.40). One of those shown includes a patented assembly incorporating water sampling by air lift – air reversal gives calibration check and filter clean.

4.8.7 Coulometric measurement of moisture in gases and liquids

Moisture from gases (or vaporized from liquids) can be absorbed by a layer of desiccant, generally phosphoric anhydride (P_2O_5), in contact with two platinum or rhodium electrodes. A d.c. voltage is applied to electrolyse the moisture, the current produced being directly proportional to the mass of moisture absorbed (Faraday's law of electrolysis). The response of such an instrument obviously depends on the flow rate of gas which is set and controlled accurately at a predetermined rate so that the current measuring meter can be calibrated in vppm moisture. Details are given in Chapter 6.

4.9 Further reading

Bailey, P. L., *Analysis with Ion-selective Electrodes*, Heyden, (1976)

Bates, R. G., *The Determination of pH* (2nd edn), Wiley Interscience, (1973)

Durst, R. A. (ed.), *Ion Selective Electrodes*, National Bureau of Standards Special Publication 314, Dept. of Commerce, Washington DC, (1969)

Eisenman, G., *Glass Electrodes for Hydrogen and Other Cations*, Edward Arnold, London/Marcel Dekker, New York, (1967)

Freiser, H. (ed.), *Ion-selective Electrodes in Analytical Chemistry* Vol. I, Plenum Press, New York, (1978)

Ives, G. J. and Janz, D. J. G., *Reference Electrodes, Theory and Practice*, Wiley Interscience, (1961)

Midgley, D. and Torrance, K., *Potentiometric Water Analysis*, Wiley Interscience, (1978)

Perrin, D. D. and Dempsey, B., *Buffers for pH and Metal Ion Control*, Chapman and Hall, London, (1974)

Sawyer, D. T. and Roberts, J. L. Jr., *Experimental Electrochemistry for Chemists*, Wiley Interscience, (1974)

5 Chemical analysis – gas analysis

C. K. LAIRD

5.1 Introduction

The ability to analyse one or more components of a gas mixture depends on the availability of suitable detectors which are responsive to the components of interest in the mixture and which can be applied over the required concentration range. Gas detectors are now available which exploit a wide variety of physical and chemical properties of the gases detected, and the devices resulting from the application of these detection mechanisms show a corresponding variety in their selectivity and range of response. In a limited number of applications it may be possible to analyse a gas mixture merely by exposure of the sample to a detector which is specific to the species of interest, and thus obtain a direct measure of its concentration. However, in the majority of cases no sufficiently selective detector is available and the gas sample requires some pretreatment, e.g. drying or removal of interfering components, to make it suitable for the proposed detector. In these cases a gas analysis system must be used.

A block diagram of the components of a typical gas analyser is given in Figure 5.1. The sample is taken into the instrument either as a continuous stream or in discrete aliquots and is adjusted as necessary in the sampling unit to the temperature, pressure and flow-rate requirements of the remainder of the system. Any treatment of the sample, for example separation of the sample into its components, removal of interfering components or reaction with an auxiliary gas is carried out and the sample is passed to the detector. The signal from the detector is amplified if necessary and processed to display or record the concentration of the components of interest in the sample.

In many gas analysers the time lag between sampling and analysis is reduced to a minimum by taking a continuous stream of sample at a relatively high flow rate, and arranging for only a small proportion to enter the analyser, the remainder being bypassed to waste. Provision is also normally made to check the zero by passing a sample, free of the species to be analysed, to the detector, and the instrument may also include facilities for calibration by means of a

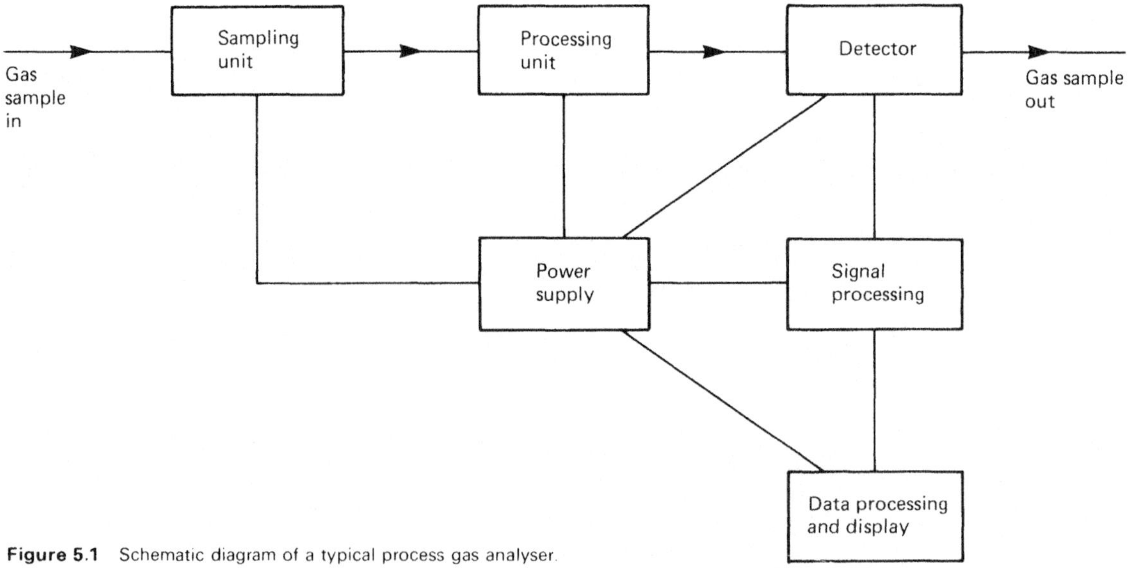

Figure 5.1 Schematic diagram of a typical process gas analyser.

'span' switch which feeds a sample of known concentration to the analyser.

For certain applications there may be a choice between the use of a highly selective detector, with relatively little pretreatment of the sample, or use of a detector which responds to a wider range of chemical species, the sample being separated into its components before it reaches the detector. In the special case of gas chromatography the sample is separated on the basis of the different times taken by each component to pass through a tube or column packed with adsorbent. The outlet gas stream may then be passed through a single detector, or through more than one detector in series or switched between detectors to analyse several components of the original sample mixture. By choice of columns, operating conditions and detectors, a gas-chromatographic analysis system may be built up individually tailored to analyse several different preselected components in a single aliquot taken from a gas sample. Because of its importance in process analysis, gas chromatography is given particularly detailed treatment.

In addition to the analysis techniques described in this chapter, a number of spectroscopic methods are given under that heading in Chapter 3, while some electrochemical methods are outlined in Chapter 4.

5.2 Separation of gaseous mixtures

Although detectors have been developed which are specific to particular gases or groups of gases, for example flammable gases or total hydrocarbons there is often a need to separate the sample into its components, or to remove interfering species, before the sample is passed to the detector. A non-specific detector, such as a katharometer, may also be used to measure one component of a gas mixture by measuring the change in detector response which occurs when the component of interest is removed from the gas mixture.

Methods for separating gaseous mixtures may be grouped under three main headings.

Chemical reaction A simple example of chemical separation is the use of desiccants to remove water from a gas stream. The percentage of carbon dioxide in blast furnace gas may be determined by measuring the thermal conductivity of the gas before and after selective removal of the carbon dioxide by passing the gas through soda-lime. Similarly the percentage of ammonia gas, in a mixture of nitrogen, hydrogen and ammonia may be measured by absorbing the ammonia in dilute sulphuric acid or a suitable solid absorbent.

Physical methods The most powerful physical technique for separation of gases is mass spectrometry, described in Chapter 3 – though only minute quantities can be handled in that way. Gases may also be separated by diffusion, for example, hydrogen may be removed from a gas stream by allowing it to diffuse through a heated tube of gold– or silver–palladium alloy.

Physico-chemical methods: chromatography Gas chromatography is one of the most powerful techniques for separation of mixtures of gases or (in their vapour phase) volatile liquids. It is relatively simple and widely applicable. Mixtures of permanent gases, such as oxygen, nitrogen, hydrogen, carbon monoxide, and carbon dioxide can easily be separated, and when applied to liquids, mixtures such as benzene and cyclohexane can be separated even though their boiling points differ by only 0.6 K. Separation of such mixtures by other techniques such as fractional distillation would be extremely difficult.

5.2.1 Gas chromatography

Chromatography is a physical or physico-chemical technique for the separation of mixtures into their components on the basis of their molecular distribution between two immiscible phases. One phase is normally stationary and is in a finely divided state to provide a large surface area relative to volume. The second phase is mobile and transports the components of the mixture over the stationary phase. The various types of chromatography are classified according to the particular mobile and stationary phases employed in each (see Chapter 2). In gas chromatography the mobile phase is a gas, known as the carrier gas, and the stationary phase is either a granular solid (gas–solid chromatography) or a granular solid coated with a thin film of non-volatile liquid (gas–liquid chromatography). In gas–solid chromatography the separation is effected on the basis of the different adsorption characteristics of the components of the mixture on the solid phase, while in gas–liquid chromatography the separation mechanism involves the distribution of the components of the mixture between the gas and stationary liquid phases. Because the components of the mixture are transported in the gaseous phase, gas chromatography is limited to separation of mixtures whose components have significant vapour pressures, and this normally means gaseous mixtures or mixtures of liquids with boiling points below approximately 450 K.

The apparatus for gas chromatography, known as the gas chromatograph, consists of a tube or column to contain the stationary phase, and itself contained in an environment whose temperature can be held at a constant known value or heated and cooled at controlled rates. The column may be uniformly packed with the granular stationary phase (packed column chromatography) and this is normally used in process instruments. However, it has been found that columns of the highest separating performance are obtained if the column is in the form of a capillary tube, with the solid or liquid stationary phase coated on its inner walls (capillary chromatography). The carrier-gas mobile phase is passed continuously through the column at a constant controlled and known rate. A facility for introduction of known volumes of the mixture to be separated into the carrier-gas stream is provided in the carrier-gas line upstream of the column, and a suitable detector, responsive to changes in the composition of the gas passing through it, is connected to the downstream end of the column.

To analyse a sample, an aliquot of suitable known volume is introduced into the carrier-gas stream, and the output of the detector is continuously monitored. Due to their interaction with the stationary phase, the components of the sample pass through the column at different rates. The processes affecting the separation are complex, but in general, in gas–solid chromatography the component which is least strongly adsorbed is eluted first, while in gas–liquid chromatography the dominant process is the solubility of the components in the liquid stationary phase. Thus the separation achieved depends on the nature of the sample and stationary phase, on the length and temperature of the column, and on the flow rate of the carrier gas, and these conditions must be optimized for a particular analysis.

The composition of the gas passing through the detector alternates between pure carrier gas, and mixtures of the carrier gas with each of the components of the sample. The output record of the detector, known as the chromatogram, is a series of deflections or peaks, spaced in time and each related to a component of the mixture analysed.

A typical chromatogram of a mixture containing five components, is shown in Figure 5.2. The first 'peak' (A) at the beginning of the chromatogram is a pressure wave or unresolved peak caused by momentary changes in carrier-gas flow and pressure during the injection of the sample. The recording of the chromatogram provides a visual record of the analysis, but for qualitative analysis each peak must be identified on the basis of the time each component takes to pass through the column by use of single pure compounds or mixtures of known composition. For quantitative analysis the apparatus must be calibrated by use of standard gas mixtures or solutions to relate the detector response to the concentration of the determinand in the initial mixture.

A significant advantage of gas chromatography is that several components of a sample may be analysed essentially simultaneously in a single aliquot extracted from a process stream. However, sampling is on a regular discrete basis rather than continuous, so that the chromatograph gives a series of spot analyses of a sample stream, at times corresponding to the time of sample injection into the instrument. Before a new sample can be analysed, it is necessary to be certain that all the components of the previous sample have

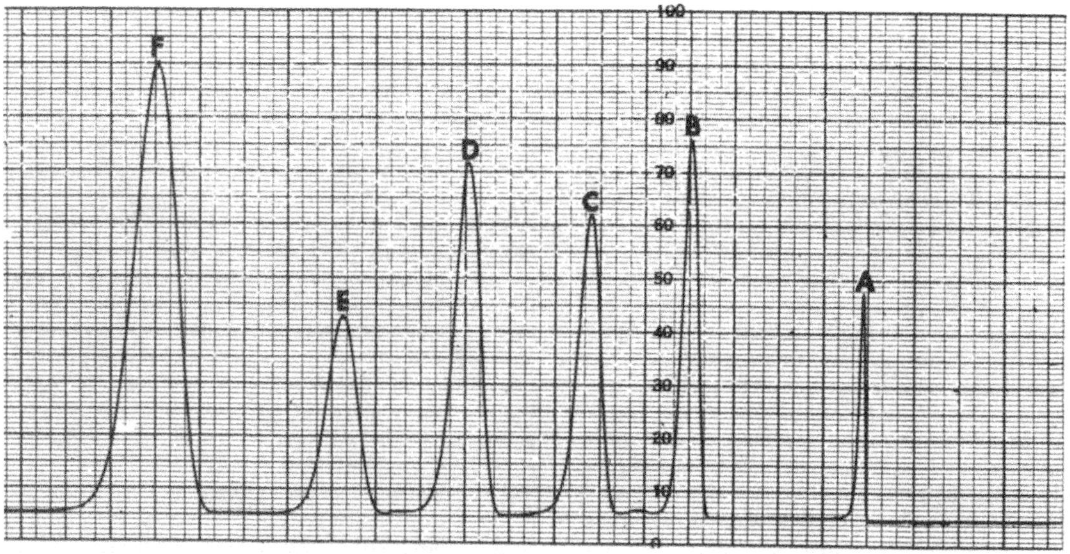

Figure 5.2 Chromatogram of a sample containing five components.

been eluted from the column. It is therefore advantageous to arrange the analytical conditions so that the sample is eluted as quickly as possible, consistent with adequate resolution of the peaks of interest.

5.3 Detectors

5.3.1 Thermal conductivity detector (TCD)

The thermal conductivity detector is among the most commonly used gas detection devices. It measures the change in thermal conductivity of a gas mixture, caused by changes in the concentration of the species it is desired to detect.

All matter is made up of molecules which are in constant rapid motion. Heat is the energy possessed by a body by virtue of the motion of the molecules of which it is composed. Raising the temperature of the body increases the energy of the molecules by increasing the velocity of the molecular motion.

In solids the molecules do not alter their position relative to one another but vibrate about a mean position, while in a liquid the molecules vibrate about mean positions, but may also move from one part of the liquid to another. In a gas the molecular motion is almost entirely translational: the molecules move from one part of the gas to another, only impeded by frequent intermolecular collisions and collisions with the walls of the vessel. The collisions with the walls produce the pressure of the gas on the walls. In a so-called 'perfect gas' the molecules are regarded as being perfectly elastic so no energy is dissipated by the intermolecular collisions.

Consideration of the properties of a gas which follow as a consequence of the motion of its molecules is the basis of the kinetic theory. Using this theory Maxwell gave a theoretical verification of laws which had previously been established experimentally. These included Avogadro's law, Dalton's law of partial pressures and Graham's law of diffusion.

Since heat is the energy of motion of the gas molecules, transfer of heat, or thermal conductivity, can also be treated by the kinetic theory. It can be shown that the thermal conductivity K of component S is given by

$$K_S = \tfrac{1}{2}\rho\tilde{v}\lambda C_v$$

where ρ is the gas density, $\tilde{v}$ is the mean molecular velocity, λ is the mean free path, and C_v is the specific heat at constant volume. Thus thermal conductivity depends on molecular size, mass and temperature.

The quantity $\tilde{v}\lambda$ is the diffusion coefficient D of the gas, and the thermal conductivity can be written

$$K_S = \tfrac{1}{2}D\rho C_v$$

According to this treatment, the thermal conductivity of the gas is independent of pressure. This is found to be true over a wide range of pressures, provided that the pressure does not become so high that the gas may no longer be regarded as being a perfect gas. At very low pressures, the conductivity of the gas is proportional to its pressure, and this is the basis of the operation of the Knudsen hot-wire manometer or Pirani gauge (see Chapter 10 of Volume 1).

It can be shown that the conductivity K_T of a pure gas at absolute temperature T, varies with temperature according to the equation

$$K_T = K_0\left[b + \frac{273}{b} + T\right]\left[\frac{T}{273}\right]^{3/2}$$

where K_0 is the thermal conductivity at 0 °C and b is a constant.

The relative thermal conductivities of some gases, relative to air as 1.00, are given in Table 5.1.

Table 5.1 Relative thermal conductivities of some common gases

Gas	*Conductivity*
Air	1.00
Oxygen	1.01
Nitrogen	1.00
Hydrogen	4.66
Chlorine	0.32
Carbon monoxide	0.96
Carbon dioxide	0.59
Sulphur dioxide	0.32
Water vapour	1.30
Helium	4.34

It can be shown that the conductivity of a binary mixture of gases is given by

$$K = \frac{K_1}{1 + A\left(\dfrac{1-x_1}{x_1}\right)} + \frac{K_2}{1 + B\left(\dfrac{x_1}{1-x_1}\right)}$$

where A and B are constants known as the Wasiljewa constants, K_1 and K_2 are the conductivities of the pure gases, and x_1 is the molar fraction of component 1.

In gas analysis, conductivities of pure gases are of limited value, and it is much more important to know how the conductivity of a mixture varies with the proportion of the constituent gases. However, as shown above, the relationship between the conductivity of a mixture of gases and the proportion of the constituents is complicated. When collisions occur between molecules of different gases the mathematics of the collisions are no longer simple, and the relationship between conductivity and the proportions of the constituents depends upon the molecular and physical constants of the gases, and on the intermolecular forces during a collision. In practice thermal conductivity instruments are therefore calibrated by establishing the required composition–conductivity curves experimentally.

Several forms of gas sensor based on thermal conductivity have been developed. The majority use the hot-wire method of measuring changes in conductivity, with the hot-wire sensors arranged in a Wheatstone bridge circuit.

5.3.1.1 Katharometer

A wire, heated electrically and maintained at constant temperature, is fixed along the axis of a cylindrical hole bored in a metal block which is also maintained at a constant temperature. The cylindrical hole is filled with the gas under test. The temperature of the wire reaches an equilibrium value when the rate of loss of heat by conduction, convection and radiation is equal to the rate of production of heat by the current in the wire. In practice, conduction through the gas is the most important source of heat loss. End-cooling, convection, radiation and thermal diffusion effects, though measurable, account for so small a part (less

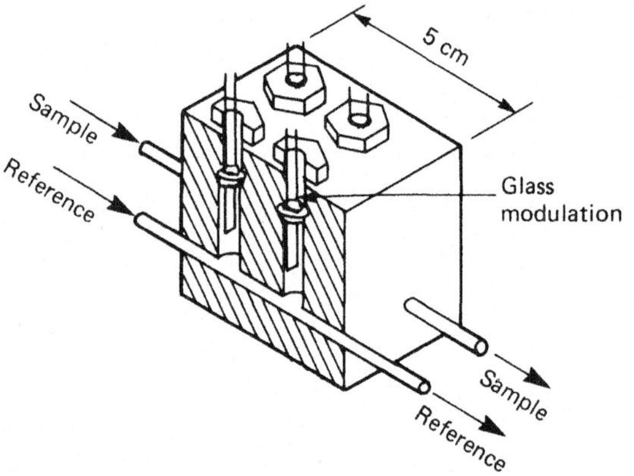

Figure 5.3 Cutaway drawing of 4-filament diffusion katharometer cell.

than 1 per cent each) of the total loss that they can satisfactorily be taken care of in the calibration. Most instruments are designed to operate with the wire mounted vertically, to minimize losses by convection. Convective losses also increase with the pressure of the gas, so the pressure should be controlled for accurate conductivity measurements in dense gases. The heat loss from the wire depends on the flow rate of gas in the sensor. In some instruments errors due to changes in gas flow are minimized because the gas does not flow through the cell but enters by diffusion, but otherwise the gas flow rate must be carefully controlled.

The resistance of the wire depends on its temperature; thus by measuring the resistance of the wire, its temperature may be found, and the wire is effectively used as a resistance thermometer. The electrical energy supplied to the wire to maintain the excess temperature is a measure of the total heat loss by conduction, convection and radiation. To measure the effects due to changes in the conductivity of the gas only, the resistance of the hot wire in a cell containing the gas to be tested is compared with the resistance of an exactly similar wire in a similar cell containing a standard gas. This differential arrangement also lessens the effects of changes in the heating current and the ambient temperature conditions. In order to increase the sensitivity two measuring and two reference cells are often used, and this arrangement is usually referred to as a 'katharometer'.

In the katharometer four filaments with precisely matched thermal and electrical characteristics are mounted in a massive metal block, drilled to form cells and gas paths. A cutaway drawing of a 4-filament cell is shown in Figure 5.3. Depending on the specific purpose, the filaments may be made of tungsten, tungsten–rhenium alloy, platinum or other alloys. For measurements in highly reactive gases gold-sheathed tungsten filaments may be used. The filaments are connected in a Wheatstone bridge circuit, which may be supplied from either a regulated-voltage or regulated-current power supply. The circuit for a constant-voltage detector is shown in Figure 5.4. The

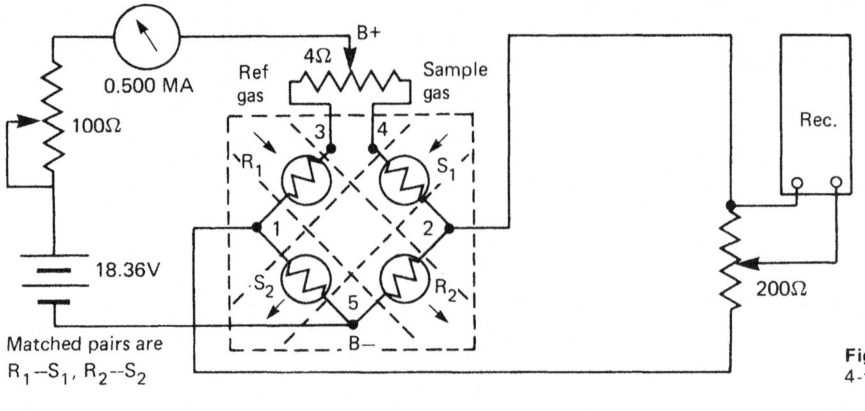

Figure 5.4 Circuit for 4-filament katharometer cell.

detector is balanced with the same gas in the reference and sample cells. If a gas of different thermal conductivity enters the sample cell, the rate of loss of heat from the sample filaments is altered, so changing their temperature and hence resistance. The change in resistance unbalances the bridge and the out-of-balance voltage is recorded as a measure of the change in gas concentration. The katharometer can be calibrated by any binary gas mixture, or for a gas mixture which may be regarded as binary, e.g. carbon dioxide in air.

A theory of the operation of the katharometer bridge follows. This is simplified but is insufficiently rigid for calibrations to be calculated. Small variations in the behaviour of individual filaments also mean that each bridge must be calibrated using mixtures of the gas the instrument is to measure.

Assume that the four arms of the bridge (Figure 5.4) have the same initial resistance R_1 when the bridge current is flowing and the same gas mixture is in the reference and sample cells. Let R_0 be resistance of filament at ambient temperature, R_1 working resistance (i.e. resistance when a current I flows), I current through one filament (i.e. half bridge current), and T wire temperature above ambient.

Then, at equilibrium, energy input is equal to heat loss

$$I^2R_1 = K_1T \tag{5.1}$$

where K_1 is a constant proportional to the thermal conductivity of the gas as most of the heat loss is by conduction through the gas. A simple expression for the working resistance is

$$R_1 = R_0(1+\alpha T) \tag{5.2}$$

where α is the temperature coefficient of resistance of the filament material. Then, from equations (5.1) and (5.2):

$$I^2R_1R_0\alpha = K_1(R_1 - R_0) \tag{5.3}$$

Then

$$R_1 = \frac{K_1R_0}{(K_1 - R_0I^2\alpha)}$$

$$= R_0 + \frac{K_1R_0}{(K_1 - R_0I^2\alpha)} - R_0$$

$$= R_0 + \frac{K_1R_0 - K_1R_0 + I^2R_0^2}{(K_1 - I^2R_0\alpha)}$$

$$= R_0 + \frac{I^2R_0}{(K_1 - I^2R_0\alpha)} \tag{5.4}$$

From equation (5.3), if $R_1 - R_0$ is small compared with R_1, then K_1 must be large compared with $I^2R_0\alpha$ and the term $I^2R_0\alpha$ can be ignored. Then

$$R_1 = R_0 + (I^2R_0^2\alpha/K_1) \tag{5.5}$$

If the two measurement filaments have a total resistance of R_1 and the reference filaments of R_2, the output voltage of the bridge E is given by

$$E = I(R_1 - R_2) \tag{5.6}$$

Combining equations (5.5) and (5.6):

$$E = I^3R_0^2\alpha[(1/K_1) - (1/K_2)] \tag{5.7}$$

where K_1 and K_2 are proportional to the conductivities of the gases in each pair of cells.

Equation (5.7) shows that the output is proportional to the cube of the bridge current but in practice the index is usually between $I^{2.5}$ and I^3. For accurate quantitative readings the bridge current must be kept constant.

This equation also shows that the output is proportional to the difference between the reciprocals of the thermal conductivities of the gases in each pair of cells. This is usually correct for small differences in thermal conductivity but does not hold for large differences.

These conditions show that the katharometer has maximum sensitivity when it is used to measure the concentration of binary or pseudo-binary gas mixtures whose components have widely different thermal conductivities and when the bridge current is as high as possible. The maximum bridge current is limited by the need to avoid overheating and distortion of the filaments, and bridge currents can be highest when a gas of high thermal conductivity is in the cell. When the katharometer is used as the detector in gas chromatography, hydrogen or helium, which have higher thermal conductivities than other common gases, is often used as the carrier gas, and automatic circuits may be fitted to reduce the current to the bridge to prevent overheating.

For maximum sensitivity, especially when it is necessary to operate the detector at low temperatures, the hot-wire filaments may be replaced by thermistors. A thermistor is a thermally sensitive resistor having a high negative coefficient of resistance, see Chapter 1. In the same manner as with hot wires, the resistance of the conductor is changed (in this case lowered) by the passage of current. Thermistor katharometers usually have one sensing and one reference element, the other resistors in the Wheatstone bridge being external resistors.

Except in the case of thermally unstable substances the katharometer is non-destructive, and it responds universally to all substances. The sensitivity is less than that of the ionization detectors, but is adequate for many applications. The detector is basically simple, and responds linearly to concentration changes over a wide range. It is used in gas chromatography and in a variety of custom-designed process analysers.

5.3.2 Flame ionization detector (FID)

An extensive group of gas detectors is based on devices in which changes in ionization current inside a chamber are measured. The ionization process occurs when a particle of high energy collides with a target particle which is thus ionized. The collision produces positive ions and secondary electrons which may be moved towards electrodes by application of an electric field, giving a measurable current, known as the ionization current, in the external circuit.

The FID utilizes the fact that, while a hydrogen–oxygen flame contains relatively few ions (10^7 ions cm^{-3}), it does contain highly energetic atoms. When trace amounts of organic compounds are added to the flame the number of ions increases (to approximately 10^{11} ions cm^{-3}) and a measurable ionization current is produced. It is assumed that the main reaction in the flame is

$$CH + O \rightarrow CHO + e$$

However, the FID gives a small response to substances that do not contain hydrogen, such as CCl_4 and CS_2. Hence it is probable that the reaction above is preceded by hydrogenation to form CH_4 or CH_3 in the reducing part of the flame. In addition to the ionization reactions, recombination also occurs, and the response of the FID is determined by the net overall ionization reaction process.

A schematic diagram of an FID is shown in Figure 5.5 and a cross-sectional view of a typical detector is shown in Figure 5.6. The sample gas, or effluent from a gas-chromatographic column, is fed into a hydrogen–air flame. The jet itself serves as one electrode and a second electrode is placed above the flame. A potential is applied across these electrodes. When sample molecules enter the flame, ionization occurs yielding a current which, after suitable amplification, may be displayed on a strip chart recorder.

The FID is a mass-sensitive, rather than concentration-sensitive, detector. This means that it does not respond to the concentration of a component entering it, but rather produces a signal which is proportional to the amount of organic material entering it per unit time. The ion current is effectively proportional to the number of carbon atoms present in the flame, and the sensitivity of the detector may be expressed as the mass of carbon passing through the flame per second required to give a detectable signal. A typical figure is 10^{-11} g C/sec.

The FID is sensitive to practically all organic substances, but is insensitive to inorganic gases and

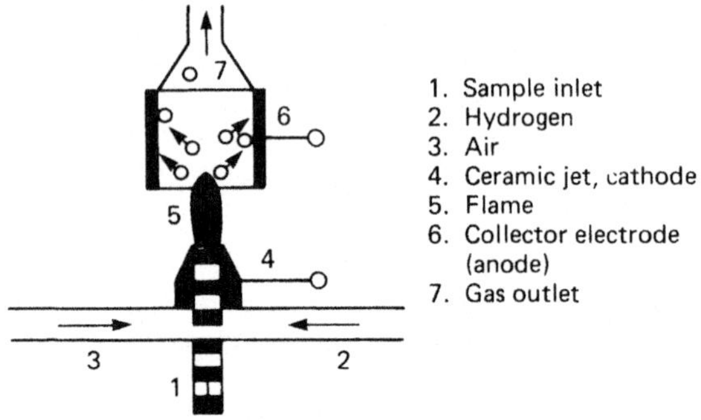

Figure 5.5 Flame ionization detector – schematic.

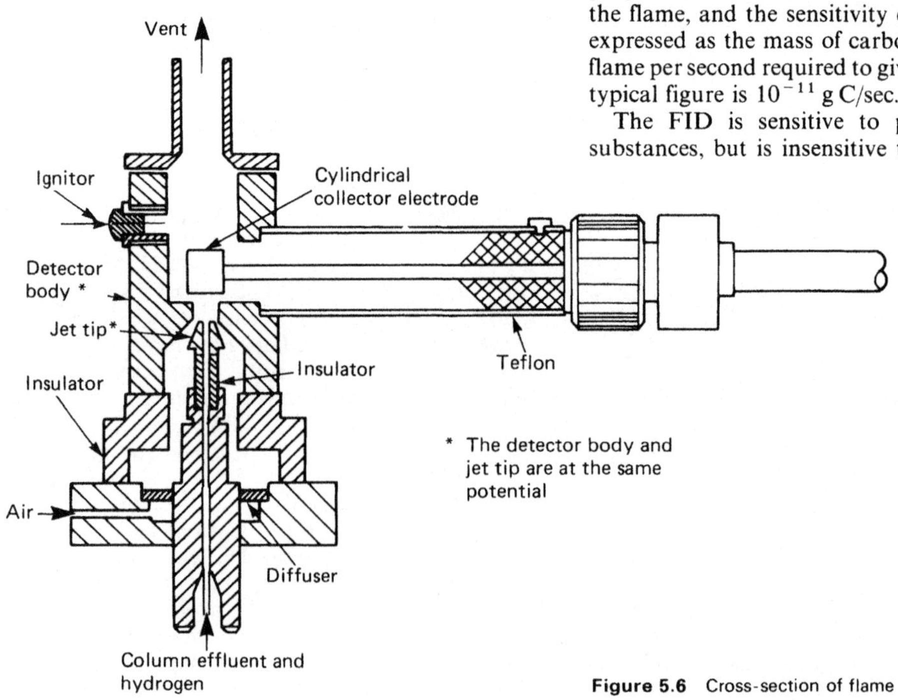

Figure 5.6 Cross-section of flame ionization detector.

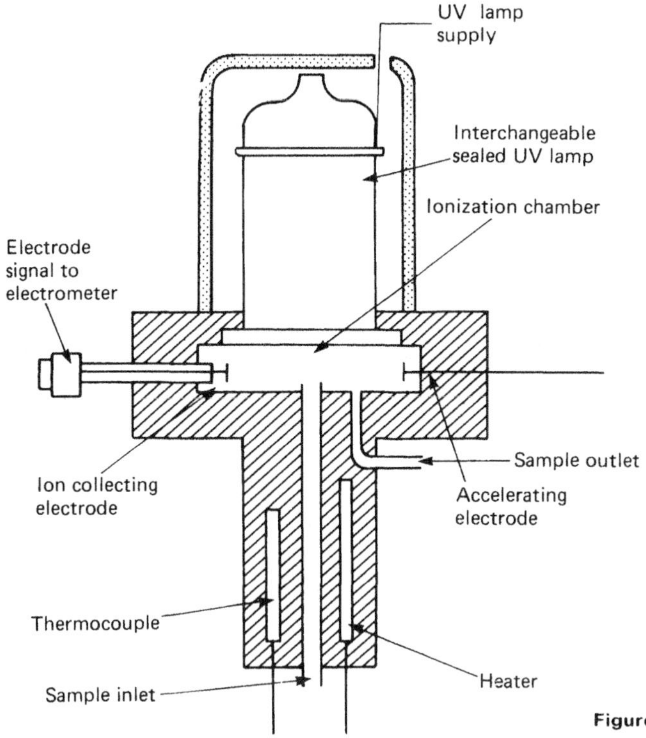

Figure 5.7 Photo-ionization detector.

water. It has a high sensitivity, good stability, wide range of linear response and low effective volume. It is widely used as a gas-chromatographic detector, and in total hydrocarbon analysers.

5.3.3 Photo-ionization detector (PID)

The photo-ionization detector (Figure 5.7) has some similarities to the flame ionization detector, and like the FID, it responds to a wide range of organic and also to some inorganic molecules. An interchangeable sealed lamp produces monochromatic radiation in the UV region. Molecules having ionization potentials less than the energy of the radiation may be ionized on passing through the beam. In practice, molecules with ionization potentials just above the photon energy of the incident beam may also be ionized, due to a proportion being in excited vibrational states. The ions formed are driven to a collector electrode by an electric field and the ion current is measured by an electrometer amplifier.

The flame in the FID is a high energy ionization source and produces highly fragmented ions from the molecules detected. The UV lamp in the PID is of lower quantum energy leading to the predominant formation of molecular ions. The response of the PID is therefore determined mainly by the ionization potential of the molecule, rather than the number of carbon atoms it contains. In addition the ionization energy in the PID may be selected by choice of the wavelength of the UV source, and the detector may be made selective in its response. The selectivity obtainable by use of three different UV lamps is shown in Figure 5.8. The ionization potentials of N_2, He, CH_3CN, CO and CO_2 are above the energy of all the lamps, and the PID does not respond to these gases.

The PID is highly sensitive, typically to picogram levels of organic compounds, and has a wide linear range. It may be used for direct measurements in gas streams or as a gas-chromatographic detector. When used as a detector in gas chromatography any of the commonly used carrier gases is suitable. Some gases, such as CO_2, absorb UV radiation and their presence may reduce the sensitivity of the detector.

5.3.4 Helium ionization detector

Monatomic gases, such as helium or argon, can be raised to excited atomic states by collision with energetic electrons emitted from a β-source. The metastable atomic states are themselves highly energetic and lose their energy by collision with other atomic or molecular species. If the helium contains a small concentration of a gas whose ionization potential is less than the excitation of the metastable helium atoms, ions will be formed in the collision, so

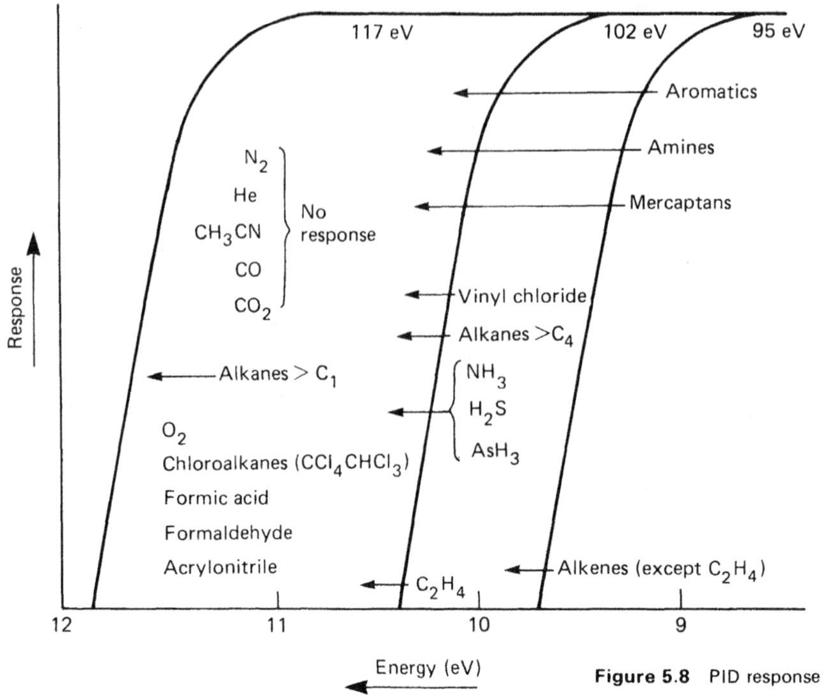

Figure 5.8 PID response for the various ultraviolet lamps.

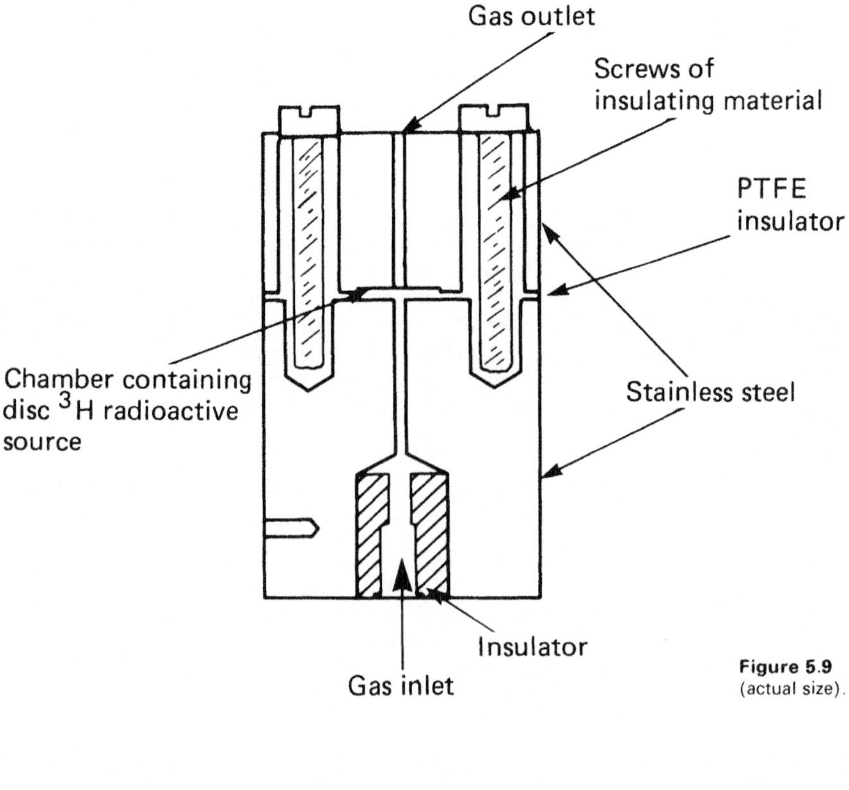

Figure 5.9 Helium ionization detector (actual size).

A – Inlet for carrier gas and anode
B – Diffuser – made of 100 mesh brass gauze
C – Source of ionizing radiation
D – Gas outlet and cathode

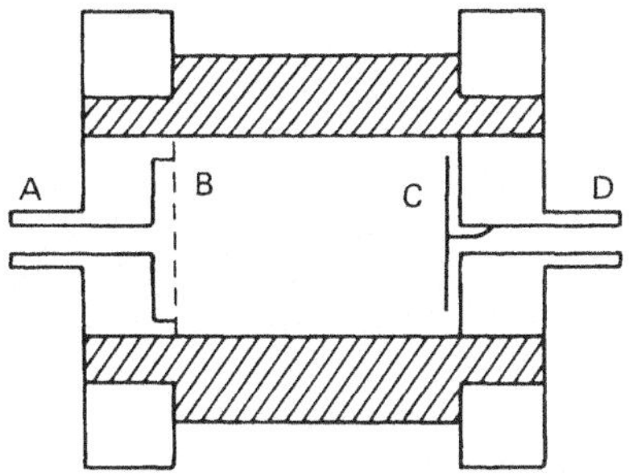

Brass

PTFE

Figure 5.10 Electron capture detector.

increasing the current-carrying capacity of the gas. This is the basis of the helium ionization detector.

The main reactions taking place can be represented as

$$He + e \rightarrow He^* + e$$

$$He^* + M \rightarrow M^+ + He + e$$

where M is the gas molecule forming ions. However, other collisions can occur, for example between metastable and ground-state helium atoms, or between metastable atoms, which may also result in ion formation.

The helium ionization detector (Figure 5.9) typically consists of a cylindrical chamber, approximately 1 cm in diameter and a few millimetres long, containing a β-emitting radioactive source. The ends of the cylinder are separated by an insulator and form electrodes. The detector is used as part of a gas-chromatographic system, with helium as the carrier gas.

It can be shown that the ionization mechanism described above depends on the number of atoms formed in metastable states. It can also be shown that the probability of formation of metastable states depends on the primary electron energy and on the intensity of the applied electric field. The reaction exhibits the highest cross-section for electrons with an energy of about 20 eV, and a field strength of 500 V/cm torr. Tritium (^3H) sources of 10–10 GBq or ^{63}Ni β-sources of 400–800 MBq activity are usually used, but the free path of the β-particles is very short, and the performance of the detector is strongly dependent on its geometry.

The helium ionization detector is used in gas chromatography, when its ability to measure trace levels of permanent gases is useful. However, the carrier gas supply must be rigorously purified.

5.3.5 Electron capture detector

The electron capture detector (Figure 5.10) consists of a cell containing a β-emitting radioactive source, purged with an inert gas. Electrons emitted by the radioactive source are slowed to thermal velocities by collision with the gas molecules, and are eventually collected by a suitable electrode, giving rise to a standing current in the cell. If a gas with greater electron affinity is introduced to the cell, some of the electrons are 'captured' forming negative ions, and the current in the cell is reduced. This effect is thc basis of the electron capture detector. The reduction in current is due both to the difference in mobility between electrons and negative ions, and to differences in the rates of recombination of the ionic species and electrons.

The radioactive source may be tritium or ^{63}Ni, with ^{63}Ni usually being preferred since it allows the detector to be operated at higher temperatures, thus lessening the effects of contamination. A potential is applied between the electrodes which is just great enough to collect the free electrons. Originally, the detector was operated under d.c. conditions, potentials up to 5 volts being used, but under some conditions space charge effects produced anomalous results. Present detectors use a pulsed supply, typically 25 to 50 volts, 1 microsecond pulses at intervals of 5 to 500 microseconds. Either the pulse interval is selected and the change in detector current monitored, or a feedback system maintains a constant current and the pulse interval is monitored.

The electron capture detector is extremely sensitive to electronegative species, particularly halogenated compounds and oxygen. To obtain maximum sensitivity for a given compound, the choice of carrier gas, pulse interval or detector current and detector temperature must be optimized.

The electron capture detector is most often used in gas chromatography, with argon, argon–methane mixture or nitrogen as carrier gas, but it is also used in leak or tracer detectors. The extreme sensitivity of the ECD to halogenated compounds is useful, but high purity carrier gas and high stability columns are required to prevent contamination. Under optimum conditions, 1 part in 10^{12} of halogenated compounds such as Freons, can be determined.

5.3.6 Flame photometric detector (FPD)

Most organic and other volatile compounds containing sulphur or phosphorus produce chemiluminescent species when burned in a hydrogen-rich flame. In a flame photometric detector (Figure 5.11) the sample gas passes into a fuel-rich H_2/O_2 or H_2/air mixture which produces simple molecular species and excites them to higher electronic states. These excited species subsequently return to their ground states and emit characteristic molecular band spectra. This emission is monitored by a photomultiplier tube through a suitable filter, thus making the detector selective to either sulphur or phosphorus. It may also be sensitive to other elements, including halogens and nitrogen.

The FPD is most commonly used as a detector for sulphur-containing species. In this application, the response is based on the formation of excited S_2 molecules, S_2^*, and their subsequent chemiluminescent emission. The original sulphur-containing molecules are decomposed in the hot inner zone of the flame, and sulphur atoms are formed which combine to form S_2^* in the cooler outer cone of the flame. The exact mechanism of the reaction is uncertain, but it is believed that the excitation energy for the $S_2 \rightarrow S_2^*$ transition may come from the formation of molecular hydrogen or water in the flame, according to the reactions

$$H + H + S_2 \rightarrow S_2^* + H_2 \ (4.5\ \text{eV})$$

$$H + OH + S_2 \rightarrow S_2^* + H_2O \ (5.1\ \text{eV})$$

As the excited S_2 molecule reverts to the ground state it emits a series of bands in the range 300–450 nm, with the most intense bands at 384.0 and 394.1 nm. The 384.0 nm emission is monitored by the photomultiplier tube.

The FPD is highly selective and sensitive, but the response is not linearly proportional to the mass-flow rate of the sulphur compound. Instead, the relationship is given by:

$$I_{S_2} = I_0[S]^n$$

where I_{S_2} is the observed intensity of the emission (photomultiplier tube output), [S] is the mass-flow rate of sulphur atoms (effectively the concentration of the sulphur compound) and n is a constant, found to be between 1.5 and 2, depending on flame conditions. Commercial analysers employing the FPD often incorporate a linearizing circuit to give an output which is directly proportional to sulphur mass-flow. The detector response is limited to two or three orders of magnitude.

The FPD is highly selective, sensitive (10^{-11} g) and relatively simple, but has an extremely non-linear response. It is used in gas chromatography and in sulphur analysers.

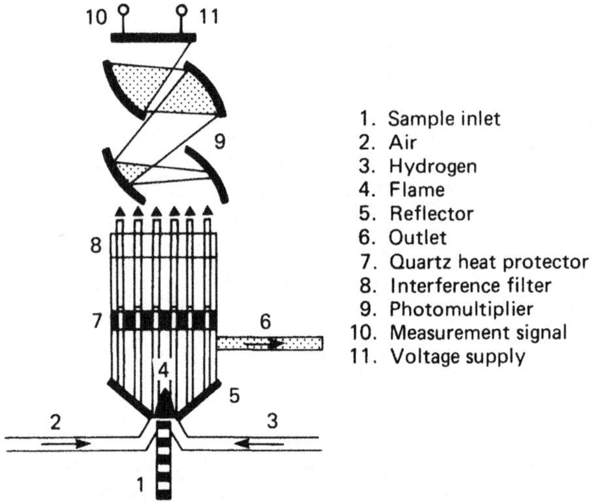

Figure 5.11 Flame photometric detector.

5.3.7 Ultrasonic detector

The velocity of sound in a gas is inversely proportional to the square root of its molecular weight. By measuring the speed of sound in a binary gas mixture, its composition can be deduced, and this technique is the basis of the ultrasonic detector (Figure 5.12). A

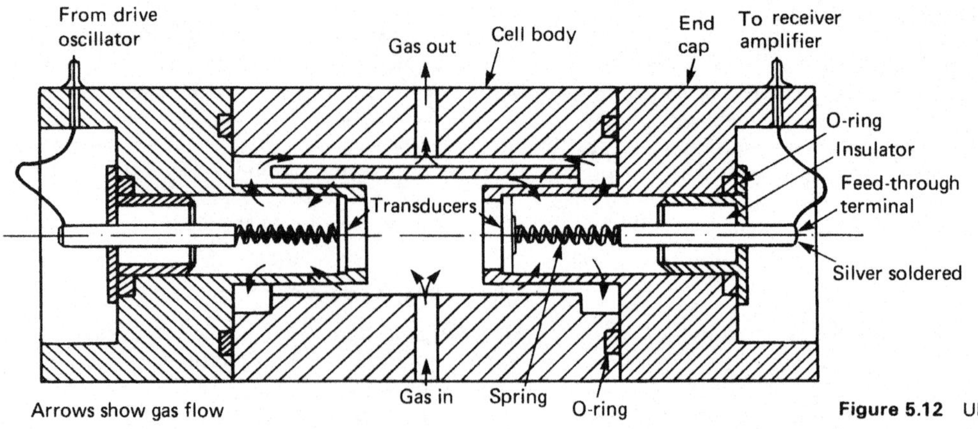

Figure 5.12 Ultrasonic detector.

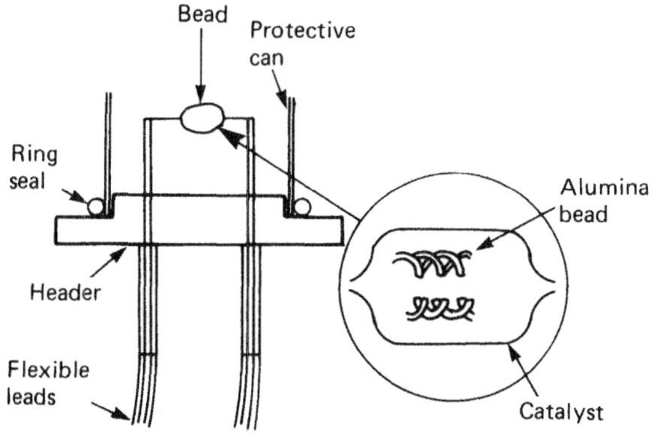

Figure 5.13 Catalytic gas-sensing element.

quartz crystal transducer located at one end of the sample cell sound tube acts as the emitter and an identical crystal located at the other end of the sound tube acts as the receiver. To obtain efficient transfer of sound energy between the gas and the transducers, the detector must be operated at above atmospheric pressure, and the gas in the cell is typically regulated to 1 to 7 bar gauge depending on the gas. The phase shift of the sound signal traversing the cell between the emitter and receiver is compared to a reference signal to determine the change in speed of sound in the detector.

The detector is most often used in gas chromatography. It has a universal response, and the output signal is proportional to the difference in molecular weight between the gaseous species forming the binary mixture. When used as a gas-chromatographic detector it has good sensitivity (10^{-9}–10^{-10} g) and a wide linear dynamic range (10^6), and allows a wide choice of carrier gas. However, precise temperature control is required, and the electronic circuitry is complex. It may be a useful alternative where flames cannot be used, or where a katharometer would not respond to all components in a mixture.

5.3.8 Catalytic detector (pellistor)

Catalytic gas detectors operate by measuring the heat output resulting from the catalytic oxidation of flammable gas molecules to carbon dioxide and water vapour at a solid surface. By use of a catalyst, the temperature at which the oxidation takes place is much reduced compared with gas phase oxidation. The catalyst may be incorporated into a solid state sensor containing an electrical heater and temperature-sensing device. A stream of sample gas is fed over the sensor, and flammable gases in the sample are continuously oxidized, releasing heat and raising the temperature of the sensor. Temperature variations in the sensor are monitored to give a continuous record of the flammable-gas concentration in the sample.

The most suitable metals for promoting the oxidation of molecules containing C—H bonds, such as methane and other organic species, are those in Group 8 of the Periodic Table, particularly platinum and palladium. The temperature sensor is usually a platinum resistance thermometer, wound in a coil and also used as the electrical heater for the sensor. The resistance is measured by connecting the sensor as one arm of a Wheatstone bridge and measuring the out-of-balance voltage across the bridge.

The construction of a typical catalytic sensing element is shown in Figure 5.13. A coil of 50 μm platinum wire is mounted on two wire supports which also act as electrical connections. The coil is embedded in porous ceramic material, usually alumina, to form a bead about 1 mm long. The catalyst material is impregnated on the outside of the bead. This type of catalytic sensor is often called a 'pellistor'. The choice of catalyst, and of the treatment of the outside of the bead, for example by inclusion of a diffusion layer, influences the overall sensitivity of the sensor, and the relative sensitivity to different gases. The sensitivity and selectivity are also influenced by the choice of catalyst and by the temperature at which the sensor is operated. Palladium and its oxides are the most widely used catalysts; they have the advantage that they are much more active than platinum, enabling the sensor to be operated at the lowest possible temperature. The sensor is mounted in a protective open-topped can as shown, so that the gas flow to the sensor is largely diffusion-controlled.

The Wheatstone bridge network commonly used with a catalytic sensor is shown in Figure 5.14. The sensing element forms one arm of the bridge, and the second arm is occupied by a compensator element.

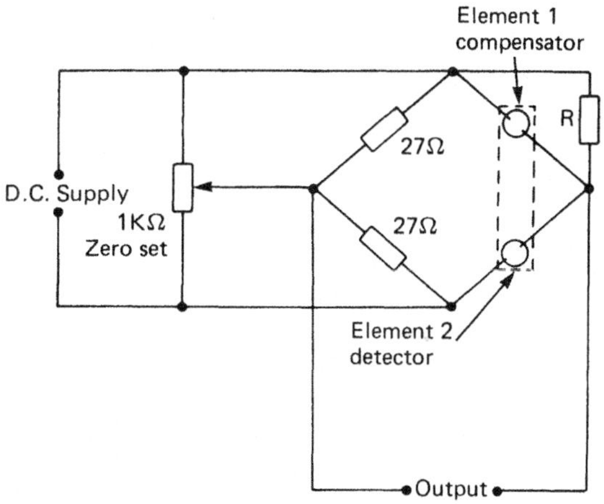

Figure 5.14 Wheatstone bridge network used with catalytic detector.

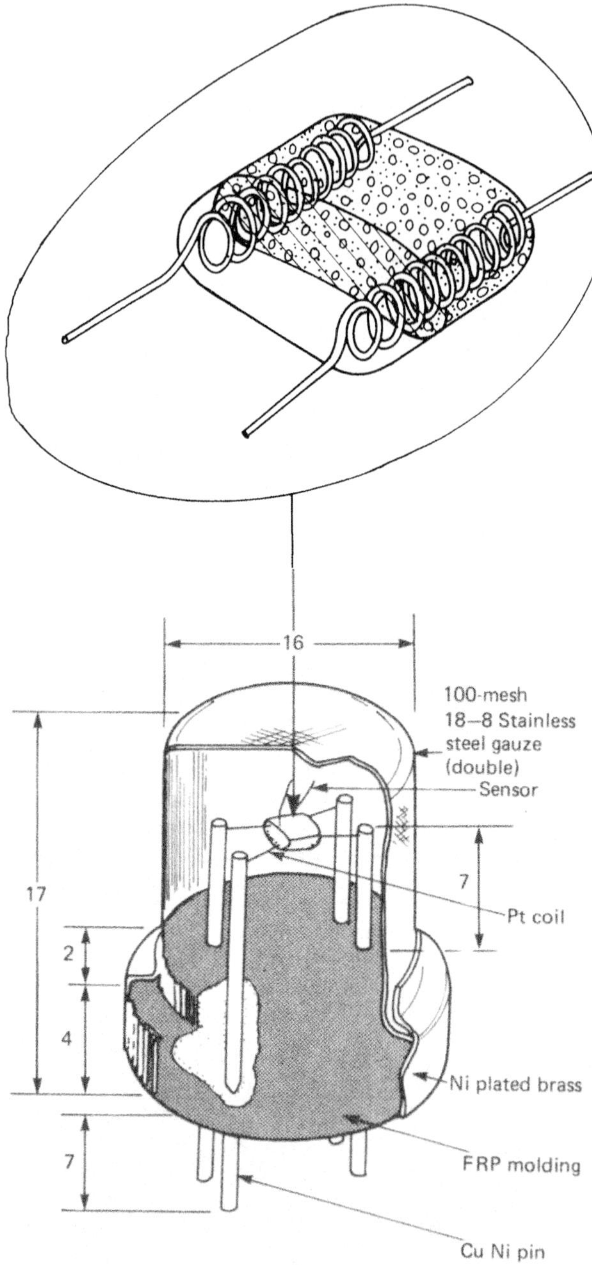

Figure 5.15 Semiconductor sensor.

This is a ceramic bead element, identical in construction to the sensor, but without the catalytic coating. The sensor and compensator are mounted close together in a suitable housing so that both are exposed to the same sample gas. The pellistor or catalytic sensor is the basis of the majority of portable flammable-gas detectors.

5.3.9 Semiconductor detector

The electrical conductivity of many metal oxide semiconductors, particularly those of the transition and heavy metals, such as tin, zinc and nickel, is changed when a gas molecule is adsorbed on the semiconductor surface. Adsorption involves the formation of bonds between the gas molecule and the semiconductor, by transfer of electrical charge. This charge transfer changes the electronic structure of the semiconductor, changing its conductivity. The conductivity changes are related to the number of gas molecules adsorbed on the surface, and hence to the concentration of the adsorbed species in the surrounding atmosphere.

A typical semiconductor detector is shown in Figure 5.15. The semiconducting material is formed as a bead, about 2–3 mm in diameter, between two small coils of platinum wire. One of the coils is used as a heater, to raise the temperature of the bead so that the gas molecules it is desired to detect are reversibly adsorbed on the surface, and the resistance of the bead is measured by measuring the resistance between the two coils. The bead is mounted in a stainless-steel gauze enclosure (Figure 5.15) to ensure that molecules diffuse to the semiconductor surface, thus ensuring that the device is as free as possible from the effects of changes in the flow rate of the sample gas.

Semiconductor detectors are mainly used as low-cost devices for detection of flammable gases. A

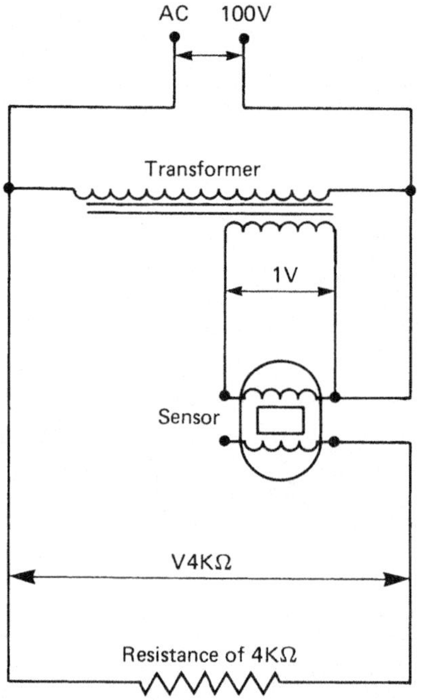

Voltage measured at the end of 4KΩ's resistance (Output)

Figure 5.16 Measuring circuit for semiconductor sensor.

Table 5.2 Properties and applications of gas detectors

Detector	*Applicability*	*Selectivity*	*Carrier or bulk gas*	*Lower limit of detection (grams)*	*Linear range*	*Typical applications*
Thermal conductivity	Universal	non-selective	He, H_2	10^{-6}–10^{-7}	10^4	Analysis of binary or pseudo-binary mixtures; gas chromatography
Flame ionization	Organic compounds	non-selective	N_2	10^{-11}	10^6	Gas chromatography; hydrocarbon analysers
Photo-ionization	Organic compounds except low molecular weight hydrocarbons	limited	N_2	10^{-11}–10^{-12}	10^7	Gas chromatography
Helium ionization	Trace levels of permanent gases	non-selective	He	10^{-11}	10^4	Gas chromatography
Electron capture	Halogenated and oxygenated compounds	response is highly compound-dependent	Ar, N_2, $N_2 + 10\%\ CH_4$	10^{-12}–10^{-13}	10^3	Gas chromatography, tracer gas detectors, explosive detectors
Flame photometric	Sulphur and phosphorus compounds	selective to compounds of S or P	N_2, He	10^{-11}	5×10^2 (S) 10^3 (P)	Gas chromatography, sulphur analysers
Ultrasonic detector	Universal	non-selective, mainly low molecular weight	H_2, He, Ar, N_2, CO_2	10^{-9}–10^{-10}	10^6	Gas chromatography
Catalytic (pellistor)	Flammable gases	selective to flammable gases	Air	*		Flammable gas detectors
Semiconductor	Flammable gases, other gases	limited	Air	*		Low-cost flammable gas detectors

* The performance of these detectors depends on the individual design and application.

suitable power-supply and measuring circuit is shown in Figure 5.16. The main defect of the devices at present is their lack of selectivity.

5.3.10 Properties and applications of gas detectors

The properties and applications of the most commonly used gas detectors are summarized in Table 5.2.

5.4 Process chromatography

On-line or process gas chromatographs are instruments which incorporate facilities to carry out automatically the analytical procedure for chromatographic separation, detection and measurement of predetermined constituents of gaseous mixtures. Samples are taken from process streams and are presented, in a controlled manner and under known conditions, to the gas chromatograph. Successive analyses may be made, on a regular timed basis, on aliquots of sample taken from a single stream, or by use of suitable stream-switching valves, a single process chromatograph may carry out automatic sequential analyses on process streams originating from several different parts of the plant.

The main components of a typical process chromatograph system are shown in Figure 5.17. These components are: a supply of carrier gas to transport the sample through the column and detector, a valve for introduction of known quantities of sample, a chromatographic column to separate the sample into its components, a detector and associated amplifier to sense and measure the components of the sample in the carrier-gas stream, a programmer to actuate the operations required during the analytical sequence and to control the apparatus, and a display or data-processing device to record the results of the analyses.

5.4.1 Sampling system

The sampling system must present a homogeneous and representative sample of the gas or liquid to be analysed, to the gas chromatograph. In process chromatography a continuous stream of the sample is taken, usually by means of a fast bypass loop, and treated as necessary for example by drying, filtering or adjustment of temperature or pressure. Discrete volumes of the treated sample stream are periodically injected into the carrier gas stream of the chromatograph by means of a gas (or liquid) sampling valve. The chromatograph is normally supplied with sample from the point or points to be monitored by use of permanently installed sampling lines. However, where

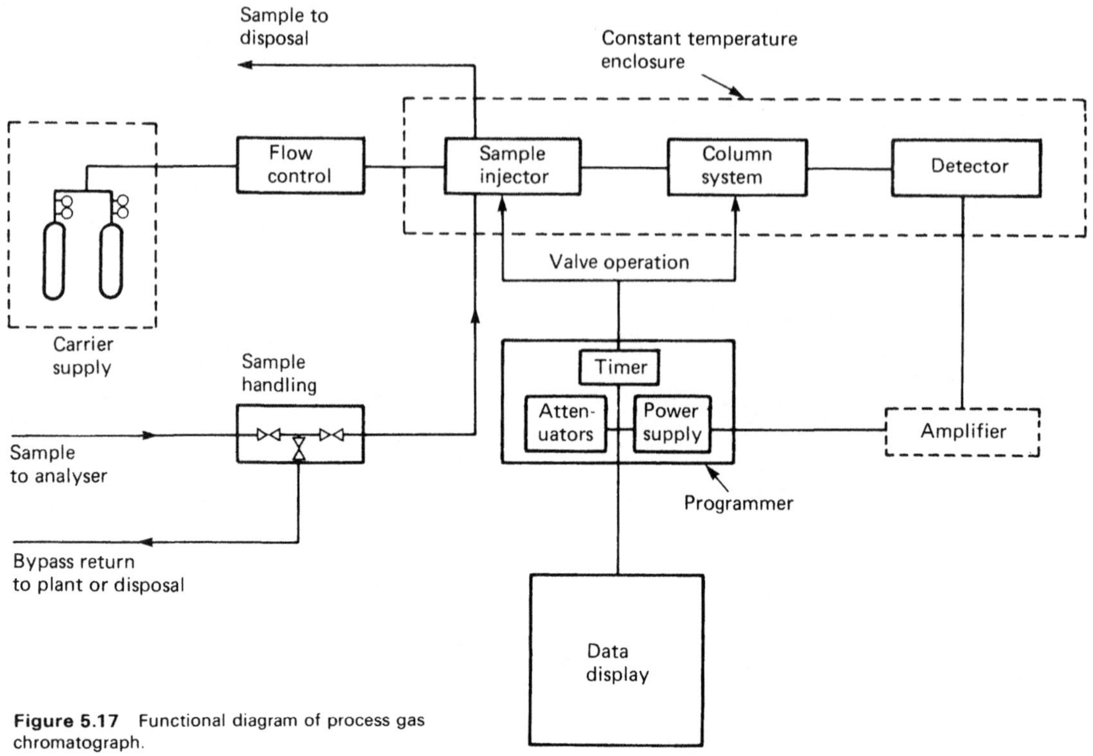

Figure 5.17 Functional diagram of process gas chromatograph.

the frequency of analysis does not justify the installation of special lines, samples may be collected in suitable containers for subsequent analysis. Gas samples may be collected under pressure in metal (usually stainless steel) cylinders or at atmospheric pressure in gas pipettes, gas sampling syringes or plastic bags. For analysis of gases at very low concentrations such as the determination of pollutants in ambient air, the pre-column or adsorption tube concentration technique is often used. The sample is drawn or allowed to diffuse through a tube containing a granular solid packing to selectively adsorb the components of interest. The tube is subsequently connected across the sample loop ports of the gas sampling valve on the chromatograph and heated to desorb the compounds to be analysed into the carrier-gas stream.

It is essential that the sample size should be constant for each analysis, and that it is introduced into the carrier gas stream rapidly as a well-defined slug. The sample should also be allowed to flow continuously through the sampling system to minimize transportation lag. Chromatographic sampling or injection valves are specially designed changeover valves which enable a fixed volume, defined by a length of tubing (the sample loop) to be connected in either one of two gas streams with only momentary interruption of either stream. The design and operation of a typical sampling valve is shown in Figure 5.18. The inlet and outlet tubes terminate in metal (usually stainless steel) blocks with accurately machined and polished flat faces. A slider of soft plastic material, with channels or holes machined to form gas paths, is held against the polished faces and moved between definite positions to fill the loop or inject the sample. The main difference between 'gas' and 'liquid' sampling valves is in the size of sample loop. In the 'gas' sampling valve the loop is formed externally, and typically has a volume in the range 0.1–10 ml. For liquid sampling the volumes required are smaller and the loop is formed in the internal channels of the valve and may have a volume as small as 1 μl. In process chromatography, sampling valves are normally fitted with electric or pneumatic actuators so that they may be operated automatically by the programmer at predetermined times during the analytical sequence.

When it is required to change between columns or detectors during an analysis, similar types of valves are required. The number of ports, and the arrangement of the internal channels, may be tailored for the individual application. Figure 5.19 shows an arrangement where a single valve is used for sample injection and backflushing in a chromatograph with two analytical columns in series. The sample is injected

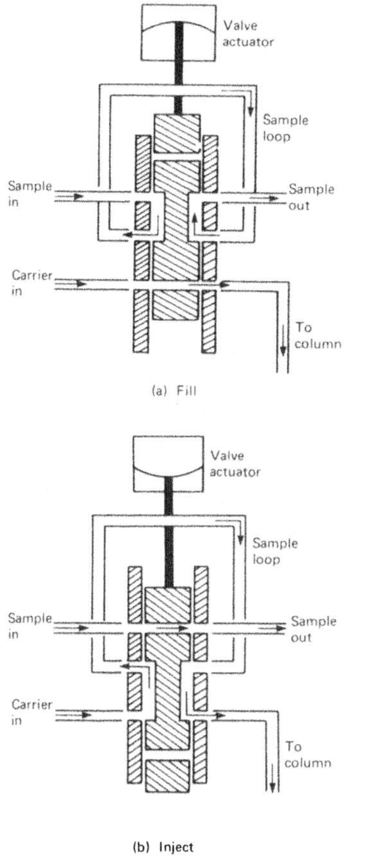

Figure 5.18 Gas-sampling valve (schematic).

onto column 1, which is chosen so that the components of interest are eluted first, and pass to column 2. At a predetermined time, the valve is switched to refill the sample loop and to reverse the flow of carrier gas to column 1, while the forward flow is maintained in column 2 to effect the final separation of the components of the sample. By this means components of no interest, such as high-boiling compounds or solvents, can be 'backflushed' to waste before they reach column 2 or the detector, thus preserving the performance of the columns and speeding the analytical procedure.

The gas sample must arrive at the sampling valve at or only slightly above atmospheric pressure, at a flow rate typically in the range 10–50 ml min^{-1}, and be free from dust, oil or abrasive particles. The sampling system may also require filters, pressure or flow controllers, pumps and shut-off valves for control and processing of the sample stream. All the components of the system must be compatible with the chemical species to be sampled, and must be capable of withstanding the range of pressures and temperatures expected.

Many applications require analysis of two or more process streams with one analyser. In these instances a sample line from each stream is piped to the analyser and sample lines are sequentially switched through solenoid valves to the sampling valve. When multistream analysis is involved, inter-sample contamination must be prevented. Contamination of samples can occur through valve leakage and inadequate flushing of common lines. To ensure adequate flushing, the capacity of common lines is kept to a minimum and the stream selection valves are timed so that while the sample from one stream is being analysed, the sample from the next stream is flowing through all common lines.

Prevention of inter-sample contamination from valve leakage is accomplished by locating valves with respect to pressure drops such that any leakage will flow to vent rather than intermix in common lines. A typical flow arrangement for gas supplies to a chromatograph for multi-stream application is shown in Figure 5.20. This is designed to ensure that the sample and other supplies are delivered at the correct flow rate

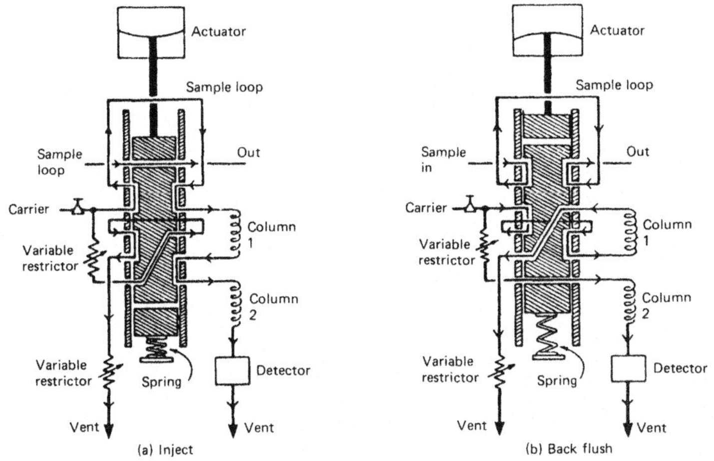

Figure 5.19 Schematic diagram of sample and backflush valve.

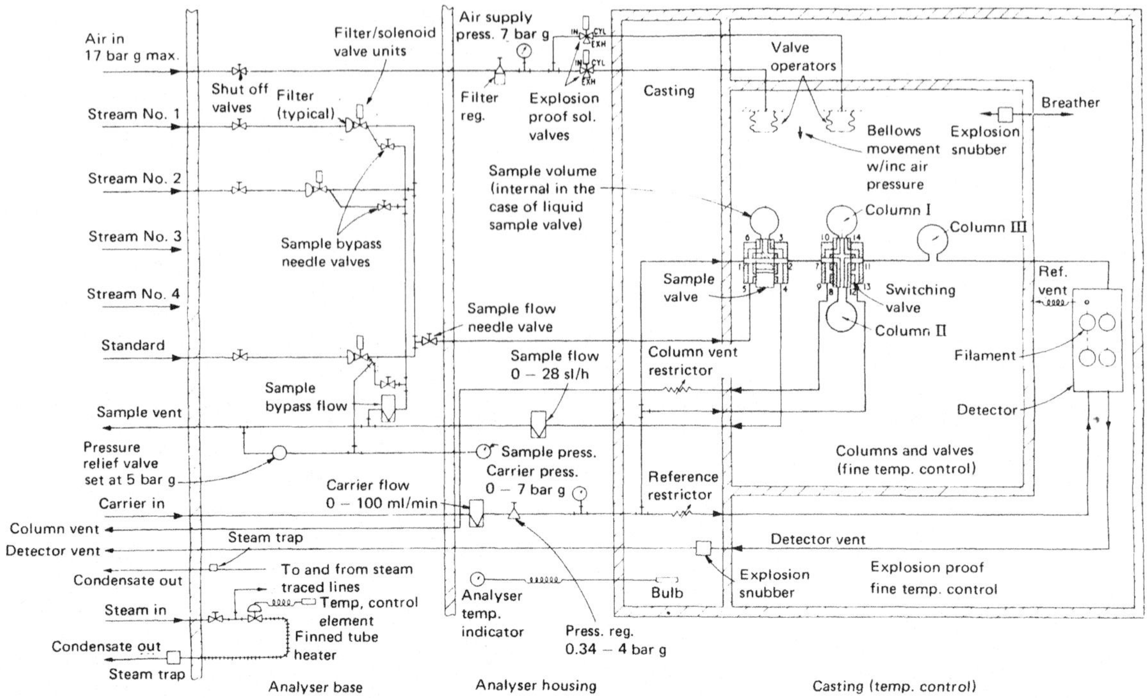

Figure 5.20 Flow diagram of multi-stream chromatograph with thermal conductivity detector (courtesy The Foxboro Company).

and pressure. A pressure-relief valve is fitted to protect the sampling valve from excessive pressure, and shut-off valves are fitted on all services except bottled gas lines.

In some applications additional conditioning of the sample is required. Typical of these would be trace-heating of sample lines to maintain a sample in a gaseous state, vaporization to change a liquid to a gas and elimination of stream contaminants by mechanical or chemical means.

5.4.2 Carrier gas

The carrier gas transports the components of the sample over the stationary phase in the chromatographic column. The carrier gas must not react with the sample, and for maximum efficiency when using long columns, it is advantageous to use a gas of low viscosity. However, the most important criterion in choosing a carrier gas is often the need to ensure compatibility with the particular detector in use.

The primary factors determining the choice of carrier gas are the effect of the gas on component resolution and detector sensitivity. The carrier gas and type of detector are chosen so that the eluted components generate large signals. For this reason, helium is generally used with thermal conductivity cells because of its high thermal conductivity. Hydrogen has a higher thermal conductivity and is less expensive than helium, but because of precautions necessary when using hydrogen, helium is preferred where suitable.

Specific properties of a particular carrier gas are exploited in other types of detectors, for example helium in the helium ionization detector. In special instances a carrier gas other than that normally associated with a particular detector may be used for other reasons. For example, to measure hydrogen in trace quantities using a thermal conductivity detector, it is necessary to use a carrier gas other than helium because both helium and hydrogen have high and similar thermal conductivities. Accordingly, argon or nitrogen is used because either has a much lower thermal conductivity than hydrogen, resulting in a larger difference in thermal conductivity and greater output.

The flow rate of carrier gas affects both the retention time of a compound in the column, and the shape of the chromatographic peak and hence the amplitude of the detector signal. It is therefore essential for the flow rate to be readily adjustable to constant known values. The gas is usually supplied from bottles, with pressure reducing valves to reduce the pressure to a level compatible with the flow control equipment, and sufficient to give the required flow rate through the column and detector.

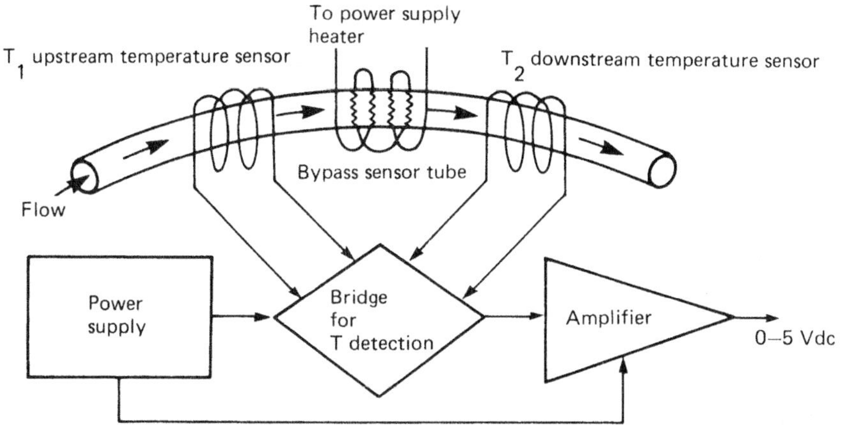

Figure 5.21 Principle of operation of electronic mass-flow controller (courtesy Brooks).

The flow rate of carrier gas may be measured and controlled either mechanically or electronically. Mechanical controllers are either precision pressure regulators which maintain a constant pressure upstream of the column and detector, or differential pressure regulators which maintain a constant pressure drop across a variable restriction. The principle of operation of one type of electronic flow controller is shown in Figure 5.21. A proportion of the gas stream is diverted via a narrow tube, fitted with an electric heating coil as shown. Sensor coils of resistance wire are wound on the tube upstream and downstream of the heating coil. Heat at a constant rate is supplied to the heating coil. Gas passing through the tube is heated by the coil, and some heat is transferred to the downstream sensor. The sensor coils are connected in a Wheatstone bridge circuit. The out-of-balance signal from the bridge, caused by the difference in temperatures and hence resistance of the upstream and downstream coils, depends on the mass-flow rate of gas through the tube, and on the specific heat of the gas. (See also Volume 1, p. 31.) The signal, suitably amplified, can be used to give a direct readout of the flow rate of gas through the tube, and can be used to control the flow by feeding the signal to open or close a solenoid valve in the main gas line downstream of the sensing device.

In cases where the carrier gas flow rate is controlled mechanically, a rotameter is provided to indicate the flow rate. However, the best indication of correct flow is often the analysis record itself, as the retention times of known components of the sample should remain constant from one injection to the next.

5.4.3 Chromatographic column

The separating columns used in process chromatographs are typically 1–2 m lengths of stainless steel tubing, 3–6 mm outer diameter, wound into a helix for convenient housing, and packed with a solid absorbent. Separation of permanent gases is normally carried out on columns packed with molecular sieve. These are synthetic zeolites, available in a range of effective pore diameters. Porous polymeric materials have been developed which are capable of separating a wide range of organic and inorganic molecules, and use of these proprietary materials gives much more predictable column performance than when liquid-coated solids are used. In addition the polymeric materials are thermally stable and do not suffer from 'bleed' or loss of the liquid stationary phase at high temperatures which can give rise to detector noise or drift in the baseline of the chromatogram.

One or more columns packed with these materials can normally be tailored to the needs of most process analyses. However, in certain cases it may be necessary to include valves to switch between columns or detectors during the analysis, or to divert the carrier gas to waste to prevent a certain component, for example a solvent present in high concentration, from reaching the detector. These switching operations are referred to as backflushing, or heart-cutting if the unwanted peak occurs in the middle of the chromatogram.

5.4.4 Controlled temperature enclosures

Many components of the gas chromatograph, including the injection valve, columns and detectors require to be kept at constant temperatures or in environments whose temperature can be altered at known rates, and separate temperature-controlled zones are usually provided in the instrument.

Two general methods are used to distribute heat to maintain the temperature-sensitive components at constant temperatures (± 0.1 K or better) and to minimize temperature gradients. One uses an air bath, and the other metal-to-metal contact (or heat sink). The former depends on circulation of heated air and the latter upon thermal contact of the temperature-sensitive elements with heated metal.

An air bath has inherently fast warm-up and comparatively high temperature gradients and offers the advantage of ready accessibility to all components within the temperature-controlled compartment. The air bath is most suitable for temperature programming and is the usual method for control of the temperature of the chromatographic column.

Metal-to-metal contact has a slower warm-up but relatively low temperature gradients. It has the disadvantage of being a greater explosion hazard, and may require the analyser to be mounted in an explosion-proof housing resulting in more limited accessibility and more difficult servicing. The detectors are often mounted in heated metal blocks for control of temperature.

The choice of the method of heating and temperature control may depend on the location where the instrument is to be used. Instruments are available with different degrees of protection against fire or explosion hazard. For operation in particularly hazardous environments, for example where there may be flammable gases, instruments are available where the operation, including temperature control, valve switching, and detector operation is entirely pneumatic, with the oven being heated by steam.

5.4.5 Detectors

A gas-chromatographic detector should have a fast response, linear output over a wide range of concentration, be reproducible and have high detection sensitivity. In addition the output from the detector must be zero when pure carrier gas from the chromatographic column is passing through the detector.

In process chromatography, the most commonly used detectors are the thermal conductivity and flame ionization types. Both have all the desirable characteristics listed above, and one or other is suitable for most commonly analysed compounds: the thermal conductivity detector is suitable for permanent gas analysis and also responds universally to other compounds while the flame ionization detector responds to almost all organic compounds. In addition these detectors can be ruggedly constructed for process use and can be used with a wide range of carrier gases. Most other detectors have disadvantages in comparison with these two, for example, fragility, non-linear response or a requirement for ultra-pure carrier-gas supplies, and although widely used in laboratory chromatographs, their application to process instruments is restricted.

The helium ionization detector may be used for permanent gas analyses at trace levels where the katharometer is insufficiently sensitive, and the ultrasonic detector may be a useful alternative in applications where a flame cannot be used or where a katharometer cannot be used for all components in a mixture. The selective sensitivity of the electron capture detector to halogenated molecules may also find occasional application. A comprehensive list of gas-detecting devices, indicating which are suitable for use in gas chromatography, is given in Table 5.2 (p. 157).

5.4.6 Programmers

Analysis of a sample by gas chromatography requires the execution of a series of operations on or by the instrument at predetermined times after the analytical sequence is initiated by injection of the sample. Other instrumental parameters must also be continuously monitored and controlled. Process gas chromatographs incorporate devices to enable the analytical sequence to be carried out automatically, and the devices necessary to automate a particular instrument are usually assembled into a single module, known as the programmer or controller.

At the most basic level the programmer may consist of mechanical or electromechanical timers, typically of the cam-timer variety, to operate relays or switches at the appropriate time to select the sample stream to be analysed, operate the injection valve and start the data recording process, combined with a facility to correct the output of the chromatograph for baseline drift. Some chromatographs with built-in microprocessors may incorporate the programmer as part of the central control and data acquisition facility, or the programmer itself may contain a microprocessor and be capable of controlling and monitoring many more of the instrumental parameters as well as acting as a data-logger to record the output of the chromatograph. Computer-type microprocessor-based integrators are available for laboratory use, and in many cases these have facilities to enable them to be used as programmers for the automation of laboratory gas chromatographs.

When the process chromatograph is operated in the automatic mode, all the time-sequenced operations are under programmer control. These will typically include operations to control the gas chromatograph and sampling system, such as sample stream selection, sample injection, column or detector switching, automatic zero and attenuation adjustment, and back-flushing. The programmer will also carry out at least some initial processing of the output data, by, for example, peak selection. It is also necessary for a process instrument to incorporate safety devices to prevent damage to itself or to the surroundings in the event of a malfunction, and also to give an indication of faults which may lead to unreliable results. Functions which may be assigned to the programmer include: fault detection and identification, alarm generation and automatic shutdown of the equipment when a fault is detected.

In addition to the automatic mode of operation the programmer must allow the equipment to be operated manually for start-up, maintenance and calibration.

5.4.7 Data-processing systems

The output from a gas chromatograph detector is usually an electrical signal, and the simplest method of data presentation is the chromatogram of the sample, obtained by direct recording of the detector output on a potentiometric recorder. However, the complexity of the chromatograms of typical mixtures analysed by chromatography means that this simple form of presentation is unsuitable for direct interpretation or display, and further processing is required. The data-processing system of a process chromatograph must be able to identify the peaks in the chromatogram corresponding to components of interest in the sample, and it must measure a suitable parameter of each peak which can be related to the concentration of that component of the sample. In addition the system should give a clear indication of faults in the equipment.

Identification of the peaks in the chromatogram is made on the basis of retention time. Provided that instrumental parameters, particularly column temperature and carrier-gas flow rate remain constant, the retention time is characteristic of a given compound on a particular column. Small changes in operating conditions may change the retention times, so the data-processing system must identify retention times in a suitable 'window' as belonging to a particular peak. In addition retention times may show a long-term drift due to column ageing; and the data-processing system may be required to compensate for this.

Relation of the output signal to the concentration of the component of interest may be made on the basis either of the height of the peak or the area under it. In both cases a calibration curve must be prepared beforehand by analysis of standard mixtures, and in the more sophisticated systems, this information can be stored and the necessary calculations carried out to give a printed output of the concentrations of the components of interest for each analysis. Automatic updating of the calibration may also be possible. The simplest data-processing systems relate peak height to concentration, but it is usually better to measure peak areas, particularly for complex chromatograms, as this gives some automatic compensation for changes in peak shape caused by adventitious changes in operating conditions. In this case the data-processing system must incorporate an integrator.

5.4.7.1 Display of chromatographic data

A refinement of the basic record of the complete chromatogram of the sample is to select and display only peaks corresponding to species of interest, each species being assigned to a separate recorder channel so that successive analyses enable changes in the concentration of each species to be seen. The peaks may be displayed directly or in bar form as shown in Figure 5.22.

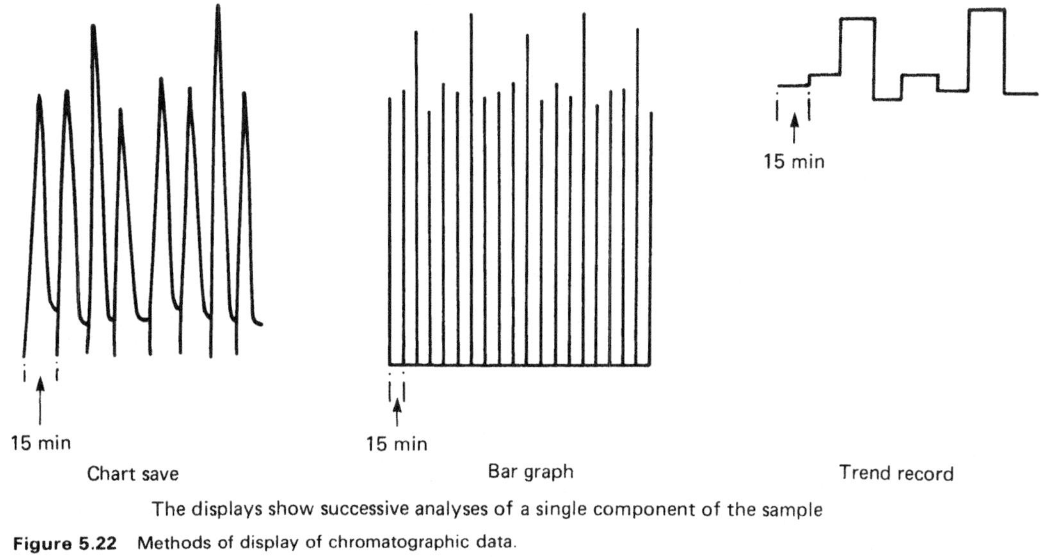

Figure 5.22 Methods of display of chromatographic data.

For trend recording a peak selector accepts the output from the process chromatograph, detects the peak height for each selected measured component and stores the data. The peak heights are transferred to a memory unit which holds the value of the height for each peak until it is updated by a further analysis. The output of this unit may be displayed as a chart record of the change in concentration of each measured species. An example of this type of output is shown in Figure 5.22.

5.4.7.2 Gas-chromatographic integrators

A variety of gas-chromatographic integrators are available to provide a measure of the areas under the peaks in a chromatogram. The area is obtained by summation of a number of individual measurements of the detector output during a peak, and the number reported by the integrator is typically the peak area expressed in millivolt-seconds. Integrators differ in the method of processing the individual readings of detector output, and in the facilities available in the instrument for further processing of the peak area values. In all instruments the analogue output signal from the gas chromatograph is first converted to digital form. In simpler integrators an upward change in the baseline level, or in the rate of baseline drift, is taken as the signal to begin the summation process which continues until the baseline level, or a defined rate of baseline drift, is regained. As the instrument has to be aware of the baseline change before it can begin integration, a proportion, usually negligibly small, of each peak is inevitably lost, the amount depending on the settings of the slope sensitivity and noise-rejection controls. This difficulty is obviated in the so-called 'computing' integrators by storing the digitized detector readings in a memory so that a complete peak, or series of merged peaks, can be stored and integrated retrospectively. Baseline assignment can then also be made retrospectively. In the most sophisticated models the memory is large enough to store data corresponding to a complete chromatogram. Use is also made of the memory to provide facilities for automatic computation of calibration curves and the integrator may then provide a printed output record giving the concentrations of each component of interest.

5.4.8 Operation of a typical process chromatograph

As an example of process chromatography, the operation of a single-stream instrument designed for high-speed on-line measurement of the concentration of a single component, or group of components, is described. The chromatograph is shown schematically in Figure 5.23, and consists of an analyser, a processor and a power unit.

The analyser unit contains those parts of the system required for sample handling and separation and detection of the components. There is a single column

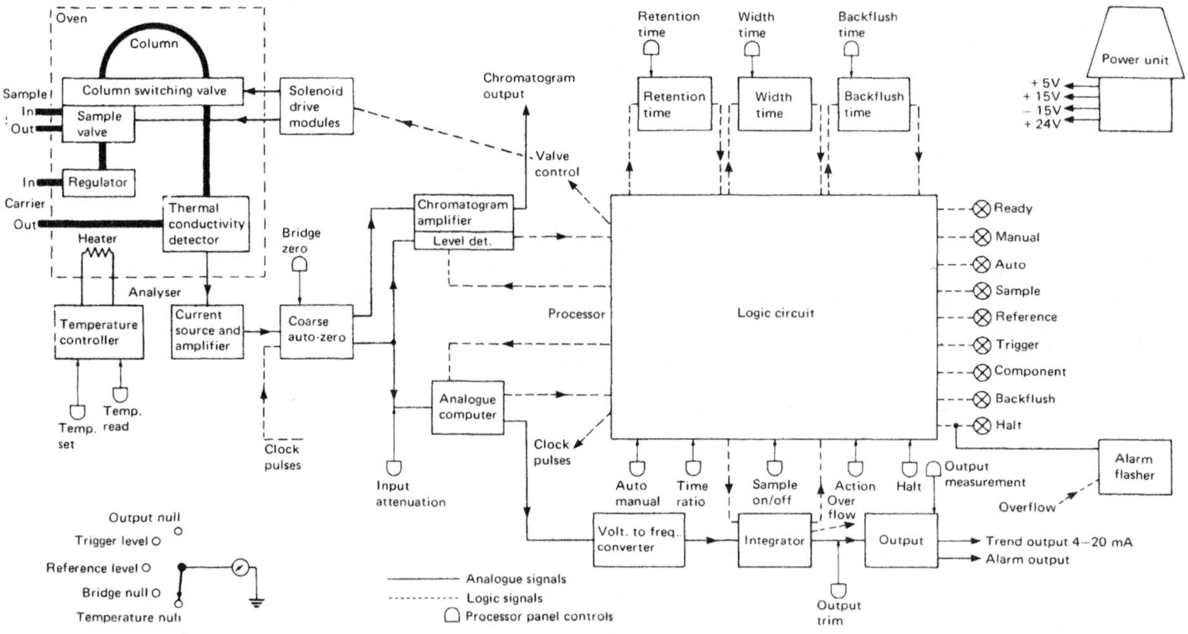

Figure 5.23 Schematic diagram of single-channel process chromatograph.

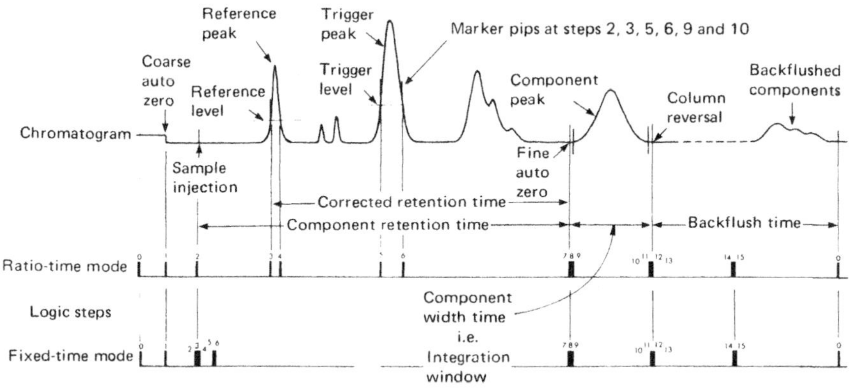

Figure 5.24 Chromatogram, showing logic and switching steps.

and thermal conductivity detector housed in a temperature-controlled zone at the top of the unit, with the associated electronics beneath. The packing and length of the small-bore column are chosen to suit the application, and the carrier-gas regulator is designed for high stability under low-flow conditions.

The small-volume thermal conductivity type detector uses thermistor elements to produce the output signals with high speed and stability. The electronic circuit modules mounted in the lower half of the main case control the oven temperature, power the detector and amplify its output, and provide power pulses to operate the valve solenoids.

The processor contains the electronic circuits which control the sequential operation of the total system. It times the operation of the simple injection and column switching valves, selects and integrates a chromatographic peak, and updates the trend output signal.

The power unit provides the low voltage regulated supplies for the analyser and the processor and may be mounted up to 3 metres from the processor.

A typical chromatogram of a sample analysed by the instrument is shown in Figure 5.24, annotated to show the various switching and logic steps during the analysis.

The operation of the chromatograph can be either on a fixed-time or ratio-time basis. In fixed-time operation the sample injection is the start of the time cycle. At preset times the 'integration window' is opened and closed, to coincide with the start and finish of the emergence of the components from the column. While the window is open the detector signal is integrated to give a measure of the concentration of the component. Other operations such as column switching and automatic zeroing are similarly timed from the sample injection. For fixed-time operation to be reliable, pressure and flow rate of carrier gas, and temperature and quantity of stationary phase in the column must be closely controlled.

Many of the problems associated with fixed-time operation may be avoided by use of ratio-time operation. In this mode of operation the retention time of components is measured from an early reference peak (corrected retention time, see Figure 5.24) instead of from the time of sample injection. The ratio of two corrected retention times (retention ratio) is less affected by changes in the critical column parameters. The corrected retention time for an early trigger peak is used to predict the time of emergence of the component of interest, that is, the integration window.

For the system to be able to operate in the ratio mode, it is necessary to have two specific peaks in the chromatogram in advance of the peak of the component of interest.

Reference peak The reference peak is due to the first component eluted from the column, with a very low retention time (such as air), and is used as the start point for the ratio timing. If a suitable component is not consistently present in the process sample, one can be injected into the column at the same time as the sample, by using the second loop of the sample valve.

Trigger peak The trigger peak must appear on the chromatogram between the reference and component peaks. It must be self-evident by virtue of size, and it must be consistent in height and width. As with the reference peak it can be from a component of the process sample, or injected separately. Alternatively it can be a negative peak derived by using a doped carrier gas. The logic circuits measure the time between reference and trigger peaks and use this, together with the preset ratio value, to compute the time for the start of the integration window. Similarly the trigger peak width is used to define the width of the window. At the start of integration the value of the signal level is stored. The integrator then measures the area under the component peak for the period of the window opening. At this point the signal level is again measured and compared with the stored start value to determine whether any baseline shift has occurred. The integration is corrected for any baseline shifts.

The final value of the integration is stored and used to give an output signal which represents the concentration of the component. As this signal is updated after each analysis the output shows the trend of the concentration.

After the completion of integration, the column is backflushed in order to remove any later components, the duration of the backflushing being ratioed from the analysis time. Alternatively, for those applications requiring a measurement such as 'total heavies', the peak of the total backflushed components can be integrated.

There are some applications where the ratio-time mode cannot be used. Typically, the measurement of a very early component, such as hydrogen, precludes the existence of earlier reference and trigger peaks. Operation of the various functions is then programmed using the fixed-time mode. Selection of the required mode is made using a switch on the processor.

Manual operation This mode of operation, selected by the 'auto/manual' switch on the front panel, provides a single analysis which is followed by column backflushing and the normal 'halt' condition. Single analyses are initiated by operation of the 'action' push-button, provided that the previous analysis has been completed. This mode of operation is used during initial programming or servicing.

5.5 Special gas analysers

5.5.1 Paramagnetic oxygen analysers

Many process analysers for oxygen make use of the fact that oxygen, alone among common gases, is paramagnetic.

5.5.1.1 Basic principles

The strength of a magnet is expressed as its magnetic moment. When a material, such as a piece of soft iron, is placed in a magnetic field, it becomes magnetized by induction and the magnetic moment of the material divided by its volume is known as the intensity of magnetization. The ratio of the intensity of magnetization to the intensity of the magnetic field is called the volume susceptibility k of the material. All materials show some magnetic effect when placed in a magnetic field, but apart from elements such as iron, nickel and cobalt and alloys such as steel, all known as ferromagnetics, the effect is very small, and intense magnetic fields are required to make it measurable.

Substances which are magnetized in the opposite direction to that of the applied field (so that k is negative) are called diamagnetics. Most substances are diamagnetic and the value of the susceptibility is usually very small. The most strongly diamagnetic substance is bismuth.

The magnetic properties of a substance can be related to its electronic structure. In the oxygen molecule, two of the electrons in the outer shell are unpaired. Because of this the magnetic moment of the molecule is not neutralized as is the commoner case, and the permanent magnetic moment is the origin of oxygen's paramagnetic properties.

A ferro- or paramagnetic substance when placed in a magnetic field in a vacuum or less strongly paramagnetic medium tries to move from the weaker to the stronger parts of the field. A diamagnetic material, in a magnetic field in a vacuum or medium of algebraically greater susceptibility tries – although the effect is very small – to move from the stronger to the weaker parts of the field. Thus when a rod of ferromagnetic or paramagnetic substance is suspended between the poles of a magnet it will set with its length along the direction of the magnetic field. A rod of bismuth, on the other hand, placed between the poles of a powerful electromagnet will set at right angles to the field.

It has been shown experimentally that for paramagnetic substances the susceptibility is independent of the strength of the magnetizing field but decreases with increase of temperature acording to the Curie–Weiss law:

$$\text{atomic susceptibility} = \frac{\text{relative atomic mass}}{\text{density}} \times \text{volume susceptibility}$$

$$= C/(T-\theta)$$

where T is the absolute temperature and C and θ are constants.

The susceptibilities of ferromagnetic materials vary with the strength of the applied field, and above a certain temperature (called the Curie temperature and characteristic of the individual material) ferromagnetics lose their ability to retain a permanent magnetic field and show paramagnetic behaviour. The Curie temperature of iron is 1000 K.

The susceptibility of diamagnetic substances is almost independent of the magnetizing field and the temperature.

The paramagnetic properties of oxygen are exploited in process analysers in two main ways; the so-called 'magnetic wind' or thermal magnetic instruments, and magnetodynamic instruments.

5.5.1.2 Magnetic wind instruments

The magnetic wind analyser, originally introduced by Hartmann and Braun, depends on the fact that oxygen, as a paramagnetic substance, tends to move

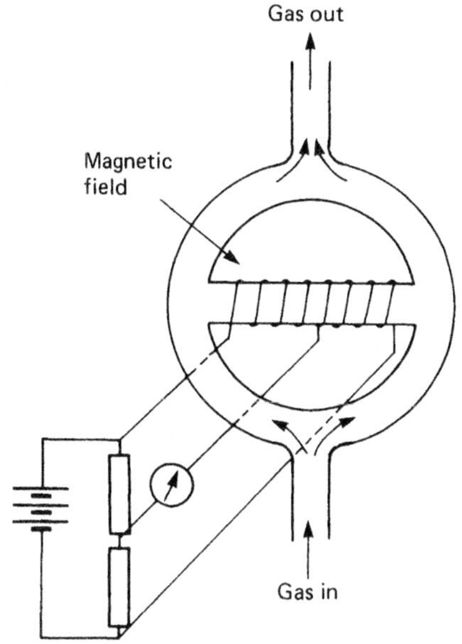

Figure 5.25 Magnetic wind oxygen analyser (courtesy Taylor Analytics).

from the weaker to the stronger part of a magnetic field, and that the paramagnetism of oxygen decreases as the temperature is raised.

$$\frac{\text{volume susceptibility}}{\text{density}} = \frac{C}{(T-\theta)} \text{ (Curie–Weiss law)}$$

$$\text{i.e. volume susceptibility} = \frac{C}{(T-\theta)} \times \text{density}$$

But for a gas, the density is proportional to $1/T$ where T is the absolute temperature. Thus

$$\text{volume susceptibility} = \frac{C}{(T^2-\theta T)}$$

The principle of the magnetic wind instrument is shown in Figure 5.25. The measuring cell consists of a circular annulus with a horizontal bypass tube on the outside of which are wound two identical platinum heating coils. These two coils form two arms of a Wheatstone bridge circuit, the bridge being completed by two external resistances. The coils are heated by means of the bridge current, supplied by a d.c. source of about 12 V. The winding on the left is placed between the poles of a very powerful magnet. When a gas sample containing oxygen enters the cell, the oxygen tends to flow into the bypass tube. Here it is heated so that its magnetic susceptibility is reduced. The heated gas is pushed along the cross-tube by other cold gas entering at the left. This gas flow cools the filaments, the left coil more than the right, and so changes their resistance, as in the flow controller mentioned in Section 5.4.2. The change in resistance unbalances the Wheatstone bridge and the out-of-balance e.m.f. is measured to give a signal which is proportional to the oxygen content of the gas.

This type of oxygen analyser is simple and reasonably robust, but it is subject to a number of errors. The instrument is temperature-sensitive: an increase in temperature causes a decrease in the out-of-balance e.m.f. of about 1 per cent per kelvin. This can be automatically compensated by a resistance thermometer placed in the gas stream near the cell. The calibration depends on the pressure of the gas in the cell.

Another error arises from the fact that the analyser basically depends on the thermal conductivity of the gas passing through the cross-tube. Any change in the composition of the gas mixed with the oxygen changes the thermal balance and so gives an error signal. This is known as the carrier-gas effect.

To a first approximation the out-of-balance e.m.f. is given by

$$e = kC_o$$

where e is the e.m.f., C_o is the oxygen concentration and k is a factor which varies with the composition of the carrier gas, and depends on the ratio of the volumetric specific heat to the viscosity of the carrier gas. For a binary mixture of oxygen with one other gas, k is a constant, and the out-of-balance e.m.f. is directly proportional to the oxygen concentration. For ternary or more complex mixtures, the value of k is constant only if the composition of the carrier gas remains constant.

Values of k for a number of common gases are given in Table 5.3 for an e.m.f. measured in volts and oxygen concentration measured in volume per cent. The value of k for a mixture can be calculated by summing the partial products:

$$k = (C_A k_A + C_B k_B)/100$$

where C_A and C_B are the percentage concentrations of

Table 5.3 k values for common gases

Gas	*k*	*Gas*	*k*
Ammonia	2.21	Nitrogen	1.00
Argon	0.59	Nitric oxide	0.94
Carbon dioxide	1.54	Nitrous oxide	1.53
Carbon monoxide	1.01	Oxygen	0.87
Chlorine	1.52	Sulphur dioxide	1.96
Helium	0.59	Water vapour	1.14
Hydrogen	1.11		

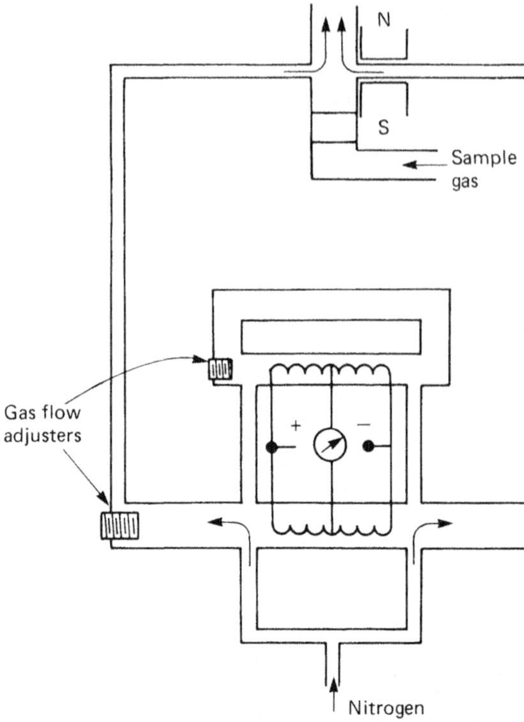

Figure 5.26 Quincke oxygen analyser (courtesy Taylor Analytics).

components A and B and k_A and k_B are the corresponding values of k.

Convective flow or misalignment of the sensor may also change the thermal balance and cause errors. In the case of flammable gases, errors may be caused if they can burn at the temperature in the cross-tube. This type of analyser is therefore usually considered to be unsuitable for oxygen measurements in hydrocarbon vapours.

5.5.1.3 Quincke analyser

The Quincke analyser is shown in Figure 5.26. A continuous stream of nitrogen enters the cell and is divided into two streams which flow over the arms of filaments of a Wheatstone bridge circuit. The flows are adjusted to balance the bridge to give zero output. One of the nitrogen streams passes the poles of a strong magnet while the other stream passes through a similar volume but without the magnetic field.

The sample gas enters the cell as shown and is mixed with the nitrogen streams immediately downstream of the magnetic field. Oxygen in the sample gas tends to be drawn into the magnetic field, causing a pressure difference in the arms of the cell and changing the flow-pattern of the nitrogen over the arms of the Wheatstone bridge. The out-of-balance e.m.f. is proportional to the oxygen concentration of the sample gas.

Because the sample gas does not come into contact with the heated filaments, the Quincke cell does not suffer from the majority of the errors present in magnetic wind instruments, but it does require a separate supply of nitrogen.

5.5.1.4 Magnetodynamic instruments

Magnetic wind instruments are susceptible to hydrocarbon vapours and to any change in the carrier gas producing a change in its thermal conductivity. These difficulties led to the development by Pauling of a measuring cell based on Faraday's work on determination of magnetic susceptibility by measuring the force acting on a diamagnetic body in a non-uniform magnetic field.

5.5.1.5 Magnetodynamic oxygen analyser

In the Pauling cell, two spheres of glass or quartz, filled with nitrogen, which is diamagnetic, are mounted at the ends of a bar to form a dumb-bell. The dumb-bell is mounted horizontally on a vertical torsion suspension, and is placed between the specially-shaped poles of a powerful permanent magnet. The gas to be measured surrounds the dumb-bell. If oxygen is present it is drawn into the field and so displaces the spheres of the dumb-bell which are repelled from the strongest parts of the field, so rotating the suspension until the torque produced is equal to the deflecting couple on the spheres, see Figure 5.27. If the oxygen content of the gas in the cell changes, there will be a change in the force acting on the spheres, which will take up a new

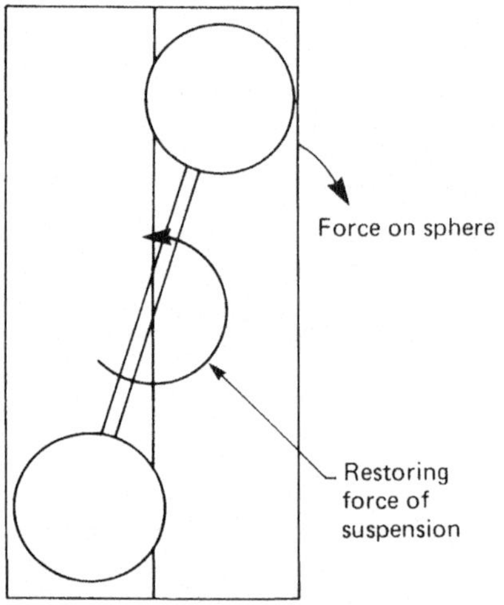

Figure 5.27 Magnetodynamic oxygen measuring cell (courtesy Taylor Analytics).

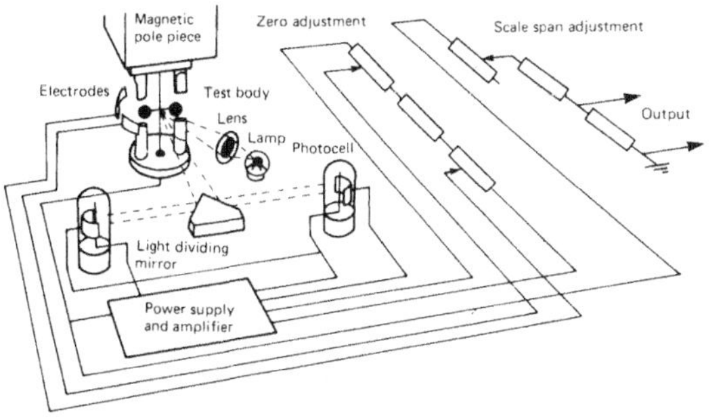

Figure 5.28 Bendix oxygen analyser.

position. The magnitude of the force on the dumb-bell may be measured in a number of ways, but a small mirror is commonly attached to the middle of the arm, and the deflection measured by focusing a beam of light on the mirror. The deflection may either be measured directly, or a force balance system may be used whereby the deflection of the dumb-bell is detected but an opposing force is applied to restore it to the null position.

Two different designs of oxygen analyser, based on the magnetodynamic principle, are shown in Figures 5.28 and 5.29. In the Bendix instrument the suspension is a quartz fibre and the restoring force is produced electrostatically by the electrodes adjacent to the dumb-bell. One electrode is held above ground potential and the other below ground potential by the amplifier controlled from the matched photocells upon which the light from the mirror falls. In the Servomex instrument (Figure 5.29) the suspension is platinum, and the restoring force is produced electrically in a single turn of platinum wire connected to the rest of the electronics through the platinum suspension. Electromagnetic feedback is used to maintain the dumb-bell in the zero position, and the current required to do this is a measure of the oxygen content of the gas.

The deflecting couple applied to the dumb-bell by the magnetic field depends on the magnetic susceptibility of the surrounding gas. The magnetic susceptibilities of all common gases at 20 °C are very small (nitrogen, -0.54×10^{-8}; hydrogen, -2.49×10^{-8}; carbon dixode, -0.59×10^{-8}) compared to that of oxygen ($+133.6 \times 10^{-8}$) and the susceptibility of the gas will depend almost entirely on the concentration of oxygen. This type of analyser is not influenced by the thermal conductivity of the gas, and is unaffected by hydrocarbons. However, the susceptibility of oxygen varies considerably with temperature. This may be overcome by maintaining the instrument at a constant temperature above ambient, or the temperature of the measuring cell may be detected and the appropriate temperature correction applied electronically. The reading also depends on the pressure of gas in the cell.

This type of analyser is suitable for measuring the oxygen content of hydrocarbon gases, but paramagnetic gases interfere and must be removed. The most important of these is nitric oxide (susceptibility $+59.3 \times 10^{-8}$), but nitrogen peroxide and chlorine dioxide are also paramagnetic. If the concentration of these gases in the sample is reasonably constant, the instrument may be zeroed on a gas sample washed in acid chromous chloride, and the oxygen measured in the usual way.

5.5.2 Ozone analyser

Continuous analysers for ozone are based on the chemiluminescent flameless reaction of ozone with

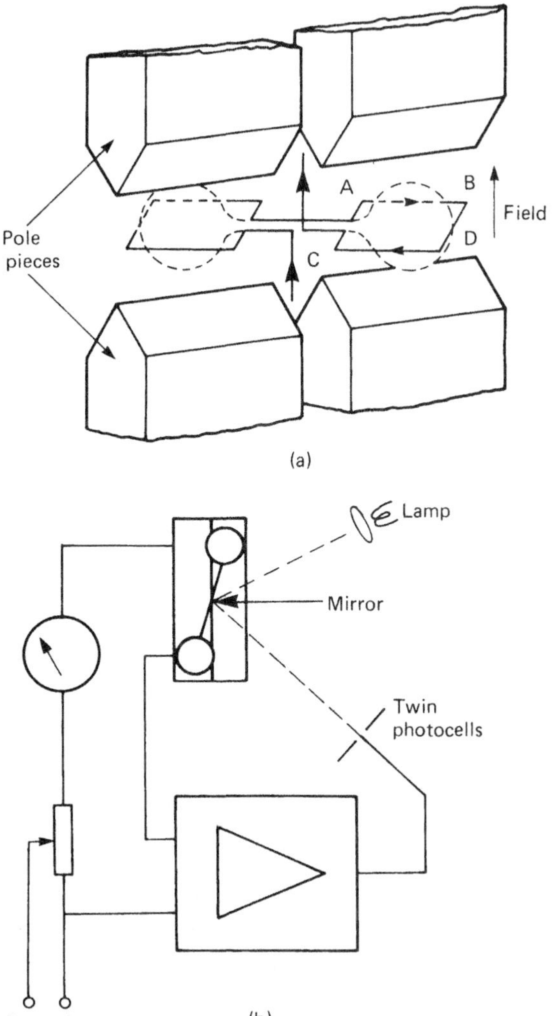

Figure 5.29 Servomex oxygen analyser (courtesy Taylor Analytics): (a) measuring cell, (b) electronic circuit.

ethylene. The light emission from the reaction, centred at 430 nm, is measured by a photomultiplier and the resulting amplified signal is a measure of the concentration of ozone in the sample stream. The flow diagram, and functional block diagram of a typical portable ozone analyser are given in Figure 5.30. The chemiluminescent light emission from the reaction chamber is a direct function of the ambient temperature, and therefore the temperature is regulated to 50 °C. The photomultiplier is contained in a thermoelectrically cooled housing maintained at 25 °C to ensure that short- and long-term drift is minimized. The instrument is capable of measuring ozone levels in the range 0.1 to 1000 ppb.

5.5.3 Oxides of nitrogen analyser

Analysers for oxides of nitrogen – NO, NO_x (total oxides of nitrogen), NO_2 – are based on the chemiluminescent reaction of nitric oxide (NO) and ozone to produce nitrogen dioxide (NO_2). About 10 per cent of the NO_2 is produced in an electronically excited state, and undergoes a transition to the ground state, emitting light in the wavelength range 590–2600 nm:

$$NO + O_3 \rightarrow NO_2^* + O_2$$

$$NO_2^* \rightarrow NO_2 + h\nu$$

The intensity of the light emission is proportional to the mass-flow rate of NO through the reaction chamber and is measured by a photomultiplier tube.

Analysis of total oxides of nitrogen (NO_x) in the sample is achieved by passing the gases through a stainless steel tube at 600–800 °C. Under these conditions, most nitrogen compounds (but not N_2O) are converted to NO which is then measured as above. Nitrogen dioxide (NO_2) may be measured directly by passing the air sample over a molybdenum catalyst to reduce it to NO, which is again measured as above, or the NO_2 concentration may be obtained by automatic electronic subtraction of the NO concentration from the NO_x value.

The flow system of a nitrogen oxides analyser is shown in Figure 5.31. Ozone is generated from ambient air by the action of UV light

$$3O_2 \xrightarrow{h\nu} 2O_3$$

and a controlled flow rate of ozonized air is passed to the reaction chamber for reaction with NO in the air sample, which is passed through the chamber at a controlled flow of 1 l min^{-1}. By selection of a switch to operate the appropriate solenoid valves, a span gas may be directed to the reaction chamber, or a zero calibration may be carried out by shutting off the flow of ozonized air to the reactor. The three-way solenoid valve downstream of the converter is switched to permit NO analysis when bypassing the converter, and NO_x analysis when the sample is passed through the converter. The analyser can measure ozone in air in the range 5 ppb to 25 ppm, with a precision of $\pm 1\%$.

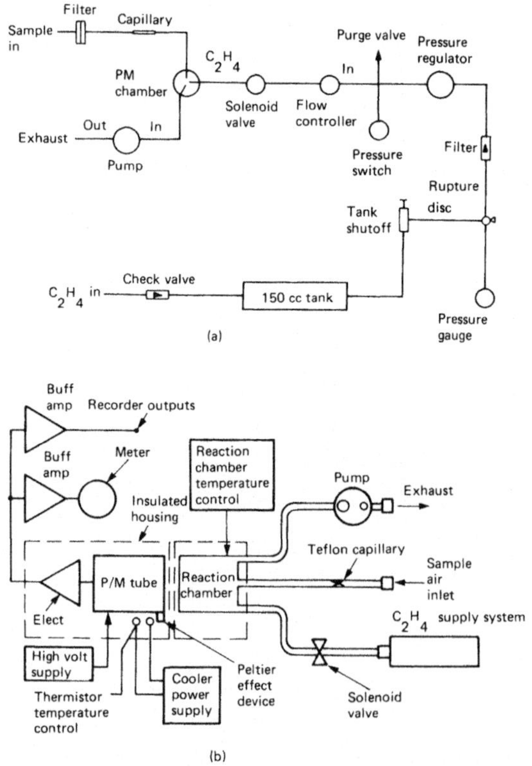

Figure 5.30 Ozone analyser (courtesy Columbia Scientific Industries Corp.): (a) flow diagram, (b) functional block diagram.

5.5.4 Summary of special gas analysers

The operating principles of analysers for the most commonly measured gases are given in Table 5.4.

5.6 Calibration of gas analysers

None of the commonly-used gas detectors is absolute; that is, they are devices where the output signal from the detector for the gas mixture under test is compared with that for mixtures of the bulk gas containing known concentrations of the determinand. The use of standard gas mixtures is analogous to the use of standard solutions in solution chemistry, but their preparation and handling present some peculiar problems. As in solution chemistry, the calibration gas

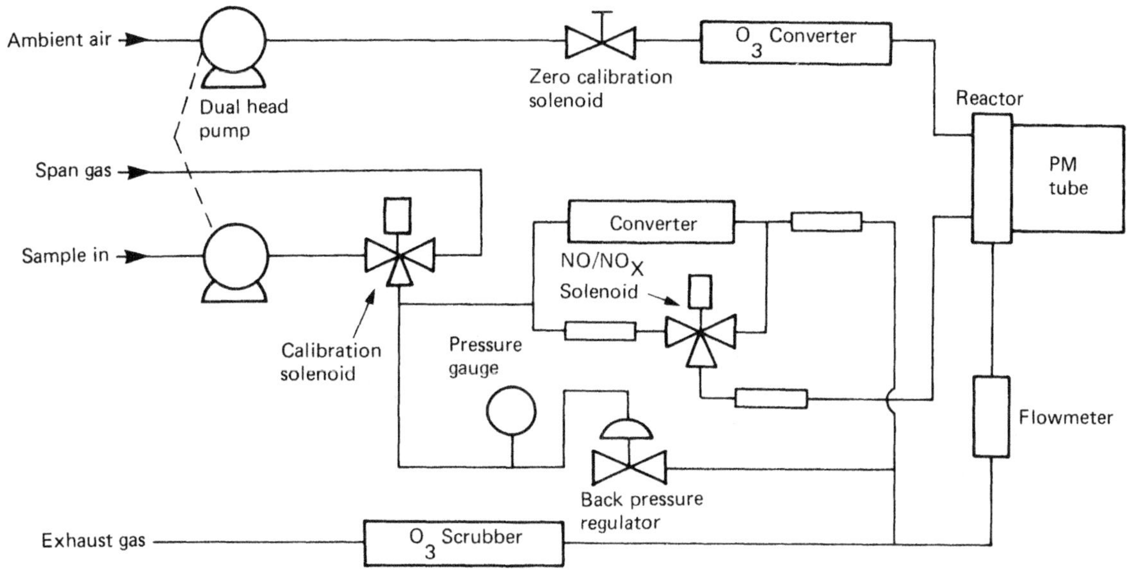

Figure 5.31 Oxides of nitrogen analyser (courtesy Beckman).

Table 5.4 Measurement principles of special gas analysers

Gas	*Measurement principle*
Oxygen	Paramagnetism Electrochemical sensor Fuel cell
Ozone	Chemiluminescence Electrochemical sensor
Nitrogen oxides	Chemiluminescence
Carbon dioxide	Infrared spectrometry
Carbon monoxide	Infrared spectrometry Electrochemical sensor
Sulphur oxides	Flame photometry
Hydrocarbons	Flame ionization detector Infrared spectrometry Catalytic detector
Flammable gases	Catalytic detector Semiconductor detector
Hydrogen sulphide	Semiconductor detector Flame photometry Electrochemical sensor

mixtures should reflect, as closely as possible, the composition of the samples it is desired to measure. Ideally a number of standard mixtures, whose concentration covers the range of samples to be measured, should be used to establish the response curve of the instrument or detector. However, for routine calibration where the response curve has previously been established or is well known, it is usual to calibrate gas analysers by use of a 'zero' gas mixture which is free of the determinand and establishes the zero of the instrument, and one or more 'span' gases containing concentrations of the determinand close to those it is desired to measure.

The accuracy to which a gas mixture can be prepared depends on the number and nature of the components, and on their concentrations. For gas mixtures prepared under pressure in cylinders, it is useful to specify two parameters, the filling and analytical tolerances. The filling tolerance describes the closeness of the final mixture to its original specification, and depends mainly on the concentrations of the components. Thus, while it may be possible to fill a cylinder with a component gas at the 50 per cent level to a tolerance of ± 2.5 per cent or ± 5 per cent of the component (that is the cylinder would contain between 47.5 and 52.5 per cent of the component), at the 10 vpm level the tolerance would typically be ± 5 vpm or ± 50 per cent of the component, and the cylinder would contain between 5 and 15 vpm of the component. The analytical tolerance is the accuracy with which the final mixture can be described, and depends on the nature of the mixture and the analytical techniques employed. Accuracies achievable are typically in the range from ± 2 per cent of component or ± 0.2 vpm at the 10 vpm level to ± 1 per cent of component or ± 0.5 per cent at the 50 per cent level. However, these figures are strongly dependent on the actual gases involved, and the techniques available to analyse them.

Gas mixtures may be prepared by either static or dynamic methods. In the static method known quantities of the constituent gases are admitted to a suitable vessel and allowed to mix, while in the dynamic method streams of the gases, each flowing at a known rate, are mixed to provide a continuous stream of the sample mixture. Cylinders containing supplies of the standard mixtures prepared under pressure are usually most convenient for fixed instruments such as process gas chromatographs, while portable instruments are often calibrated by mixtures prepared dynamically. Where mixtures containing low concentrations of the constituents are needed, adsorptive effects may make the static method inapplicable, while the dynamic method becomes more complex for mixtures containing large numbers of constituents.

Before any gas mixture is prepared, its properties must be known, particularly if there is any possibility of reaction between the components, over the range of pressures and concentrations expected during the preparation.

5.6.1 Static methods

Static gas mixtures may be prepared either gravimetrically or by measurement of pressure. Since the weight of gas is usually small relative to the weight of the cylinder required to contain it, gravimetric procedures require balances which have both high capacity and high sensitivity, and the buoyancy effect of the air displaced by the cylinder may be significant. Measurement of pressure is often a more readily applicable technique.

After preparation gas mixtures must be adequately mixed to ensure homogeneity, usually by prolonged continuous rolling of the cylinder. Once mixed, they should remain homogeneous over long periods of time. Any concentration changes are likely to be due to adsorption on the cylinder walls. This is most likely to happen with mixtures containing vapours near their critical pressures, and use of such mixtures should be avoided if possible.

5.6.2 Dynamic methods

5.6.2.1 Gas flow mixing

Gas mixtures of known concentration may be prepared by mixing streams of two or more components, each of which is flowing at a known rate. The concentration of one gas in the others may be varied by adjustment of the relative flow rates, but the range of concentration available is limited by the range of flows which can be measured with sufficient accuracy. Electronic mass-flow controllers are a convenient method of flow measurement and control.

5.6.2.2 Diffusion-tube and permeation-tube calibrators

Standard gas mixtures may be prepared by allowing the compound or compounds of interest to diffuse through a narrow orifice or to permeate through a membrane, into a stream of the base gas which is flowing over the calibration source at a controlled and known rate.

Typical designs of diffusion and permeation tubes are shown in Figure 5.32. In both cases there is a reservoir of the sample, either a volatile liquid or a liquefied gas under pressure, to provide an essentially constant pressure, the saturation vapour pressure, upstream of the diffusion tube or permeation membrane. After an initial induction period it is found that, provided the tube is kept at constant temperature, the permeation or diffusion rate is constant as long as there is liquid in the reservoir. The tube can then be calibrated gravimetrically to find the diffusion or permeation rate of the sample. The concentration of the sample in the gas stream is then given by

$$C = RK/F$$

where C is the exit gas concentration, R is the diffusion or permeation rate, K is the reciprocal density of the sample vapour, and F is the gas flow rate over the calibration device. The diffusion or permeation rate depends on the temperature of the tube, and on the molecular weight and vapour pressure of the sample. Additionally, the diffusion rate depends on the length

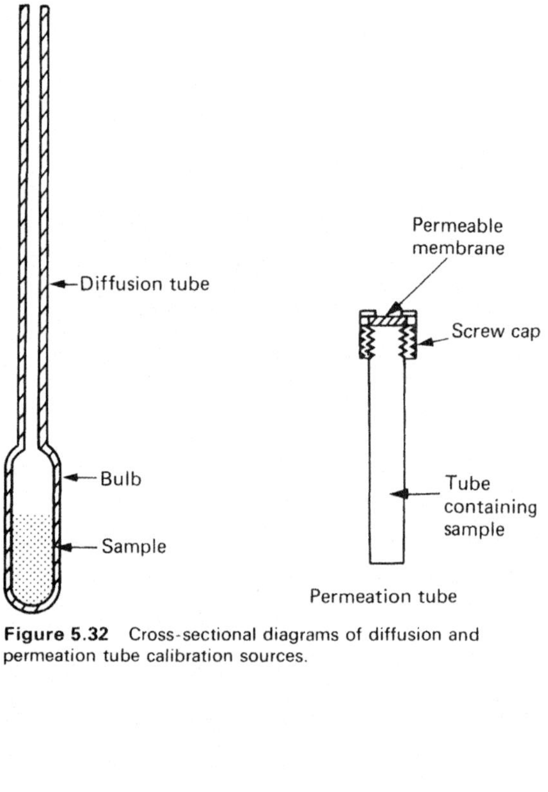

Figure 5.32 Cross-sectional diagrams of diffusion and permeation tube calibration sources.

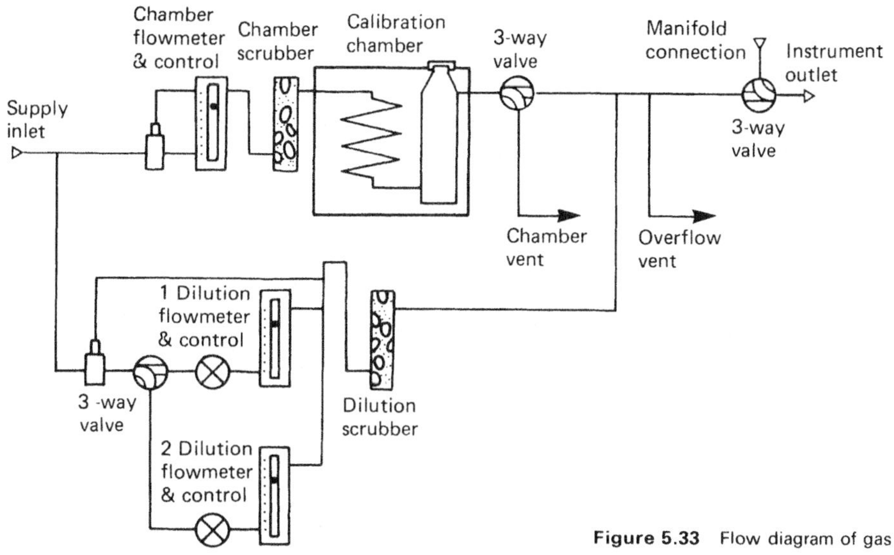

Figure 5.33 Flow diagram of gas calibrator.

and inner diameter of the capillary tube and the permeation rate depends on the nature, area and thickness of the permeation membrane. Data are available for a large number of organic and inorganic vapours to allow tubes to be designed with the required diffusion or permeation rate, and the exact rate for each tube is then established empirically.

The temperature-dependence of diffusion or permeation means that the tubes must be carefully thermostated for accurate calibrations. The empirical equation for the temperature-dependence of permeation rate is:

$$\log \frac{R_2}{R_1} = 2950\left(\frac{1}{T_1} - \frac{1}{T_2}\right)$$

where R_1 is permeation rate at T_1 K and R_2 is permeation rate at T_2 K. The permeation rate changes by approximately 10 per cent for every 1 K change in temperature. Thus the temperature of the permeation tube must be controlled to within 0.1 K or better if 1 per cent accuracy in the permeation rate, and thus the concentration that is being developed, is to be achieved.

The flow diagram of a typical calibrator for use with diffusion or permeation tubes is shown in Figure 5.33. The gas supply is scrubbed before passing through a thermostated coil and over the calibration source or sources in the calibration chamber. Secondary streams of purified gas may be added to the effluent gas stream to adjust the final concentration to the range required.

The diffusion or permeation technique is especially useful for generating standard mixtures at low concentrations, for example of organic compounds in air for calibration of environmental monitors, air pollution monitors etc., and the calibrator can be made portable for field use. The range of compounds which can be used is limited by their saturation vapour pressure; if this is too low, the diffusion or permeation rates, and hence the concentrations available, are very small, while compounds with high saturation vapour pressures present problems in construction and filling of the calibration tubes.

5.6.2.3 Exponential dilution

In the exponential dilution technique a volume of gas contained in a vessel, in which there is perfect and instantaneous mixing, is diluted by passing a stream of a second gas through the vessel at a constant flow rate. It can be shown that, under these conditions, the concentration of any gaseous species in the vessel, and hence the instantaneous concentration in the effluent stream of diluent gas, decays according to the law:

$$C = C_0 \exp\left(-\frac{Ut}{V}\right)$$

where C is the concentration of the diluted species at time t, C_0 is the initial concentration, U is the flow rate of diluent gas, and V is the volume of the vessel.

The vessel may either be filled with the gaseous species to be analysed, in which case the concentration decays from an initial value of 100 per cent, or the vessel may be filled with the diluent gas and a known volume of the gas of interest may be injected into the diluent gas just upstream of the dilution vessel at the start of the experiment. In either case the concentration of the species of interest in the effluent gas stream may be calculated at any time after the start of the dilution.

The exponential dilution vessel is typically a spherical or cylindrical glass vessel of 250–500 ml capacity, fitted with inlet and outlet tubes, and a septum cap or gas sampling valve for introduction of the gas to be diluted. The vessel must be fitted with a stirrer, usually magnetically driven, and baffles to ensure that mixing is as rapid and homogeneous as possible. The diluent gas flows through the vessel at a constant known flow rate, usually in the range 20–30 ml min^{-1}. For a vessel of the dimensions suggested above, this gives a tenfold dilution in approximately 30 minutes.

The exponential dilution technique is a valuable calibration method especially suitable for use at very low concentrations. It is also valuable for studying or verifying the response of a detector over a range of concentrations. However it should be noted that strict adherence to a known exponential law for the decay of concentrations in the vessel depends on the attainment of theoretically perfect experimental conditions which cannot be achieved in practice.

Changes in the flow rate of the diluent gas or in the temperature or pressure of the gas in the dilution vessel and imperfect or non-instantaneous mixing in the vessel lead to unpredictable deviations from the exponential decay law. Deviations also occur if the determinand is lost from the system by adsorption on the walls of the vessel. Since the technique involves extrapolation from the known initial concentration of the determinand in the diluting gas, any deviations are likely to become more important at the later stages of the dilution. If possible it is therefore advisable to restrict the range of the dilution to two or three orders of magnitude change in concentration. Where the gas to be diluted is introduced to the dilution vessel by injection with a valve or syringe, the accuracy and precision of the entire calibration curve resulting from the dilution is limited by the accuracy and precision of the initial injection.

5.7 Further reading

Cooper, C. J. and De Rose, A. J., 'The analysis of gases by chromatography', *Pergamon Series in Analytical Chemistry*, Vol. 7, Pergamon, (1983)

Cullis, C. F. and Firth, J. G. (eds), *Detection and Measurement of Hazardous Gases*, Heinemann, (1981)

Grob, R. L. (ed.), *Modern Practice of Gas Chromatography*, Wiley, (1977)

Jeffery, P. F. and Kipping, P. J., *Gas Analysis by Gas Chromatography*, International Series of Monographs in *Analytical Chemistry*, Vol. 17, Pergamon, (1972)

Sevcik, J., *Detectors in Gas Chromatography*, *Journal of Chromatography Library*, Vol. 4, Elsevier, (1976)

Also review artices in *Analytical Chemistry*, and manufacturers' literature

6 Chemical analysis – moisture measurement

D. B. MEADOWCROFT

6.1 Introduction

The measurement and control of the moisture content of gases, liquids and solids is an integral part of many industries. Numerous techniques exist, none being universally applicable, and the instrument technologist must be able to choose the appropriate measurement technique for his application. It is particularly important to measure moisture because of its presence in the atmosphere, but it is awkward because it is a condensable vapour which will combine with many substances by either physical adsorption or chemical reaction. Moisture measurement may be needed to ensure the level remains below a prescribed value or within a specified band, and the range of concentrations involved can be from less than one part per million to percentage values.

A few examples will illustrate the range of applications:

Gases In gas-cooled nuclear reactors the moisture level of the coolant has to be within a prescribed band (e.g. 250–500 volume parts per million) or below a certain value (e.g. 10 vppm) depending on the type of reactor. Rapid detection of small increases due to leaks from the steam generators is also essential. Moisture must be excluded from semiconductor device manufacture, and glove boxes are fitted with moisture meters to give an alarm at, say, 40 vppm. Environmental control systems need moisture measurement in order to control the humidity, and even tumble driers can be fitted with sensors to automatically end the clothes drying cycle.

Liquids The requirement is usually to ensure the water contamination level is low enough. Examples are the prevention of corrosion in machinery, breakdown of transformer oil and loss of efficiency of refrigerants or solvents.

Solids Specified moisture levels are often necessary for commercial reasons. Products sold by weight (e.g. coal, ore, tobacco, textiles) can most profitably have moisture contents just below the maximum acceptable limit. Some textiles and papers must be dried to standard storage conditions to prevent deterioration caused by excessive wetness and to avoid the waste of overdrying as the moisture would be picked up again during storage. Finally, many granulated foods must have a defined moisture content.

The purpose of this chapter is to introduce the reader to the major measurement techniques which are available. The three states, gas, liquid and solid, will be treated separately. In addition, many commercial instruments measure some parameter which changes reproducibly with moisture concentration and these instruments must be regularly calibrated by the user. The chapter therefore ends with a discussion of the major calibration techniques which the average user must be willing to employ when using such instruments.

First it is necessary to clarify a further aspect of moisture measurement which can confuse the newcomer, which is to define the large number of units which are used, particularly for gases, and show how they are interrelated.

6.2 Definitions

6.2.1 Gases

Although water vapour is not an ideal gas, for most hygrometry purposes, and to gain an understanding of the units involved, it is sufficient to assume water vapour does behave ideally. The basic unit of moisture in a gas against which other units can readily be referred is *vapour pressure*, and Dalton's Law of Partial Pressures can be assumed to hold if the saturated vapour pressure is not exceeded.

In environmental applications the unit often used is *relative humidity* which is the ratio in per cent of the actual vapour pressure in a gas to the saturation vapour pressure of water at that temperature. It is therefore temperature-dependent but is independent of the pressure of the carrier gas.

For chemical measurements the concentration of moisture is usually required. The *volume concentration* is given by the vapour pressure of moisture divided by the total pressure, often multiplied by 10^6 to give

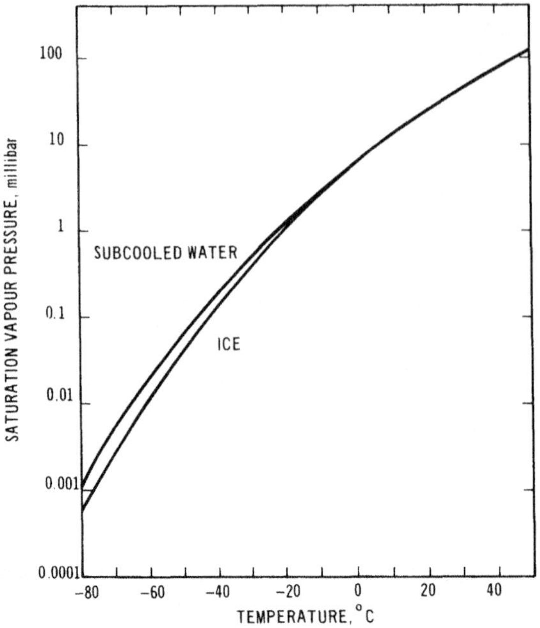

Figure 6.1 The relationship between saturation vapour pressure and dew point and frost point temperatures.

volume parts per million (vppm). The concentration by *weight* in wppm is given by the volume concentration multiplied by the molecular weight of water and divided by that of the carrier gas. Meteorologists often call the weight concentration the 'mixing ratio' and express it in g/kg.

When the prime aim is to avoid condensation the appropriate unit is the *dew point* which is the temperature at which the vapour pressure of the moisture would become saturated with respect to a plane surface. Similarly the *frost point* refers to the formation of ice. The relationship between dew and frost points and saturated vapour pressure is derived from thermodynamic and experimental work and is shown in Figure 6.1. It should be noted that below 0 °C the dew point and frost point differ. It is possible for supercooled water to exist below 0 °C, which can give some ambiguity, but this is unlikely very much below 0 °C (certainly not below –40 °C). In addition it can be seen that the saturated vapour pressure increases by an order of magnitude every 15–20 degrees so that in the range –80 °C to 50 °C dew point there is a vapour pressure change of five orders of magnitude. Table 6.1 lists the vapour pressure for dew or frost point between –90 °C and +50 °C.

Table 6.2 gives the interrelationships between these various units for some typical values.

Table 6.1 The relationship between dew/frost point and vapour pressure (μbar which is equivalent to vppm at 1 bar total pressure)

Frost point (°C)	*Saturated vapour pressure* (μbar)	*Frost point* (°C)	*Saturated vapour pressure* (μbar)	*Dew point* (°C)	*Saturated vapour pressure* (μbar)
–90	0.10	–40	128	0	6110
–80	0.55	–36	200	4	8120
–75	1.22	–32	308	8	10 700
–70	2.62	–28	467	12	14 000
–65	5.41	–24	700	16	19 200
–60	10.8	–20	1030	20	23 400
–56	18.4	–16	1510	25	31 700
–52	30.7	–12	2170	30	41 800
–48	50.2	–8	3100	40	73 000
–44	81.0	–4	4370	50	120 000

Table 6.2 Some examples of the relationships between the various units for moisture in gases

Dew/frost point (°C)	*Vapour pressure* (μbar *or* vppm at 1 bar)	*RH at 20 °C ambient* (%)	*Mixing ratio in air* (g/kg)
–70	2.5	0.01	1.5×10^{-3}
–45	72	0.3	0.045
–20	1030	4.4	0.64
0	6110	26	3.8
10	12 300	53	7.6
20	23 400	100	14.5

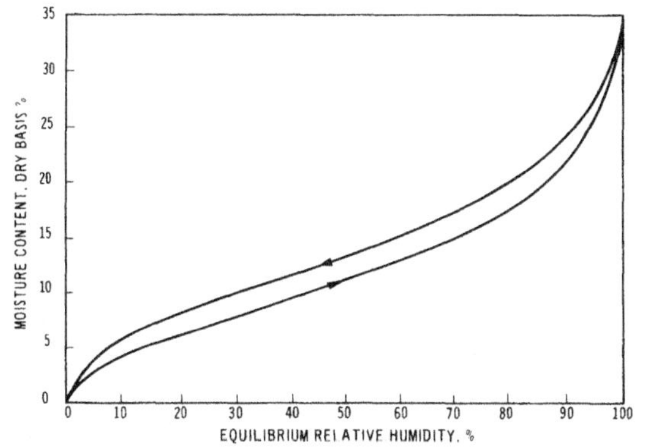

Figure 6.2 The relationship between the moisture content of a substance and the equilibrium relative humidity of the surrounding gas, for the example of wool.

6.2.2 Liquids and solids

Generally measurements are made in terms of *concentration*, either as a percentage of the total wet weight of the sample (e.g. in the ceramics industry for clay) or of the dry weight (e.g. in the textile industry where the moisture concentration is called 'regain'). In addition if a liquid or solid is in equilibrium with the gas surrounding it, the *equilibrium relative humidity* of the gas can be related to the moisture content of the solid or liquid by experimentally derived isotherms (e.g. Figure 6.2), or by Henry's law for appropriate non-saturated liquids. For liquids which obey Henry's law the partial vapour pressure of the moisture P is related to the concentration of water dissolved in the liquid by $W=KP$ where K is Henry's law constant. K can be derived from the known saturation values of the particular liquid, i.e. $K=W_s/P_s$ where W_s and P_s are respectively saturation concentration and saturation vapour pressure at a given temperature.

6.3 Measurement techniques

Techniques which allow automatic operation have the important advantage that they can be used for process control. We therefore concentrate our attention here on such techniques. Again, those available for gases, liquids and solids will be discussed separately.

6.3.1 Gases

There is a huge choice of techniques for the measurement of moisture in gases, reflecting the large number of ways in which its presence is manifested. The techniques range from measuring the extension of hair in simple wall-mounted room monitors to sophisticated electronic instruments. To some extent the choice of technique depends on the property required – dew point, concentration or relative humidity. Only the major techniques are discussed here. More extensive treatments are given in the bibliography.

6.3.1.1 Dew point instruments

The determination of the temperature at which moisture condenses on a plane mirror can be readily estimated (Figure 6.3) using a small mirror whose temperature can be controlled by a built-in heater and thermoelectric cooler. The temperature is measured by a thermocouple or platinum resistance thermometer just behind the mirror surface and the onset of dew is detected by the change of reflectivity measured by a lamp and photocell. A feedback circuit between the cell output and the heater/cooler circuit enables the dew point temperature to be followed automatically. Systematic errors can be very small and such intruments are used as secondary standards, yet with little loss of sophistication they can be priced competitively for laboratory and plant use. Mirror contamination can be a problem in dirty gases and in some instruments the mirror is periodically heated to reduce the effect of contamination. Condensable carrier gases which condense at similar temperatures to the moisture invalidate the technique. It is an ideal method if the dew point itself is required, but if another unit is to be derived from it accurate temperature measurements are essential because of the rapid change in vapour pressure with dew point temperature (see Section 6.2.1).

6.3.1.2 Coulometric instruments

The gas is passed at a constant rate through a sampling tube in which the moisture is absorbed onto a film of partially hydrated phosphoric anhydride (P_2O_5) coated on two platinum electrodes (Figure 6.4). A d.c. voltage is applied across the electrodes to decompose the water, the charge produced by the electrolysis being directly proportional to the mass of water absorbed (Faraday's law). Thus the current depends

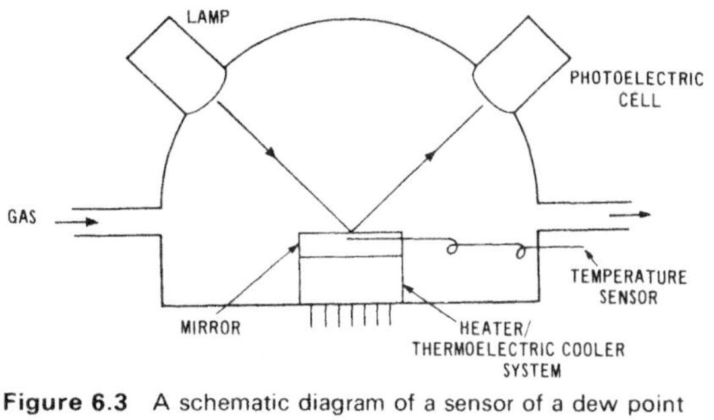

Figure 6.3 A schematic diagram of a sensor of a dew point mirror instrument.

on the flow rate, which must be set and controlled accurately at a predetermined rate (usually 100 ml min^{-1}) so that the current meter can be calibrated directly in ppm. Several points are worth making:

(a) The maximum moisture concentration measurable by this technique is in the range 1000–3000 vppm but care must be taken to ensure surges of moisture level do not wash off the P_2O_5.
(b) There is generally a zero leakage current equivalent to a few ppm. To allow for this error, when necessary, the current should be measured at two flow rates and the difference normalized to the flow for 100 ml min^{-1}.
(c) Platinum electrodes are not suitable for use in gases containing significant amounts of hydrogen. The platinum can catalyse the recombination of the electrolysed oxygen and this water is also electrolysed giving inaccurate measurements. Gold or rhodium elements reduce this effect.
(d) In the absence of recombination and gas leaks the response of a coulometric instrument can be regarded as absolute for many purposes.
(e) Cells which work at pressure can be obtained. This can increase the sensitivity at low moisture levels as it is possible to use a flow rate of 100 ml min^{-1} at the measuring pressure, which does not increase the velocity of gas along the element and hence does not impair the absorption efficiency of the P_2O_5.

6.3.1.3 Infrared instruments

Water vapour absorbs in the 1–2 μm infrared range, and infrared analysers (see Chapter 3) can be successfully used as moisture meters. For concentrations in the vppm range the path length has to be very long and high sample flow rates of several litres per minute can be necessary to reduce the consequent slow response time. Both single-beam instruments, in which the zero baseline is determined by measuring the absorption at a nearby non-absorbing wavelength, and double-beam instruments, in which a sealed parallel cell is used as reference, can be used. Single-beam instruments are less affected by deposits on the cell windows and give better calibration stability in polluted gases.

6.3.1.4 Electrical sensor instruments

There are many substances whose electrical impedance changes with the surrounding moisture level. If this absorption process is sufficiently reproducible on a thin film the impedance, measured at either an audio frequency or a radio frequency, can be calibrated in terms of moisture concentration or relative humidity. Materials used in commercial instruments include polymers, tantalum oxide, silicon oxide, chromium oxide, aluminium oxide, lithium chloride mixed with plastic, and carbon-loaded plastics which change length and hence resistance with moisture level. Many such instruments are available commercially, particularly using an anodized aluminium oxide layer which has a very narrow columnar pore structure (Figure 6.5), but ageing and other deterioration processes can occur so that regular calibration is essential. A major advantage of such sensors is that as no imposed gas flow is necessary they can simply be placed in the gas to be measured – for example an environmental chamber. In addition they can be used at high pressure, they have a wide response range (typically 50 °C to −80 °C dew point for a single

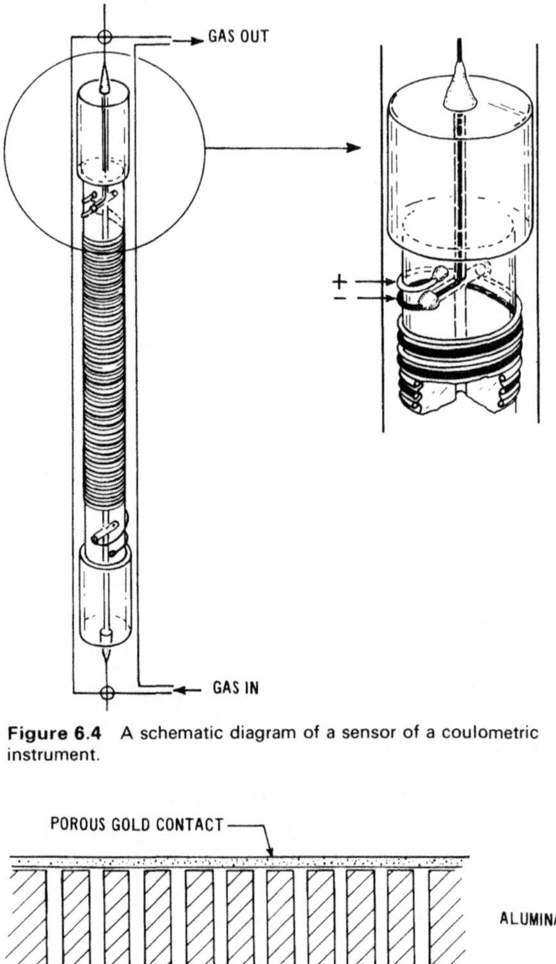

Figure 6.4 A schematic diagram of a sensor of a coulometric instrument.

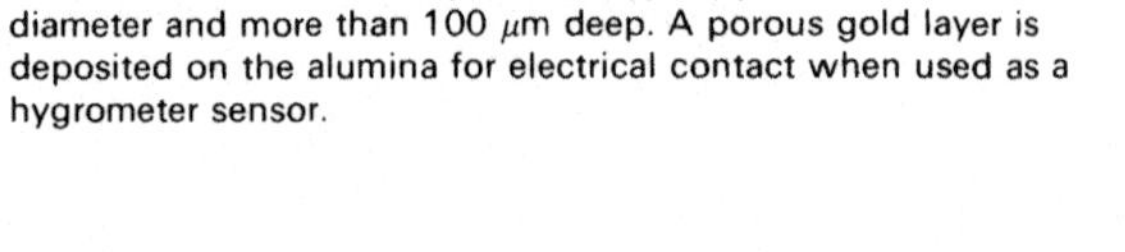

Figure 6.5 An idealized representation of the pore structure of anodized alumina. The pores are typically less than 20 nm in diameter and more than 100 μm deep. A porous gold layer is deposited on the alumina for electrical contact when used as a hygrometer sensor.

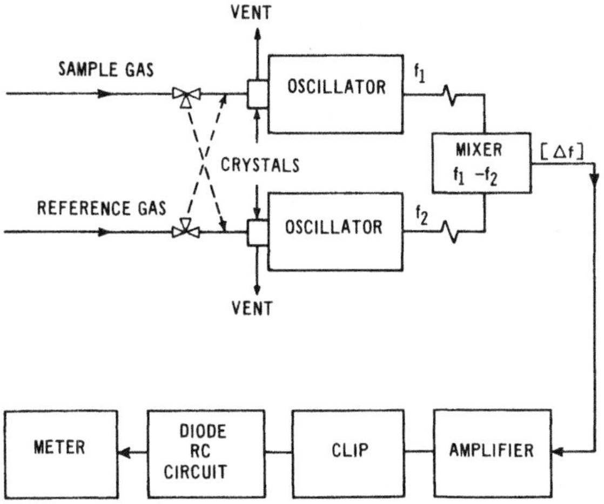

Figure 6.6 A block diagram of the arrangement of a piezoelectric humidity instrument (courtesy Du Pont Instruments (UK) Ltd.).

aluminium oxide sensor), have a rapid response, and are generally not expensive. These advantages often outweigh any problems of drift and stability, and the requirement for regular calibration, but they must be used with care.

6.3.1.5 Quartz crystal oscillator instrument

The oscillation frequency of a quartz crystal coated with hygroscopic material is a very sensitive detector of the weight of absorbed water because very small changes in frequency can be measured. In practice as shown in Figure 6.6, two quartz crystal oscillators are used, and the wet and a dry gas are passed across them alternately, usually for 30 seconds at a time. The frequency of crystal oscillation is about $9 \,.\, 10^6$ Hz, and that of the crystal exposed to the wet gas will be lowered and that of the crystal exposed to the dry gas will rise. The resultant audio frequency difference is extracted, amplified and converted to voltage to give a meter response whose maximum value on each 30-second cycle is a measure of the moisture level. The range of applicable concentrations is 1–3000 vppm and at lower levels the fact that the value after a certain time is measured rather than an equilibrium value means that the instrument can have a more rapid response than alternative methods (sample lines, however, often determine response time). Because the crystals see the sample gas for equal times contamination of the two crystals should be similar, and the frequency difference little affected, resulting in stability. However, regular calibration is still necessary and the complexity of the instrument makes it expensive.

6.3.1.6 Automatic psychrometers

The measurement of the temperature difference between a dry thermometer bulb and one surrounded by a wet muslin bag fed by a wick is the classical meteorological humidity measurement. This is called psychrometry, and automated instruments are available. The rate of evaporation depends on the gas flow as well as on the relative humidity, but generally a flow rate greater than $3\,\mathrm{m\,s^{-1}}$ gives a constant temperature depression. It is most useful at high relative humidities with accurate temperature measurements.

6.3.2 Liquids

6.3.2.1 Karl Fischer titration

The Karl Fischer reagent contains iodine, sulphur dioxide and pyridine (C_5H_5N) in methanol; the iodine reacts quantitatively with water as follows:

$$[3C_5H_5N + I_2 + SO_2] + H_2O \rightarrow 2C_5H_5NHI + C_5H_5NSO_3$$

$$C_5H_5NSO_3 + CH_3OH \rightarrow C_5H_5NHSO_4CH_3$$

If a sample containing water is titrated with this reagent the end-point at which all the H_2O has been reacted is indicated by a brown colour showing the presence of free iodine. This is the basic standard technique, and is incorporated into many commercial instruments, with varying levels of automation. In process instruments the end-point is determined electrometrically by amperometric, potentiometric, or coulometric methods (see Chapter 3). In the amperometric method two platinum electrodes are polarized and when free iodine appears they are depolarized and the resultant current is measured to define the end-point. Potentiometrically, the potential of an indicator electrode is monitored against a calomel electrode and the end-point is characterized by a sudden change in potential. Coulometrically, iodine is generated by a constant electrolysing current from a modified reagent and the time taken to reach the end-point gives the mass of water in the sample. This last technique lends itself to automatic operation, with samples injected sequentially or, in one instrument, the moisture in a sample flow is measured continuously by mixing with standardized reagent, and the electrolysis current is a measure of the mass flow of water.

6.3.2.2 Infrared instruments

The same comments apply as for gases (Section 6.3.1.3), but sample cell lengths are usually shorter, in the range 1–100 mm. It is an attractive method for on-line analysis but care must be taken that other components in the liquid do not interfere with the

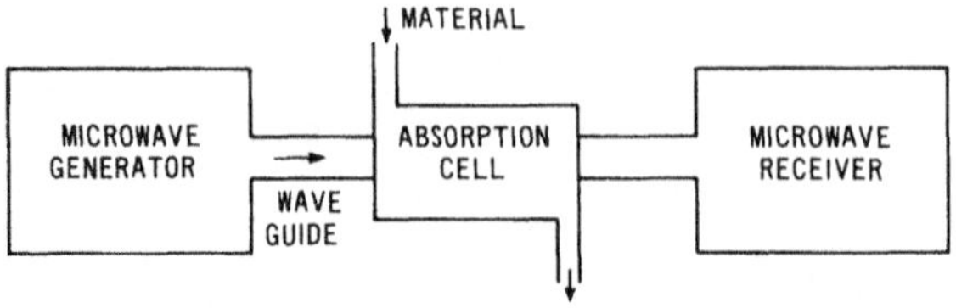

Figure 6.7 The basic concept for measuring moisture by microwave absorption.

measurement. Single-beam instruments are most often used.

6.3.2.3 Vapour pressure methods

As discussed in Section 6.2.2 the equilibrium relative humidity above a liquid can be used to determine the moisture content in the liquid. Either the relative humidity in a closed volume above the liquid can be measured, or a sensor which responds to the moisture vapour pressure in the liquid can be immersed in the liquid. The aluminium oxide sensor (Section 6.3.1.4) can be used, either above the liquid because it does not require a gas flow rate, or within the liquid because though the aluminium oxide pores will adsorb water molecules they will not adsorb the liquid molecules. These techniques are not appropriate if suspended free water is present in the liquid.

One manufacturer has developed a system in which the sensor is a moisture-permeable plastic tube which is immersed in the liquid. A fixed quantity of initially dry gas is circulated through the tube and the moisture in the gas is measured by an optical dew point meter. When equilibrium is reached the dew point measured equals that of the moisture in the liquid.

6.3.2.4 Microwave instruments

The water molecule has a dipole moment with rotational vibration frequencies which give absorption in the microwave S-band (2.6–3.95 GHz) and X-band (8.2–12.4 GHz) suitable for moisture measurement (Figure 6.7). The S-band needs path lengths four times longer than the X-band for a given attenuation and therefore the microwave band as well as cell dimensions can be chosen to give a suitable attenuation. Electronic developments are causing increased interest in this technique.

6.3.2.5 Turbidity/nephelometer

Undissolved water must be detected in aviation fuel during transfer. After thorough mixing the fuel is divided into two flows – one is heated to dissolve all the water before it passes into a reference cell; the other passes directly into the working cell. Light beams split from a single source pass through the cells and suspended water droplets in the cell scatter the light and a differential output is obtained from the matched photoelectric detectors on the two cells. 0 to 40 ppm moisture can be detected at fuel temperatures of −30 to 40 °C.

6.3.3 Solids

The range of solids in which moisture must be measured commercially is wide and many techniques are limited to specific materials and industries. In this book just some of the major methods are discussed.

6.3.3.1 Equilibrium relative humidity

The moisture level of the air immediately above a solid can be used to measure its moisture content. Electrical probes as discussed in Section 6.3.1.4 are generally used, and if appropriate can be placed above a moving conveyor. If a material is being dried its temperature is related to its equilibrium relative humidity and a temperature measurement can be used to assess the extent of drying.

6.3.3.2 Electrical impedance

Moisture can produce a marked increase in the electrical conductivity of a material and, because of water's high dielectric constant, capacitance measurements can also be valuable. Electrical resistance measurements of moisture in timber and plaster are generally made using a pair of sharp pointed probes (Figure 6.8(a)) which are pushed into the material, the meter on the instrument being calibrated directly in percentage moisture. For on-line measurements of granular materials electrodes can be rollers, plates (Figure 6.8(b)), or skids but uniform density is essential. A difficulty with this and other on-line methods which

Figure 6.8 Two techniques for electrical measurements of moisture in solids: (a) pointed probes for insertion in wood, plaster, etc. to measure resistance, (b) capacitance plates to measure moisture in flowing powder or granules.

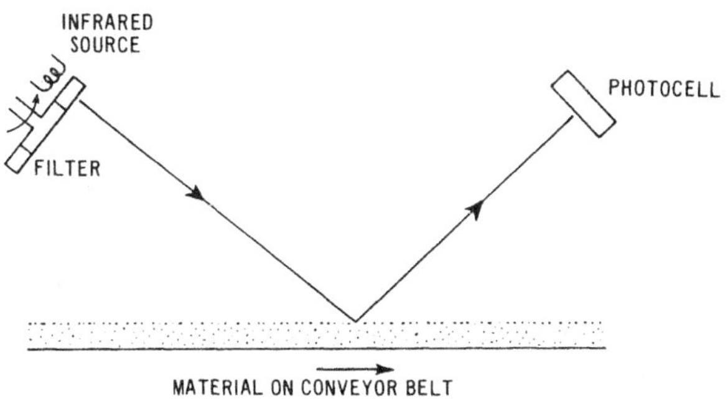

Figure 6.9 The principle of infrared reflectance used to measure moisture in a solid on a conveyor.

require contact between the sensor and the material is that hard materials will cause rapid erosion of the sensor.

6.3.3.3 Microwave instruments

Most comments appropriate to liquids also apply to solids, but, as above, constant packing density is necessary. For sheet materials, such as paper or cloth, measurement is simple, the sheet passing through a slot in the waveguide. For granular materials uniform material density is achieved by design of the flow path; alternatively extruders or compactors can be useful.

6.3.3.4 Infrared instruments

The basic difference from measurements in gases and liquids is that, for solids, reflectance methods (see Figure 6.9) are usually used rather than transmission methods. Single-beam operation is used with a rotating absorption and reference frequency filter to give regular zero readings. The calibration of a reflectance method can be substantially independent of the packing density as it measures only the surface concentration. For material on a conveyor belt a plough is often used in front of the sensing position to ensure a measurement more typical of the bulk. The method is not suitable for poorly reflecting materials, e.g. carbon and some metal powders.

6.3.3.5 Neutron moderation

Hydrogen nuclei slow down ('moderate') fast neutrons and therefore if a fast neutron source is placed over a moist material with a slow neutron detector adjacent, the detector output can be used to indicate the moisture concentration. The concentration of any other hydrogen atoms in the material and its packing density must be known. This technique is described in Chapter 4 of Volume 3. Nuclear magnetic resonance can also be used to detect hydrogen nuclei as a means of measuring moisture content.

6.4 Calibration

It will be seen from the above sections that many moisture measurement techniques are not absolute and must be calibrated, generally at very regular intervals. It must first be emphasized that the absolute accuracy of moisture measurement, particularly in gases, is not usually high. Though it is possible to calibrate moisture detectors for liquids or solids to 0.1 to 1.0 per cent, such accuracies are the exception rather than the rule for gases. Figure 6.10 shows the accuracies of some of the techniques discussed in this chapter compared with the absolute gravimetric standard of the US National Bureau of Standards.

6.4.1 Gases

First of all the difficulties of making accurate moisture measurements must be stressed. This is particularly so at low levels, say less than 100 vppm because as all materials absorb moisture to some extent sample lines must come to equilibrium as well as the detector. At low moisture levels this can take hours particularly at low flow rates. A rapid-flow bypass line can be valuable. Patience is mandatory and if possible the outputs of the instruments should be recorded to establish when stable conditions are achieved. Many plastics are permeable to moisture and must never be used. At high moisture levels copper, Teflon, Viton, glass or quartz can be satisfactorily used but at low levels stainless steel is essential. Finally, at high moisture levels it must be remembered that the sample lines and detectors must be at least 10 kelvins hotter than the dew point of the gas.

There are two basic calibration methods which can, with advantage, be combined. Either a sample gas is passed through a reference hygrometer and the

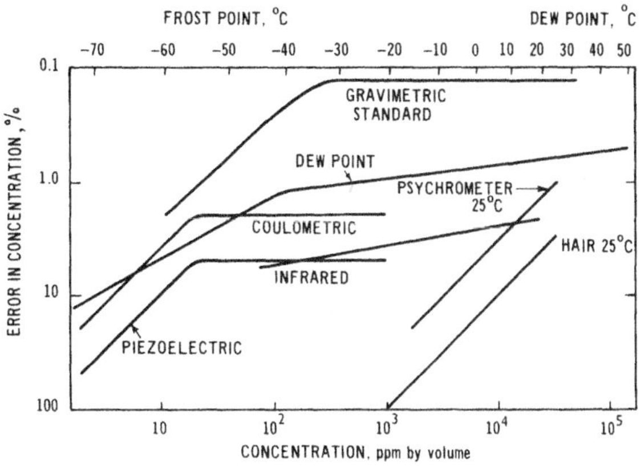

Figure 6.10 The accuracy of some of the major techniques for measuring moisture in gases, after Wexler (1970).

instrument under test, or a gas of known humidity is generated and passed through the instrument under test. Obviously it is ideal to double-check the calibration by using a known humidity and a reference hygrometer.

The most suitable reference hygrometer is the dew point meter, which can be readily obtained with certified calibration traceable to a standard instrument. For many applications less sophisticated dew point instruments would be adequate and coulometric analysers are possible for low moisture levels. At high levels gravimetric methods can be used but they are slow and tedious and difficult to make accurate.

There are a range of possible humidity sources, some of which are available commercially, and the choice depends on the facilities available and the application:

(a) A plastic tube, permeable to moisture, held in a thermostatically controlled water bath, will give a constant humidity for a given flow rate. Some manufacturers sell such tubes precalibrated for use as humidity sources but obviously the method is not absolute and the permeation characteristics of the tubes may change with time.
(b) Gas cylinders can be purchased with a predetermined moisture level which does not significantly drift because of the internal surface treatment of the cylinder. However to prevent condensation in the cylinder the maximum moisture level is limited to about 500 vppm even with a cylinder pressure of only 10 bar. They are most suitable for spot checks of instruments on site.
(c) If an inert gas containing a known concentration of hydrogen is passed through a bed of copper oxide heated to $\sim 350\,^{\circ}C$ the hydrogen is converted to water vapour. This method relies on the measurement and stability of the hydrogen content which is better than for moisture. The generated humidity is also independent of flow rate.
(d) Water can be continuously injected into a gas stream using either an electrically driven syringe pump or a peristaltic pump. The injection point should be heated to ensure rapid evaporation. The method can be used very successfully, syringes in particular allowing a very wide range to be covered.
(e) If a single humidity level can be generated, a range can be obtained using a flow mixing system, but to achieve sufficient accuracy mass flow meters will probably be necessary.

6.4.2 Liquids

The basic absolute method is that of the Karl Fischer titration which was described in Section 6.3.2.1.

6.4.3 Solids

There are several methods which allow the absolute moisture level of a solid to be determined, but, for all of them, samples of the specific substance being measured by the process technique must be used. The most common technique is of course to weigh a sample, dry it, and then weigh again. Drying temperature and time depend on the material; if necessary the temperature must be limited to avoid decomposition, loss of volatile components or absorption of gases from the atmosphere. Balances can be obtained with a built-in heater, which give a direct reading of moisture content for a fixed initial sample weight. Other favoured techniques include measuring the water vapour given off by absorbing it in a desiccant to avoid the effects of volatiles; the Karl Fischer method again; or mixing the substance with calcium carbide in a closed bomb and measuring the pressure of acetylene produced. The method must be carefully chosen to suit the substance and process technique being used. Finally it is worth noting that rather than an absolute calibration, calibration directly in terms of the desired quality of the substance in the manufacturing process may be the most appropriate.

6.5 Bibliography

Mitchell, J. and Smith, D:, *Aquametry. Part 1, A Treatise on Methods for the Determination of Water*, Chemical Analysis Series No 5, Wiley, New York, (1977)

Mitchell, J. and Smith, D., *Aquametry. Part 2, The Karl Fischer Reagent*, Wiley, New York, (1980)

Verdin, A., *Gas Analysis Instrumentation*, Macmillan, London, (1973)

Wexler, A., 'Electric hygrometers', National Bureau of Standards Circular No 586, (1957)

Wexler, A. (ed.), *Humidity and Moisture* (3 volumes), papers presented at a conference, Reinhold, New York, (1965)

Wexler, A., 'Measurement of humidity in the free atmosphere near the surface of the Earth', *Meteorological Monographs*, **11**, 262–82, (1970)

Index

Printed in Dunstable, United Kingdom

91221444R00112